LIFE OUT OF BALANCE

NEXUS

NEW HISTORIES OF SCIENCE, TECHNOLOGY, THE ENVIRONMENT, AGRICULTURE & MEDICINE

NEXUS is a book series devoted to the publication of high-quality scholarship in the history of the sciences and allied fields. Its broad reach encompasses science, technology, the environment, agriculture, and medicine, but also includes intersections with other types of knowledge, such as music, urban planning, or educational policy. Its essential concern is with the interface of nature and culture, broadly conceived, and it embraces an emerging intellectual constellation of new syntheses, methods, and approaches in the study of people and nature through time.

LIFE OUT OF BALANCE

Homeostasis and Adaptation in a Darwinian World

JOEL B. HAGEN

THE UNIVERSITY OF ALABAMA PRESS TUSCALOOSA

The University of Alabama Press
Tuscaloosa, Alabama 35487-0380
uapress.ua.edu

Typeface: Scala Pro

Cover image: Knut Schmidt-Nielsen with a kangaroo rat in his laboratory at Duke University; courtesy University Archives Photograph Collection, Arthur Rubenstein Library, Duke University, Durham, North Carolina.
Cover design: Michele Myatt Quinn

Cataloging-in-Publication data is available from the Library of Congress.
ISBN: 978-0-8173-2089-8
E-ISBN: 978-0-8173-9347-2

Contents

Figures

Foreword

Few sciences have attracted as much attention from historians in the recent decades as ecology. In exploring the nuances of ecology's origins and development, however, the concept of homeostasis, the self-regulating process by which biological systems maintain the stability necessary for survival, has been curiously overlooked despite its centrality to the modern life sciences. Joel Hagen, a pioneering scholar in the history of biology, rectifies this in *Life Out of Balance: Homeostasis and Adaptation in a Darwinian World*, tracing the history of a scientific idea. Homeostasis was introduced in its modern form by physiologists in the early twentieth century. Hagen shows how this concept gained new traction—and criticism—as the debates between geneticists and evolutionary biologists swirled towards the modern synthesis. The application of homeostasis as a framework for understanding was eventually extended beyond the self-regulation of individual organisms to ecological systems, and then to human society, economic markets, and cybernetics. Hagen is especially interested in the first phase of this expansion as lab-based medical scientists became engaged in the interactions of populations, even entire ecosystems, in natural environments; what would become physiological ecology. Their work was informed by applied research projects during World War II as well as by an abiding belief in an interdisciplinary and holistic approach to biological synthesis. The fraught expansion of the concept of homeostasis illuminates the debate within the discipline of biology distinguishing functional and evolutionary approaches to biology but also a more general and evolving discussion about the application of concepts. The threshold of applicability—too flexible to be meaningful or too rich and fruitful to be discarded—of a scientific concept is historical, by Hagen's telling. The larger questions *Life Out of Balance* explores about cross-disciplinary synthesis, scale, and intellectual cultures are located at the intersection of history of science, history of technology, and environmental history. This volume thus demonstrates the advantages of scholarly interdisciplinarity that NEXUS seeks to advance. We heartily welcome this contribution to the NEXUS Series.

Alexandra E. Hui
For the NEXUS editors

Acknowledgments

I completed the broad outlines of this book during a sabbatical leave from Radford University in 2014. The university also provided generous support for travel to far-flung archives during and after the leave. I thank my department chairs, Christine Small and Justin Anderson, for their encouragement throughout this project. I am also grateful to my dean, Orion Rogers, for his enthusiastic support for my sabbatical leave. The book could not have been completed without the generous assistance of numerous librarians and archivists from collections at Duke University, Harvard University, University of Minnesota, University of Chicago, Rochester University, Yale University, University of Alaska, University of Pennsylvania, University of Florida, University of Georgia, and University of California at San Diego. I am particularly thankful for help in locating and duplicating the illustrations in the book provided by Bert Hart, Astrid Schmidt-Nielsen, Meredith Gozo, Marianne Reed, Peregrine Wolff, Brooke Guthrie, and Becky Butler.

I thank Laura Martin for organizing a historical session at the 2015 centennial of the Ecological Society of America, where I presented an early version of material in chapters 3 and 4 of the book. Darrell Arnold encouraged me to think more deeply about the ecological uses of homeostasis within the broader context of systems theory when he invited me to contribute to his edited collection *Traditions of Systems Theory*. I presented some of my early thoughts about the work of George Bartholomew at a conference on museum collections and systematics organized by Staffan Müller-Wille and Michael Ohl at Schloss Herrenhausen. I appreciate the critical comments and suggestions from the conference participants. Parts of chapters 7 and 8 were presented at a conference on cross-species comparisons at Kings College London organized by Abigail Woods and Rachel Mason Dentinger. I appreciate the ideas raised during the discussion of the paper. I am very grateful to Betty Smocovitis for her constant encouragement during the project and for inviting me to discuss my thoughts on proximate and ultimate causation as part of a symposium series in the Biology Department at the University of Florida. In addition to Betty's many helpful suggestions for improvement, I also learned a great deal from discussions about physiological ecology with Brian McNab, Harvey Lillywhite, Robert Holt, and other faculty and students at the University of Florida. I appreciate the many email discussions I have had with Todd Sformo about Per Scholander, Laurence Irving, and the history of arctic biology in Alaska. William R. Dawson generously answered a number of my questions about George Bartholomew and the development of physiological ecology after World War II.

Thank you to everyone at the University of Alabama Press who helped bring this book to fruition. I am indebted to my acquisitions editor, Claire Lewis Evans, for shepherding my manuscript through the review and revision process. Her advice and suggestions were impeccable. Joanna Jacobs managed the production of the book and kept procedures on schedule. The copy editor is an author's best friend, and I appreciate Lisa Williams's efforts toward bringing the manuscript to final form. I am also grateful to the anonymous reviewers who made many insightful criticisms, pointed out flaws in my thinking, and made numerous positive suggestions for improvement.

I am most thankful for the continuous interest and support from members of my extended family. I am deeply grateful for the encouragement that my parents shared for my professional work. Sue, Kirsten, Chris, and Kenneth always provide inspiration for my work. A very special thanks goes to Tricia and Bert, too.

LIFE OUT OF BALANCE

Introduction

Perhaps the most striking characteristic of living things is their ability to maintain integrity and some degree of constancy in a fluctuating environment. This self-regulation, often referred to as homeostasis, is so fundamental that for some biologists it is the sine qua non of life. Indeed, for physiological ecologist J. Scott Turner, homeostasis is "biology's second law" and is irreducible to the first law of natural selection.[1] From molecular pathways to ecosystems, biological systems regulate themselves following a few fundamental rules, or, as the evolutionary developmental biologist Sean B. Carroll succinctly claims, in biology "everything is regulated."[2] Thus, life emerges out of a balance between the internal environment of the organism and the external environment within which it exists. This intimate "fit" between organism and environment is one definition of adaptation. At the same time, adaptation involves an evolutionary process shaped by natural selection. "Fitness" in modern evolutionary theory is defined as relative reproductive success among the competing—and sometimes cooperating—individuals (or genes) that make up evolving populations. Faced with a fluctuating and seemingly capricious environment, successful populations must adaptively track environmental changes if they are to avoid extinction. From this perspective life is perpetually out of balance in a never-ending process of adaptive change that some evolutionary biologists have analogized to the Red Queen's race in *Through the Looking-Glass.*[3] Here, Alice runs just to stay in place. Similarly, in an unbalanced world, species must continually evolve just to maintain a state of adaptation as the environment decays around them.

Surely, it would seem, homeostasis must be an important part of adaptation in both its physiological and evolutionary senses. Temperature regulation, water balance, closely regulated oscillations in blood glucose, and other homeostatic processes must all contribute importantly to both an organism's fit with the environment and its reproductive success. Yet these important biological concepts have an uneasy history. *Life Out of Balance* explores the persistent appeal of homeostasis in recent biological thought but also the challenges of applying it to a Darwinian world of competing organisms, evolving populations, and interacting ecosystems.

Life Out of Balance focuses on a particular period in history when new ideas of self-regulation, adaptation, and fitness became central to a variety of biological disciplines. During the mid-twentieth century, the decade preceding World War II and the two decades following the war, these ideas developed in a number of quite different contexts. The idea of homeostasis, popularized by the physiologist Walter B. Cannon during the 1930s, emerged from laboratory studies in a biomedical context. Particularly in its later cybernetic guise, heavily influenced by Norbert Wiener and other mathematical systems thinkers, homeostasis seemed to provide new ways of discussing balance and self-regulation that avoided discredited positions taken by earlier champions of vitalism and mechanism. Indeed, it provided both a common perspective and terminology for discussing complex systems whether mechanical, biological, or social. It also provided a language for discussing apparently goal-directed or teleological processes in scientifically and philosophically acceptable ways. Broadening homeostasis in this way encouraged its diffusion into areas of biology far removed from Cannon's medical studies. Not all physiologists were enamored by homeostasis, and indeed the idea faced early and sustained criticism. Nonetheless, it attracted adherents not only in physiology but also among other biologists who discussed balance and self-regulation in various living systems from populations to ecosystems. During this same historical period, evolutionary biologists consolidated Mendelian genetics and Darwinian natural selection in the so-called modern evolutionary synthesis. This theoretical synthesis not only combined what previously had been rival evolutionary explanations but also promised to provide a broad explanatory foundation for a more unified organismal biology applicable across multiple disciplinary boundaries.

Comparative physiology seemed to provide an obvious meeting place for homeostasis and the new evolutionary theory. In contrast to medical physiologists, who often reduced the external environment to disturbing influences causing deviations from normality, comparative physiologists drew on zoological traditions that emphasized both the complexity of natural environments and the ways that organisms adaptively interact with the world around them. Unlike the medically oriented focus on health and disease, comparative physiologists were often impressed by the adaptability of animals faced with multifaceted and constantly fluctuating natural environments. It is one thing to study temperature regulation in the controlled setting of the laboratory, and quite another to consider the challenges faced by a desert animal coping with extreme heat and lack of water while avoiding predators, hunting for food, and competing for mates. Working in the museum and field, as well as in the laboratory, comparative physiologists appreciated the diversity of mechanisms that organisms use to both meet and manipulate environmental variables. To be sure, the boundary between medical and comparative physiologies was never absolute, and *Life Out of Balance* highlights the careers of several medically trained researchers who became deeply involved with broader studies of adaptation and self-regulation in natural habitats. Yet the central focus of my book is how various groups of biologists creatively combined evolutionary biology, ecology, behavior, and physiology to study the interaction of the internal environments of organisms and the external environments within which they live.

During World War II, a prominent group of comparative physiologists became involved with applied research projects on survival and human performance in harsh environments. After the war, and often using military funding, these same scientists turned attention to other species using explicitly ecological and evolutionary approaches to study self-regulation and adaptation in deserts, the arctic, high altitudes, and underwater habitats. Often referred to as physiological ecology, this area of research focused on the interaction of the whole organism with its physical environment—always with an eye toward recent developments in ecology, evolutionary biology, and behavior. Although sometimes interested in centralized control by nerves and hormones, many of these physiological ecologists embraced a holistic perspective on the organism and its relationship to a seemingly capricious external environment. Homeostasis, usually in a broader sense than Cannon originally intended, was a central focus of these studies.

The novelty, as well as the long history, of these approaches was captured by C. Ladd Prosser in *Comparative Animal Physiology* (1950).[4] The book was written as an advanced textbook but also as a sourcebook for professional zoologists. As such, it attempted an overview of what its preface described as the "enormous" literature in comparative physiology, including more than 2,800 citations. Although some of this literature dealt with mammals or model organisms such as frogs that were commonly used in research laboratories,

Comparative Animal Physiology also discussed adaptations of insects and a variety of other invertebrates, as well as microscopic protozoans. Even a cursory inspection of Prosser's book would convince readers that comparative physiology was a mature area of research that had deep roots extending back to nineteenth-century zoology and physiology. What was novel in Prosser's presentation was his explicit attempt to ground comparative physiology on recent developments in evolutionary biology and ecology. To a greater extent than his predecessors and many of his contemporaries, Prosser was drawn to the modern synthesis of genetics and natural selection. In the years following the publication of his book Prosser actively engaged with leading evolutionary biologists, including Theodosius Dobzhansky, Ernst Mayr, and C. H. Waddington. *Life Out of Balance* explores the close ties that developed among physiology, ecology, and evolution during the decades following World War II, but also some contentious disagreements among specialists from different fields. Ideas of evolutionary fitness and physiological "fit" of organism and environment sometimes harmonized but often conflicted, particularly when biologists extended homeostasis to populations, communities, and ecosystems.

Perhaps some of these disagreements might have been avoided if biologists had carefully distinguished between the different types of questions studied by various disciplines in the life sciences. In 1961 the eminent evolutionary biologist Ernst Mayr drew a distinction between proximate causation used by functional biologists to answer "how" questions and ultimate causation used by evolutionary biologists to answer "why" questions.[5] Both were important, but Mayr claimed that most biologists (and most biological disciplines) focused on one or the other alternative. According to Mayr, pointless controversies arose when biologists failed to recognize the distinction and confused the two forms of causation.

A number of historians and philosophers have argued that Mayr used the dichotomies between proximate versus ultimate causation and functional versus evolutionary biology to justify and defend organismal biology against what he viewed as the aggressive rise of molecular biology and a reductionist philosophy of science that looked to physics as the model for scientific thought.[6] Yet the sharp dichotomies that Mayr drew seem curious given the emphasis that he placed on evolutionary theory as a unifying and synthesizing foundation for all of biology. The dichotomy between functional biology and evolutionary biology was particularly problematic in broad interdisciplinary fields such as physiological ecology. One might distinguish between "how" and "why" questions, and it might be true that many biologists tended to focus attention on one or the other, but it was also true that many interesting questions did not neatly fit the dichotomy. *Life Out of Balance* explores this situation by examining how Mayr and other evolutionary biologists used homeostasis to discuss speciation and other evolutionary phenomena, sometimes blurring the distinction between proximate and ultimate causation as they did so. The book also analyzes cases in which functional and evolutionary explanations overlapped in complex ways that led to disagreements among specialists from different fields. Given his attempts to carefully distinguish between proximate and ultimate causation, it is ironic that Mayr himself became embroiled in these very controversies.

The distinction between proximate and ultimate causation is historically important. Mayr's thinking on the two forms of causation developed throughout his career as a result of debates that both he and his contemporaries considered central to understanding biology.[7] Many biologists still find the distinction a useful way to categorize questions, organize disciplines, and distinguish between alternative modes of explanation. The two forms of causation continue to be widely taught in biology textbooks. For Mayr, it was a cornerstone of a new philosophy of biology that took seriously questions of organic diversity and evolutionary change. Mayr was deeply committed to organismal biology, and he was critical of any reductionism that claimed to explain biology on physical principles alone. *Life Out of Balance* extends and broadens the historical discussion of Mayr's commitment to defending organismal biology but also his deep interest in synthesizing and unifying biology, particularly through his use of homeostasis at the levels of both individual organisms and populations. At the same time, *Life Out of Balance* seriously examines the work of other prominent biologists who proposed philosophical perspectives that were less antagonistic to the physical sciences and avoided the sharp dichotomies that Mayr used to defend organismal biology. Some of these biologists had conducted applied research in the physical sciences during World War II, and these early experiences shaped the way they approached biological problems in the decades following the war. More than Mayr, they enthusiastically embraced broad interdisciplinary approaches to studying organisms, without privileging the fields of population genetics and systematics that were so central to the modern synthesis. Although less familiar to historians and philosophers of biology, these alternative philosophical perspectives also had profound effects on the development of modern studies of organisms, populations, communities, and ecosystems.

This broader context is particularly significant in light of recent claims by various biologists about homeostasis. In his widely acclaimed *The Serengeti Rules,* Sean B. Carroll makes a sweeping claim that all biological systems are regulated according to a common set of principles and that understanding self-regulation is the key to both human health and protecting the environment.[8] Carroll supports his claim with an engaging set of historical vignettes drawn from diverse fields, from genetics and cancer biology to population biology and ecosystem conservation. Yet his bold claim about self-regulation largely ignores the controversies—past and present—surrounding biological regulation. Even in medical physiology, where the idea originated, homeostasis has been challenged by a broad array of different—sometimes incompatible—interpretations. Extended to other fields such as evolutionary biology and ecology, homeostasis has been even more controversial. Critics have often flatly rejected ideas of self-regulation applied to populations, communities, or ecosystems. *Life Out of Balance* explores the sometimes fruitful, but often troubled, attempts to combine functional and evolutionary explanations at various levels of biological organization.

1

Adaptation and the Wisdom of the Body

In the spring of 1929, the publisher William W. Norton approached Walter Bradford Cannon about writing a popular book on physiological self-regulation.[1] His publishing company would later become highly successful, but at the time it was a fledgling business run out of Norton's home in New York. Norton rather brashly offered the eminent Harvard physiologist a contract with a guaranteed minimum payment of $1,000 (about $15,000 in today's dollars) even if the book did not turn a profit. Although deferential to his prominent potential author, Norton assured Cannon that he knew how to sell scientific books to the general public, and he was confident that a book on homeostasis would be popular. Much of his confidence rested on the recent commercial success of another Norton book by the geneticist and eugenicist Herbert Spencer Jennings that served as the publisher's model for Cannon's *The Wisdom of the Body*.[2]

Cannon eventually warmed to the idea, but he was initially ambivalent toward Norton's overture. He had recently coined the term "homeostasis" and outlined the concept in a lecture at Cambridge University and in a long article in the *Physiological Review*. Nonetheless, Cannon expressed doubts that he could popularize the technical details of nerves and hormones in a book that would appeal to a broad general audience. He was loath to enter a contractual agreement before starting to write. From the outset Cannon also wanted the freedom to approach other publishers if, indeed, he decided to write the book.

During the following two years, Norton assiduously wooed Cannon with promises of flexible deadlines, generous royalties, and an advertising budget—even as the country tumbled into the Great Depression. He was delighted when the physiologist finally signed the contract in May 1931.[3] By daily "pegging along" at his summer home, Cannon managed to write most of what would become *The Wisdom of the Body* during the next two months. Despite his earlier worries, Cannon's book became both commercially successful and widely influential among scientists in diverse disciplines. This outcome was due in large part to Cannon's deep insights into organic self-regulation and the clarity of his writing style. It was also due, in no small part, to Norton's subtle guidance in popularizing self-regulation, particularly in a final chapter on social homeostasis. Despite that popularity—or perhaps because of it—homeostasis also faced indifference and hostile criticism from various physiologists.

One of the nagging criticisms Cannon faced was that homeostasis was unoriginal.[4] Some critics claimed that he simply created a name for ideas formulated by Claude Bernard without properly crediting the famous French physiologist. Others suggested that Cannon had lifted the idea from one of his Harvard colleagues, the physiological chemist Lawrence J. Henderson. Still others dismissed homeostasis as a commonplace idea widely shared by Victorian thinkers and their intellectual descendants in the United States. The decision to gloss over the historical context of homeostasis may have been largely due to Norton's insistence that readers had little interest in intellectual precursors. The publisher's brash claim that Cannon had "discovered" homeostasis caused some discomfort for the Harvard physiologist; still, Norton made sure that assertion appeared prominently on the dust jacket of the book.

One might read *The Wisdom of the Body* as an expanded but simplified account of Cannon's *Physiological Review* article "Organization for Physiological Homeostasis," published three years earlier.[5] Nonetheless, the two presentations differed in significant ways. Importantly, given criticisms of inadequately crediting predecessors, Cannon had provided a rather detailed historical account of the development of self-regulation in the 1929 article. The introductory section not only paid homage to Bernard and recognized contributions from other physiologists but also provided a historical context for emphasizing what was new and different in his concept of homeostasis. Homeostasis might not have been a discovery, but as a number of historians have argued, it was far more than simply a restatement of nineteenth-century ideas of self-regulation and stability, as some of Cannon's critics claimed.[6] Cannon readily accepted the importance of Bernard's distinction between the relatively constant fluid matrix of the body (*milieu intérieur*) and the fluctuating external environment in which the organism lives. He recognized the importance of what he acknowledged was Bernard's "more precise analysis" of traditional physiological ideas about balance and self-regulation traceable to ancient Hippocratic medicine. He also accepted, without critical comment, Bernard's famous dictum that maintaining the fixity of the *milieu intérieur* was the condition of "free and independent life" and that all vital mechanisms contributed toward this end. Indeed, the free and independent life emerged out of the balance between internal and external environments. What set Cannon's ideas apart, and what was captured by the term "homeostasis," was the idea that organisms were open systems that maintained steady states through the complex integration of physiological organs and organ systems. This systems thinking, the fundamentals of which Cannon attributed to Alfred Lotka, differentiated homeostasis from equilibrium as understood in the physical sciences. The stability of living systems was in sharp contrast to their unstable chemical constituents, but maintaining this stability came at a steep energetic cost. Compared to chemical equilibria, homeostasis was more dynamic, because living systems face the constant challenge of environmental disturbances. Indeed, Cannon emphasized the broad, but energetically costly, "margins of safety" built into most homeostatic processes. Finally, a constant equilibrium was not possible in a developing organism, and the precision with which regulation occurred often differed throughout the stages of the life cycle—particularly early and late in life. Cannon claimed that his dynamic, whole-organism perspective also set homeostasis apart from more limited ideas of physiological equilibrium, including the way his friend and colleague Henderson had described the blood as a complex physicochemical system that maintained constancy of respiratory functions. For Cannon, all of the organs of the body were coordinated toward resisting disturbing forces from the environment and maintaining the integrity of the body's fluid matrix within broad limits.

Cannon's emphasis on integration was important but ambiguous. Historian Garland Allen characterized Cannon's perspective as a form of "holistic materialism."[7] For Allen, Cannon's homeostasis exemplified a new way of looking at physiological function that made nineteenth-century controversies between vitalism and mechanism obsolete. In his 1929 article, Cannon emphasized that homeostasis was maintained by the "integrated cooperation" of multiple organs and the "interplay of these organs in the organism as a whole."[8] This physiological holism seemed to rule out simple mechanistic explanations. According to Cannon, "The factors which operate in the body to maintain uniformity are often so peculiarly physiological that any hint of immediate explanation in terms of relatively simple mechanics seems misleading."[9] Even though the details of self-regulation were often unknown, Cannon argued that this holistic perspective played an important heuristic role in suggesting patterns of integration.[10] The results of Cannon's own experiments on the sympathetic nervous system and the hormones secreted by the adrenal gland seemed to be a particularly apt example of the direction that future research on homeostasis might take. He held out the hope that more quantitative measures of neuroendocrine control would lead to a better understanding of the organism's "fitness," its response to stress, and developmental changes in homeostasis throughout the life cycle.[11]

Ironically, this perspective also served as the basis for criticism that his account was not holistic at all, but rather a form of mechanistic reductionism that was reactive, rather the proactive.[12] The seemingly mechanistic examples used in Cannon's 1929 article fueled this suspicion, despite his claims for integration. The simplified presentation of discrete homeostatic responses in *The Wisdom of the Body* further highlighted the accusation of mechanistic reductionism by his critics.

Cannon's diagrammatic representation of homeostatic regulation of blood glucose (figure 1) captures the tension between holistic materialism and a simpler mechanistic perspective. As the centerpiece of his 1929 article, the general scheme for this important homeostatic process was clear, even though there remained important questions and uncertainties about the details. The body maintains blood glucose within the "common variation" of 70–130 mg % whether vigorously exercising or digesting a rich meal. In Cannon's time the central role of insulin in lowering blood sugar was well understood, although there was debate about the possibility that the hormone might also act to release sugar into the blood. Cannon was skeptical of this claim for the dual action of insulin, primarily because his own work on adrenal hormones (later identified as epinephrine) and the sympathetic nervous system implicated these as counterbalancing the effects of insulin.[13] Despite the uncertainties, Cannon presented a compelling scheme for automatic control based on the antagonistic effects of two sets of nerves and hormones. Because these agents had opposing effects of elevating and depressing glucose in the blood, they acted together as a coordinated system of checks and balances to maintain blood sugar within a "normal" range of values. Above the normal range, there was a safety margin—or "margin of economy"—involving elimination of excess sugar by the kidneys during urine production. This overflow resulted in a waste of energy, illustrating Cannon's claim that homeostasis favored stability over energetic efficiency.

Despite Cannon's emphasis on integration at the organismal level, one could also interpret the diagram and accompanying discussion of blood

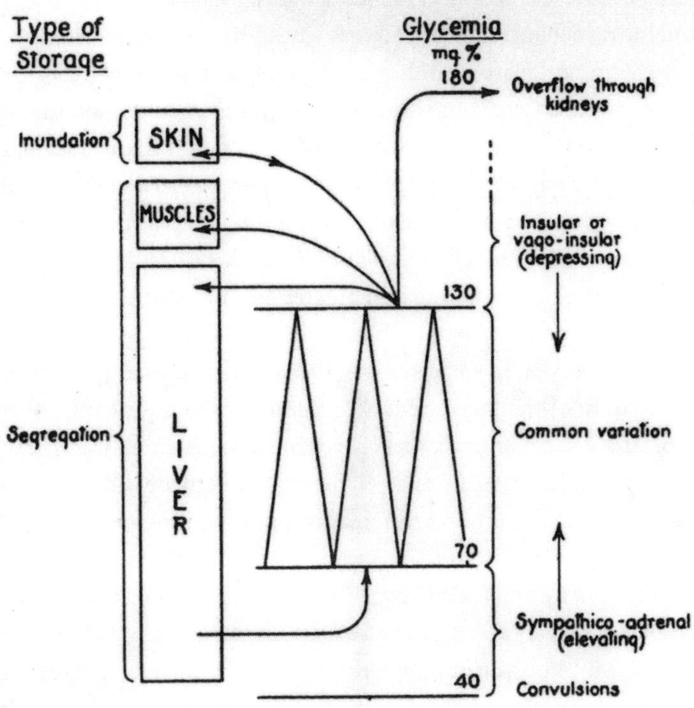

Figure 1. Diagrammatic representation of blood glucose regulation. Walter B. Cannon, "Organization for Physiological Homeostasis," Physiological Reviews 9 (1929): 399–431, 410.

glucose regulation as a "mechanism" analogous to a thermostat or other single-purpose automatic control system devised by engineers. Indeed, Cannon invoked the analogy of a thermostat in his discussion of homeostatic temperature regulation in his 1929 article.[14] His diagram of glucose regulation highlighted internal control through nerves and hormones but largely ignored the complexities of the external environment, which he reduced to "disturbing influences" in his accompanying text. As later critic Robert Perlman contends, the tradition of medical physiology that Cannon inherited from Bernard largely disregarded the external environment, particularly in its ecological and evolutionary contexts, resulting in an "impoverished concept of organisms."[15]

The organization of *The Wisdom of the Body* further encouraged a reductionist interpretation of homeostasis. Each chapter of the book was devoted to the automatic control of a particular vital constituent of the organism:

water, sugar, pH, temperature, etc. This was, perhaps, a logical way to guide the general reader through the complexity of living systems while emphasizing the common features of homeostatic regulation. Nonetheless, this arrangement opened Cannon to criticism for presenting a compartmentalized model of the body narrowly focused on control, rather than considering the integrated organism interacting with its complex external environment. The popular success of the book meant that later biologists, including physiologists, usually turned to *The Wisdom of the Body* rather than Cannon's earlier review article, which had placed greater emphasis on integration from a whole-organism perspective.

An equally persistent criticism was that Cannon presented an overly optimistic account of the body's ability to maintain stability. It might seem an odd criticism, given Cannon's earlier pathbreaking work on traumatic shock and his wartime experiences as a physician in a battlefield hospital. Yet it is undeniably true that Cannon's focus was on normal function when he wrote *The Wisdom of the Body*. Before settling on the final title, which borrowed from an earlier lecture by the British physiologist Ernest Starling, Cannon and Norton discussed several other possibilities. Cannon initially favored "Factors in Stabilization," but Norton pressed for a more human-centered title: "Organization for Stability in Human Beings." While writing the book, Cannon suggested a variety of alternatives for Norton's consideration, including "The Self-Regulating Stability of Man," "Organized Stabilization in Man," and "Automatic Human Stabilization." Other possibilities that Norton preferred were "How We Stay Normal" or "Keeping Normal," which the publisher and author agreed concisely captured the main idea of the book.[16] *The Wisdom of the Body* turned out to be a more euphonious title, although it, too, evinced an optimistic perspective. Although Cannon briefly mentioned traumatic shock and other pathological conditions, normality and stability remained the overarching themes of the book.

Cannon was a committed evolutionist, but it is perhaps unsurprising that as a medical researcher, he was unaware of the robust evolutionary theory developing from the combination of natural selection and Mendelian genetics at the time he was writing his book. Cannon's prescient recognition of fight or flight responses pointed to the importance of competition and predation in the life of an animal but hardly captured the richer sense of Darwinian struggle for existence. Nor was he interested in variation both within and among species. Cannon embedded homeostasis within a linear evolutionary perspective that emphasized progress from lower to higher animals, rather than a richly branching tree of life. The frog's inability to control water loss from its moist skin prevented it from living for extended periods on dry land, and its failure to maintain constant body temperature condemned it to what Cannon characterized as a "sluggish numbness" in the muddy bottom of the pond during the winter. Although reptiles had evolved defenses against water loss, only birds and mammals had "acquired the freedom" to live in cold temperatures.[17] Ironically, after outlining this progressive evolutionary perspective on homeostasis, Cannon suggested that self-regulation might be a common property of any complex organized system—living or nonliving. Thus, understanding the highly evolved homeostatic processes in mammals provided a vantage point for understanding more rudimentary self-regulation in both lower animals and human societies.[18]

Cannon had suggested this broader perspective on homeostasis in his 1929 article without elaboration, but he more fully developed the idea of social homeostasis in *The Wisdom of the Body*. Moving from the comfort of his specialty of mammalian physiology to a broader discussion of stabilizing mechanisms in American society made Cannon nervous. He equivocated on including the chapter and referred to it as an "epilogue." Norton was insistent on including the chapter, which he viewed not simply as an addendum but as the culmination of Cannon's argument. Norton's approach to popular science writing involved using biology to support progressive political and social views. In their correspondence one senses a real tension between Norton's enthusiasm for sweeping social generalizations based on biology and Cannon's hesitancy about going out on a limb in a field outside his professional training. In the end Norton won the day, but only after Cannon had the chapter thoroughly vetted by several Harvard sociologists and economists.[19]

Cannon admitted that previous, largely discredited comparisons between human societies and organisms rested on naive analogies. Equating organs with factories or white blood cells with police officers contributed little to understanding physiology or society. By contrast, Cannon was confident that broad organic claims about self-regulation might be illuminating, even if only by emphasizing what sociologists already knew. Progressive evolution as a process of attaining freedom from the disruptive forces of the external environment applied in both cases. In human terms, this freedom involved balancing security, social stability, and liberty. In his later presidential address on social homeostasis to the American Association for the Advancement of Science, Cannon stated, "A social order ensuring not security alone, not freedom alone, but both security and freedom, is the goal."[20]

Cannon wrote *The Wisdom of the Body* during the depths of the Depression, and he delivered his presidential address less than a year before the United States entered World War II.[21] This tense and turbulent time called forth emergency responses to promote social and economic stability caused by extreme disturbances. Social homeostasis relied upon governmental programs that might seem uneconomical under normal circumstances. These emergency measures were buffered by "large margins of safety" analogous to the range of blood sugar levels tolerated by the body as a result of disturbances and the wasteful spillover of excessive glucose in the kidneys when blood sugar rose to dangerous levels. Indeed, Cannon contrasted the dire conditions that the nation faced with a more normal state in which supply and demand maintained harmonious internal relations with a minimum of government regulation and a high degree of individual autonomy. Yet, despite some optimism about the future of democracy, Cannon warned that society lacked the "delicate indicators" found in living systems to warn of disturbances and call forth homeostatic responses. Lacking such control systems, societies were always in danger of disturbances leading to dangerous oscillations. Cannon gave muted praise to the social experimentation of the New Deal that held some promise for a limited form of social control and regulation. Although social evolution had not progressed very far compared with organic evolution, Cannon optimistically concluded, "Control to a greater degree than now prevails might lead, however, to more extensive liberty of action, as control by stop-and-go signals has eliminated traffic jams and promoted free movement, or as control of infections has enormously expanded the liberties of mankind by providing safe food and drink and by isolating the carriers of disease. More control is tolerable if it results in greater human freedom."[22]

Despite his initial hesitancy, Cannon's foray into social homeostasis was well received. The *New York Times* reported that an audience of five hundred alumni applauded his presentation of the idea during a speech at the Massachusetts Institute of Technology in 1933. The newspaper dubbed Cannon's social homeostasis a new "biocracy" that rivaled the currently fashionable technocracy movement, which advocated government led by engineers and scientists. Others presented the two movements as complementary. *Publishers Weekly* reported that *The Wisdom of the Body* had been included on the technocracy reading list compiled by Howard Scott, a leading figure in the technocracy movement. Norton was thrilled with the publicity and envisioned Cannon writing another book completely devoted to biocracy.[23] Although Cannon made it clear that he had no interest in such a venture, his AAAS presidential address on social homeostasis eight years later provides ample evidence of his continued interest in expanding the concept of homeostasis beyond mammalian physiology.

The success of social homeostasis pointed to another ambiguity in Cannon's thinking about self-regulation. Aside from its final chapter, *The Wisdom of the Body* was almost entirely a discussion of mammalian physiology within a medical context. From Cannon's progressive evolutionary perspective, frogs, other lower animals, and human societies had only the most rudimentary mechanisms for self-regulation. Nonetheless, even in his initial presentation of homeostasis in 1929, Cannon suggested that homeostasis might be a characteristic of any organized system. The MIT mathematician Norbert Wiener greatly expanded this possibility in his cybernetics—a science inspired, in part, by Cannon's homeostasis.[24] Wiener drew analogies between the automatic antiaircraft guns that he helped design during World War II and physiological control systems. Together with one of Cannon's colleagues, the Mexican physiologist Arturo Rosenblueth, Wiener studied parallels between pathological nervous disorders in humans and malfunctions that prevented automated guns from tracking their targets. In return, ideas of information and programs, negative and positive feedback, effectors and sensors became widely popular in biological discussions of self-regulation. The cybernetic terms, if not necessarily the underlying mathematics, encouraged a wide range of biologists to invoke homeostasis in systems at all levels in the biological hierarchy. Less propitiously, it also encouraged analogizing organic self-regulation to simple, single-purpose control mechanisms such as thermostats—thus perpetuating a mechanistic reductionism of the external environment to simple inputs or "disturbing influences."

The Wisdom of the Body was a model of scientific popularization. Reviewers described it as science that read like poetry.[25] Cannon captured the essence of his own research on the autonomic nervous system and adrenal hormones, placed it within the broader context of self-regulation, and suggested ways to generalize homeostasis. Despite its medical focus and mechanistic implications, the book also suggested a more outward-looking focus on adaptation and integrated responses of the organism as a whole to its external environment. Particularly for ecologists and evolutionary biologists, this held out the promise that homeostasis might broadly apply to populations, communities, and ecosystems both as interacting systems and as evolving entities. This possibility was enticing but ultimately controversial.

Adaptation and Homeostasis in a Darwinian World

Among biologists drawn to Cannon's discussion of homeostasis in *The Wisdom of the Body* was the population geneticist and evolutionary theorist Theodosius Dobzhansky. Although he did not explicitly discuss homeostasis in populations until the mid-1950s, Dobzhansky was already thinking broadly in terms of self-regulation twenty years earlier when he wrote his influential *Genetics and the Origin of Species* (1937). This book went through three editions and provided one of the important foundations for the developments in evolutionary theory often referred to as the modern synthesis.[26] Dobzhansky characterized the work as popularizing the new evolutionary theory in a form accessible to general biologists.[27] Widely read, Dobzhansky's book turned out to be deeply influential in consolidating the synthesis of Mendelian genetics with Darwinian natural selection and adaptation.

Genetics and the Origin of Species popularized the idea of the genotype as a complex, interacting system in dynamic equilibrium with the fluctuating external environment. The alternative view—what Dobzhansky's ally Ernst Mayr derisively referred to as "beanbag genetics"—emphasized the autonomy and independent action of individual genes.[28] Even before formally adopting the terminology of homeostasis, Dobzhansky presented both individual organisms and populations as self-regulating, physiological entities. What differentiated Dobzhansky's perspective from Cannon's formulation of homeostasis was the evolutionist's deep appreciation for diversity and his recognition that the external environments within which organisms existed were both highly complex and often unpredictable. For Dobzhansky, adaptation might mean a "harmony" between organism and environment, but it also implied a "precarious balance" or "ceaseless conflict" with constantly fluctuating external conditions.[29]

Dobzhansky introduced his discussion of adaptation by emphasizing diversity and change: "The multitude of the distinct 'kinds' or species of organisms is seemingly endless, and within a species no uniformity prevails."[30] According to Dobzhansky, there were two "methodologically distinct" ways to study this diversity. One could take an essentially historical approach to describe the diversity using the methods of morphology and systematics. Alternatively, one could approach the problem from a nomothetic (law-creating) perspective using the experimental methods of genetics and physiology. From this latter perspective, evolutionary biology was a search for the "causal mechanisms" of change and stability.[31] For Dobzhansky and many of his contemporaries, genetic interactions constituted physiology at its most fundamental level. Ideas of genetic self-regulation provided an explanation for the dynamic equilibrium between organism and environment. The interacting genome (or "germ plasm") became part and parcel of Dobzhansky's thinking about genetic systems, underlying both individual organisms and populations.

Dobzhansky's unusual background uniquely prepared him to apply both the historical and nomothetic approaches to studying adaptation and evolutionary change. Indeed, according to Garland Allen, no other biologist was more successful than Dobzhansky at bridging natural history and experimental traditions.[32] Allen claims that this unification was Dobzhansky's major contribution to evolutionary biology. Originally trained in the natural history practices of collecting and classifying butterflies and beetles as a student at the State University of Kiev, Dobzhansky continued to relish field research throughout his career. After immigrating to the United States in 1927, he mastered the new laboratory practices of cytology and experimental genetics of *Drosophila* at Columbia University, working with Thomas Hunt Morgan, the foremost geneticist of the early twentieth century. Dobzhansky's views on population genetics were later heavily influenced by theoretical considerations raised by the University of Chicago geneticist Sewall Wright, among other mathematical theoreticians. Wright was deeply interested in physiological genetics and systems thinking, both of which reinforced the idea of populations as interacting gene complexes in dynamic equilibrium with complex, shifting environments.[33] For Dobzhansky, this meant that neither individual organisms nor populations were "bundles of adaptations," but rather highly integrated entities capable of adaptively tracking environmental changes.[34]

According to Dobzhansky, adaptation, in the sense of a harmony between the organism and its environment, could be arrived at in two distinct ways. First, a genotype might be a narrow specialist producing a specific, well-adapted phenotype to the particular set of environmental conditions that it was most likely to encounter. Following an alternative adaptive strategy—one that Dobzhansky found more likely—a genotype could be an adaptable generalist with a broad "norm of reaction" capable of producing well-adapted phenotypes under a wide range of environmental conditions.[35] An ideal genotype would be capable of responding optimally to any environment that it encountered, although no organism had attained such a "paragon of adaptability." Nonetheless, the idea of an adaptable generalist implied that genotypes were capable of self-regulating developmental and physiological processes to meet many of the challenges posed by a fluctuating environment.

Although Dobzhansky characterized adaptation as a harmonious fit between organism and environment, he also described it as a series of conflicts. Because environments were in a constant state of flux—sometimes slow, other times catastrophic—adaptations were always temporary solutions to environmental challenges. Genotypes that were well adapted in the past might not be in the present or future. As a result, populations were always undergoing a process of genetic reorganization—or going extinct. Although committed to explaining the process of adaptation (and the resulting adaptive manifestations) in terms of natural selection acting on the raw material of genetic recombination and mutation, Dobzhansky admitted that important questions about this process remained unsettled. Indeed, throughout subsequent editions of *Genetics and the Origin of Species*, Dobzhansky struggled to strike a balance between what he viewed as two extreme positions on adaptation.

At one extreme, Dobzhansky pointed to the prominent population geneticist R. A. Fisher who argued that evolution was nothing but progressive adaptation driven by natural selection.[36] Opposing this extreme selectionist view, were the museum naturalists G. C. Robson and O. W. Richards, who had recently presented a skeptical critique of adaptation and natural selection in *Variation of Animals in Nature* (1936). Although not denying either adaptation or natural selection, the two British naturalists scrutinized both concepts and highlighted problems with applying them to explain variation found in natural populations. On the surface, Robson and Richards's critique might seem a minor contribution to a parochial dispute in British philosophy of biology, but it is historically important for at least two reasons.[37] First, although Dobzhansky was not entirely sympathetic to their views, he continued to refer to Robson and Richards's critique even though other prominent evolutionary biologists largely ignored it. Second, although they were not major contributors to the development of modern evolutionary theory, Robson and Richards raised important questions about adaptation, self-regulation, and natural selection that would continue to bedevil Dobzhansky and other evolutionary biologists.

Robson and Richards claimed that the common belief that all structures are adaptive was the result of "the human craving for a good story" that led evolutionary biologists to invoke natural selection even when evidence was contradictory or lacking.[38] Just as problematic was the "plasticity" of organisms, and the difficulty of unambiguously distinguishing between the effects of heredity and environment on the development of characteristics.[39] The two naturalists also noted that biologists often applied the term "adaptation" to highly specialized structures that might be useful in a particular environment but prove to be evolutionary dead ends if the environment changes. Organisms face endlessly changing environments, which are constantly deteriorating, at least in terms of present adaptations. Following a line of reasoning similar to Dobzhansky's, Robson and Richards claimed that an adaptable generalist might prove more successful than a narrow but well-adapted specialist, at least in the long run.

More fundamentally, Robson and Richards drew a philosophical distinction between the "statistical adaptation" championed by Fisher and "organismal adaptation," which they attributed to the Austrian biologist and later systems theorist Karl Ludwig von Bertalanffy and the British ecologist Charles Elton. They acknowledged Fisher's claim that most adaptations were compromises among conflicting selective pressures and that in an unstable and unfriendly world, organisms often had to "make the best of a bad job." Still, this selective tug-of-war did not mean that organisms were mere pawns in a hostile environment. Although not necessarily in opposition to natural selection, the idea of organismal adaptation focused attention upon adaptability and flexible self-regulation.[40]

Faced by hostile conditions, animals often migrated to a more favorable environment. Along with such behavioral responses to environmental change, Robson and Richards pointed out that animals also responded physiologically by modification, compensation, and self-regulation. Complete independence from the external environment was an unachievable ideal, yet all organisms had some self-regulatory powers. Birds and mammals exemplified this ability with thermoregulation analogous to central heating and air conditioning in modern buildings, but there were equally impressive ways that other organisms compensated or regulated body temperature in a fluctuating environment.[41] Insects living on inhospitable sand dunes burrowed into the soil during the heat of the day and surfaced only at night when temperatures cooled. Bees, termites, and other social insects maintained their colonies at optimum temperature through a combination of behavioral and physiological processes that could cool or heat depending on external conditions. Contrary to what they viewed as Fisher's extreme emphasis on natural selection, Robson and Richards argued that organisms were not pebbles sifted by "blind mechanical sieving," but rather active agents capable of adaptively responding to environmental challenges and modifying the conditions of existence.[42]

In her compelling study of scientific rhetoric, Leah Ceccarelli emphasized how subsequent editions of Dobzhansky's *Genetics and the Origin of Species* changed as the modern synthesis solidified.[43] The geneticist added new material, dropped older information, completely rewrote some sections, and reorganized chapters in subsequent editions. With all that change, it is surprising that in 1951 Dobzhansky continued to refer to Robson and Richards's critique and to juxtapose it with Fisher's strong selectionist position. The two contrasting positions served as bookends for the "balanced" theoretical explanation for adaptation that Dobzhansky was building during the fifteen-year span of *Genetics and the Origin of Species*. Without endorsing Robson and Richards's skepticism toward natural selection, Dobzhansky was sympathetic to their emphasis on integration, self-regulation, and adaptability. These ideas harmonized with his own growing sense that adaptation and homeostasis were characteristics not only of individual genotypes and phenotypes, but also of entire populations. By the mid-1950s Dobzhansky would draw strong parallels between Cannon's physiological homeostasis in organisms, the homeostatic norm of reaction of genotypes, and the genetic homeostasis of populations.

Even in 1937, Dobzhansky was outlining the self-regulatory mechanisms internal to populations that provided stability in a fluctuating environment. In every generation, genetic recombination associated with sexual reproduction produced an array of new genotypes. Each individual genotype had a norm of reaction, either specialized or generalized, to meet the challenges of the environment. The variation among genotypes also provided a kind of "evolutionary plasticity" that buffered the population against extreme fluctuations and provided some protection against extinction. Like Cannon's homeostasis, in which stability was more important than economy, evolutionary plasticity came at a high cost to poorly adapted individuals that inevitably arose due to the "scattering of variation" during sexual reproduction.[44] Populations were on an endless treadmill tracking environmental change. From this evolutionary perspective, life was perpetually out of balance—or, at least, in danger of becoming so. At the same time—and more optimistically—both organisms and populations were capable of maintaining at a least a tenuous balance with a fluctuating environment through the internal self-regulating mechanisms constituting the norm of reaction.

The idea of a genetic equilibrium could refer to the formal mathematical treatment of the subject exemplified by the so-called Hardy-Weinberg equation that was usually limited to simple cases of alternative alleles for a single gene. However, Dobzhansky was also interested in a more expansive, though less quantitative, notion of equilibrium involving the entire "germ plasm." He was committed to the general idea that genetic variation was a mechanism for stabilization and change in both individual organisms and populations. Genetic variation in alleles at different loci on a chromosome broadened the norms of reaction for individual genotypes, buffered populations against environmental changes, and provided the raw material for both adaptation and speciation. Referring to the famous pictorial representation of an "adaptive landscape" created by Sewall Wright, Dobzhansky wrote, "The symbolic picture of a rugged field of gene combinations strewn with peaks and valleys helps to visualize the fact that the genotype of each species represents at least a tolerably harmonious system of genes and chromosome structures."[45]

His thinking borrowed heavily from Wright's mathematical theory but also rested on Dobzhansky's broader appreciation for the complexities of natural environments.[46] The adaptive landscape was a metaphor primarily aimed at presenting Wright's mathematical theory to a nonmathematical audience, but it also encouraged thinking in ecological, as well as genetic, terms.[47] For Wright, the peaks and valleys on the landscape represented differences in fitness of various genotypes, but Dobzhansky sometimes suggested that the peaks and valleys also represented various ecological niches encountered by a species. In this broader sense, the landscape encouraged unifying Darwinian adaptation with physiological adaptation. Moving from an adaptive peak to a higher one meant traversing an adaptive valley of lowered fitness, an impossibility for natural selection. This transition ultimately might require a random event or revolutionary reorganization of the genotype, but Dobzhansky also suggested that highly adaptable genoytpes might tolerate suboptimal conditions indefinitely while "exploring" unfilled niches of the adaptive landscape. From a physiological perspective, broad phenotypic plasticity and the ability to acclimate to environmental variation could preadapt organisms and populations to new ecological niches.

Preadaptation, Adaptation, and Physiological Limits of Tolerance

Preadaptation was an idea that blurred the distinction between proximate and ultimate causation and the boundaries dividing functional and evolutionary biology. Acclimation, or reversible phenotypic changes, could not be inherited, yet these physiological responses might allow organisms and populations to exist indefinitely in marginal habitats or to expand their ranges to previously uninhabited areas. If a mutation mimicked such a physiological change, the preadaptation might become an adaptation in the Darwinian sense. Unifying these functional and evolutionary responses was of particular interest to C. Ladd Prosser, who forged a new approach to comparative physiology emphasizing ecology and the modern synthesis of Mendelian genetics and Darwinian evolutionary theory.

Colleagues remembered Prosser for his deep curiosity and the breadth of his physiological perspective.[48] As a graduate student at Johns Hopkins University, he studied amoeboid locomotion but in 1932 he completed his dissertation on the nervous system of the earthworm. He spent a year on a postdoctoral fellowship at Harvard studying the neurophysiology and behavior of crayfish, as well as interacting with leading physiologists at the university, including Cannon. Moving to England in 1934, he spent time at both Cambridge and Oxford. At Cambridge he continued his research on the nervous system of earthworms in the laboratory of Edgar Adrian, who had recently been awarded a Nobel Prize for his work on the function of neurons. Prosser also conducted research on the sympathetic nervous system of cats at Oxford with the future Nobel laureate John Eccles.

Returning to the United States, Prosser briefly joined the faculty at Clark University before moving to the Zoology Department at the University of Illinois at Champaign-Urbana in 1939. He spent summers at the Marine Biological Laboratory in Woods Hole, Massachusetts, conducting research that included work on neural transmission using the giant axon of the squid with its discoverer John Z. Young. Although he began laying the groundwork for his monumental *Comparative Animal Physiology* during the late 1930s, Prosser interrupted his physiological research to work on the Manhattan Project during World War II and briefly on radiation biology projects after the war ended.[49] *Comparative Animal Physiology*, finally published in 1950, was an important constituent of Prosser's goal of building a broadly comparative physiology based on ecological and evolutionary principles.

In the introduction to *Comparative Animal Physiology*, Prosser wrote, "Foremost among general principles which emerge from a study of comparative physiology is the functional adaptation of organisms to their environment. The distribution of a species is determined through natural selection by its limits of tolerance."[50] Combining the ideas of "functional adaptation," "limits of tolerance," and natural selection provided a conceptual framework for thinking physiologically about how organisms confronted the challenges of a heterogeneous landscape of different ecological niches, even though the genetic bases for most complex physiological processes were unknown. The physiological flexibility exhibited by all species preadapted organisms to explore and invade previously uninhabited spaces, thus potentially extending the species' geographical range and setting the stage for speciation. This emphasis on flexibility harmonized with Prosser's broad perspective on homeostasis and adaptation. Because much of his research was conducted on invertebrates, Prosser identified homeostasis not only with the internal constancy emphasized by Cannon but also with the ability of organisms to conform to environmental changes in a precise and predictable way. Conforming was often an energetically inexpensive way to maintain a dynamic equilibrium with a fluctuating environment, especially because many conformers were resilient and could quickly recover from short-term stress. Longer-term acclimatory adjustments were also homeostatic. For example, seasonal changes in temperature or other environmental variables often resulted in an initial conformational change in physiology, which gradually shifted back toward the original state. From Prosser's broader perspective, this type of adjustment or compensation, so typical of acclimation, was another manifestation of homeostasis.[51]

Broadly conceived, homeostasis played an important role in Prosser's evolutionary thinking about adaptation and speciation. Physiological modification allowed individuals to explore new habitats and push species' limits of tolerance. Under the right conditions, these preadaptations set the stage for both adaptation and speciation. Self-consciously paraphrasing Dobzhansky, Prosser characterized evolution as "the creative response of organisms to environmental opportunity."[52] He explicitly denied any neo-Lamarckian implications of inheritance of acquired traits in his claims about acclimation, and he noted the importance of distinguishing between genetic and environmental influences on the phenotype. Nevertheless, he believed that acclimation played an important, though indirect, role in adaptation and speciation. He was encouraged that geneticists and various other evolutionary biologists had proposed somewhat similar ideas of phenocopies and genetic assimilation.[53]

Prosser pointed out that the geneticist C. H. Waddington had documented cases of "anticipatory adaptations" in which phenotypic responses to environmental stresses sometimes were replaced by mutations in genes affecting the characteristic. One of Waddington's favorite examples was the calluses on the skin of ostriches where the body contacted the ground while sitting. These thickenings of the skin were protective, but unlike calluses that develop on human hands in response to abrasion, the callosities of ostriches developed when the embryonic bird was still in the egg. This was a case of what Waddington referred to as the "genetic assimilation" of an acquired trait. Although controversial among evolutionary biologists, Waddington's ideas reinforced

Prosser's ideas about the transition from preadaptation to Darwinian adaptation. For Prosser, this also seemed to provide a foundation for unifying the disciplines of physiology and genetics.

In his optimism Prosser also thought that this common ground held the promise for integrating evolutionary biology—with a prominent role for the new comparative physiology that he was promoting. More than most physiologists, Prosser actively engaged other evolutionary biologists, particularly during symposia and conferences organized around the Darwin centennial in 1957.[54] Prosser was confident that natural selection and mutation explained the functional adaptation that so interested comparative physiologists. For their part, Dobzhansky, Waddington, and other geneticists turned to homeostasis as a way to discuss self-regulation in both organisms and populations. Yet, despite this shared enthusiasm for integration (a popular goal among postwar biologists), much of my book will be devoted to analyzing the difficulties of creating a truly integrated and synthetic evolutionary biology during the decades following World War II.

Proximate and Ultimate Causation

Perhaps biologists might have avoided some of the difficulties if they had carefully distinguished between the different types of questions asked by physiologists and evolutionary biologists. Ernst Mayr made precisely this argument in a highly influential philosophical article, "Cause and Effect in Biology," published in *Science* in 1961.[55] He distinguished between "functional biologists" who asked "how" questions related to proximate causation and "evolutionary biologists" who asked historical or "why" questions about ultimate causation. According to Mayr, explaining why the warbler at his vacation home in New Hampshire migrated south on August 25 could be divided into four causal parts. A physiologist would focus on *intrinsic proximate causes* involving the internal state of the bird and levels of various hormones leading to migratory behavior or *extrinsic proximate causes* such as the temperature drop that actually triggered the departure on that particular day. By contrast, an evolutionary biologist would consider *ultimate causes* involving the genetic program underlying the bird's physiology (*genetic cause*) and the evolutionary history of insect eating in the species that required migration for survival during the winter (*ecological cause*).

A complete explanation of the migration of warblers required the integration of all four causes. Mayr referred to "integration" five times in the short article and he emphasized the importance of having a basic knowledge and appreciation for other fields. Nonetheless, he more strongly emphasized the importance of not confusing the types of causation—a confusion that often led to unnecessary and unproductive controversies. From Mayr's perspective, the growth of biological thought required a kind of division of labor along the functional—evolutionary divide. Despite points of contact and overlap, functional biology and evolutionary biology were "two largely separate fields which differ greatly in methods, *Fragestellung,* and basic concepts."[56] According to Mayr, most biologists spent their careers doing one or the other.

Ironically, Mayr's dichotomy between functional biology and evolutionary biology seemed to divide and exclude, more than to integrate biology. Several historians have convincingly argued that Mayr wrote "Cause and Effect in Biology" primarily as a defense of organismal biology against the perceived threats of an aggressive molecular biology that was claiming intellectual authority and monopolizing institutional and financial resources.[57] Separating functional biology (and functional biologists) from evolutionary biology provided an argument against reductionism and at the same time placed the type of organismal biology done by naturalists on an equal footing with experimental, laboratory sciences. However, by emphasizing the existence of two largely separate fields within which individual scientists tended to work, Mayr diverted attention from the important areas of overlap, even though acknowledging that these areas existed. Indeed, one might argue that Mayr's own research demonstrated how the most interesting questions in biology often lay at the intersections of fields and specialties. He himself explored the cross-disciplinary relations between functional and evolutionary biology by incorporating homeostasis into his discussion of adaptation and speciation. Like Dobzhansky, Mayr argued that homeostasis applied to populations as well as to organisms. As a conservative, stabilizing process, homeostasis held populations together in a tug-of-war with diversification and adaptation to local environments. To be successful, the evolution of new species was a revolutionary event requiring a rapid reintegration of species-specific homeostatic mechanisms. From this vantage point, the distinction between proximate and ultimate causation was blurred, and Mayr moved freely between functional and evolutionary uses of homeostasis. Although alluding to this broad use of homeostasis in "Cause and Effect in Biology," Mayr placed more emphasis on distinguishing between functional biology and evolutionary biology, rather than on integrating the two. That would make sense within the context of defending Mayr's approach to organismal biology against reductionism and freeing philosophy of biology from physics. Although "Cause and Effect in Biology" became the exemplar of a new philosophy of biology, it exuded a certain defensiveness and divisiveness seemingly at odds with the goal of integrating biology.

Other Approaches to Integrating Biology

Quite a different perspective on integrating biology was presented by George A. Bartholomew, a physiological ecologist at the University of California at Los Angeles. A decade and a half younger than Mayr and other prominent figures in the modern synthesis, Bartholomew's education encouraged his broad interdisciplinary approach to studying adaptation. His early training and research centered in natural history museums and field studies of avian behavior.[58] During World War II, he worked in a physics laboratory for the Naval Bureau of Ordnance. This experience introduced him to electronic instrumentation, which he later put to use developing miniature thermocouples for measuring internal body temperatures in small mammals and large insects during rest and activity. Although conducting much of his physiological ecology in the laboratory, Bartholomew maintained lifelong interests in the behavior, ecology, and natural history of the species that he studied. He considered himself an "experimental naturalist."[59]

In a review article published three years after Mayr's "Cause and Effect in Biology," Bartholomew outlined the challenges and opportunities of using an integrated approach to studying the adaptations of animals living in hot, dry environments.[60] The general topic of desert adaptation had been covered much more comprehensively in a recently published book by Knut Schmidt-Nielsen, with whom Bartholomew had a friendly correspondence. Acknowledging Schmidt-Nielsen's book provided a justification for Bartholomew to focus almost exclusively on studies that he and his students were conducting on desert adaptations of a wide array of birds, mammals, and reptiles primarily in the Southern California desert. Bartholomew also used the review article to express a broad philosophical perspective for integrating biology around the physiological ecology that he was pioneering. Although he never mentioned Mayr, Bartholomew's philosophical position challenged the older evolutionary biologist's claims about boundaries between functional and evolutionary areas of biology, as well as the threat of philosophical reductionism in biological explanation. Organismal biology was just as important to Bartholomew as it was to Mayr, but physiological ecology provided an interdisciplinary approach to integrating biology that did not rest on sharp dichotomies between fields, levels of organization, or types of causation.

According to Bartholomew, the growth of science had resulted in "extreme specialization" and "tunnel vision," which hindered communication and led scientists to ignore important developments outside of their narrow specialties. This was particularly a problem for a newly emerging physiological ecology, because it was broadly interdisciplinary, combining insights from comparative physiology, ecology, evolutionary biology, and animal behavior. Understanding how and where animals lived required a deep understanding of natural environments both physical and biological, physiological and behavioral responses to environmental variables, competition and other interspecific interactions, as well as the evolutionary history of the species being studied. In most cases animals were capable of living under a broader range of conditions than where they were actually found, so the observed geographical distribution of a species was not usually set by the physiological limits of tolerance studied by traditional comparative physiologists. In many cases the observed distribution of a species was also limited by biotic interactions such as interspecific competition or predation. Furthermore, individuals were usually not uniformly distributed within the geographical range but actively selected the most suitable microhabitats. Thus, comparative physiology needed to be informed by a deep understanding of the natural history of a species, including adaptive behaviors, the multiplicity and variability of ecological microhabitats, and phylogenetic constraints on adaptive strategies.

The problem of tunnel vision in modern biology also involved a preoccupation with phenomena at a particular level of organization, rather than considering connections among levels from molecules to ecosystems. Specialists, whether biochemists or ecologists, tended to view their own work as fundamental, and information from other areas as peripheral or irrelevant. Although this problem of disciplinary chauvinism was widely recognized, Bartholomew claimed that it was "emotionally" difficult for specialists to apply the obvious solution: "This is the idea that there are a number of levels of biological integration and that each level offers unique problems and insights, and further, that each level finds its explanations of mechanism in the levels below, and its significance in the levels above."[61] By the very nature of their work, physiological ecologists were forced to focus more broadly than on a single level or even a few adjacent levels of organization, to consider multiple levels "continuously and usually simultaneously." For Bartholomew, this was "the elementary philosophical idea" that grounded a unified study of organisms and environments integrating investigations at all levels from molecules to ecosystems.

Mayr had also emphasized the unity of biological explanation, but Bartholomew's insistence on the integration of multiple perspectives in the day-to-day practice of physiological ecology was in marked contrast to the sharp dichotomy that the prominent evolutionary biologist drew between evolutionary biology and functional biology. Mayr claimed that despite areas of contact and overlap, the dichotomy applied to both disciplines and the work of individual scientists. Scientific progress for Mayr depended upon a division of labor among specialists, but for Bartholomew progress meant individual scientists combining multiple disciplinary perspectives and levels of organization all of the time. Difficult as this synthesis might be, Bartholomew claimed that this broad approach was possible, and indeed necessary, for a physiological ecologist to explain adaptation in all its complexity. Also absent from Bartholomew's philosophical perspective on organismal biology was the defensiveness that Mayr evinced toward the reductionism of molecular biology and physics. Instead, Bartholomew emphasized the necessity of integrating levels of organization. In this sense integration provided a metaphor for approaches that focused on the organism but at the same time encouraged moving up and down biological levels from molecules and cells to populations and ecosystems.

Because physiological ecology focused on the whole organism in its environment as an inseparable system, integration also meant that adaptation always involved multiple organs and physiological processes working together. In this sense integration resonated with older ideas of self-regulation. But for Bartholomew, homeostasis was a very broad concept that involved physiological flexibility as much as internal constancy. For example, many of the small mammals he and his students studied allowed their body temperature to fluctuate widely in order to conserve energy. Conversely, some cold-blooded insects used a combination of behavior and physiology to maintain the elevated body temperature required for flight and other energetic activities. These examples of "heterothermy" contrasted with the constant temperature regulation that Cannon had equated to a thermostat, but from Bartholomew's perspective flexible body temperature was equally homeostatic in adapting an organism to its fluctuating temperature environment. Maintaining homeostasis also depended upon the organism's ability to select among environmental options. Environments were never uniform but provided an "almost infinite series" of microhabitats from which highly adaptable organisms could choose in an "intricate and precise way."[62] For Bartholomew, homeostasis, both physiological and behavioral, was the key concept for understanding the integration of organism and environment as an inseparable and integrated system in dynamic—though rarely constant—equilibrium. "It is clear," he claimed, "that the organism exists as a dynamic equilibrium and thus, as long as it is alive, it is the example par excellence of the phenomenon of homeostasis."

Bartholomew's broad interest in integrating levels of biological organization combining natural history and laboratory experimentation and explaining adaptation using an expansive concept of homeostasis was widely shared by his students as well as other biologists. Bartholomew's UCLA colleague Theodore Bullock was a much more traditional laboratory physiologist who gained considerable recognition for his many contributions to neurophysiology, including chemical transmission across synapses, the structure and function of infrared sense organs in pit vipers, electroreceptors in fish, and computational studies of neurotransmission.[63] Yet he maintained a deep interest in natural history, working on a wide variety of vertebrate and invertebrate taxa as he became a leading authority on the evolution of the brain. Although much of his work involved integration—both physiological and methodological—at levels ranging downward from the organism to the cell, Bullock stressed the importance of focusing on the whole organism interacting with its environment. This observational, natural historical approach was a rich source of interesting problems and questions to be tackled in the laboratory, but it also played an integral role in the interpretation and explanation of results from even the most reductionist neurophysiological experiments.[64] Similarly, understanding the organism in its environment—homeostasis in a broad sense—was a necessary part of integrating the perspectives from neighboring disciplines. One could artificially divide the study of adaptation and self-regulation into distinct genetic, ecological, behavioral, and physiological categories much in the way that Mayr had suggested in "Cause and Effect in Biology," but from Bullock's broad physiological perspective the most interesting (and challenging) questions emerged at the interstices among these fields.[65] Scientific progress was not to be made through an incremental step-by-step process within fields or levels of organization but by a "simultaneous attack" at all levels. This observation echoed Bartholomew's claim that physiological ecologists needed to consider multiple levels of organization both continuously and simultaneously.

Integrating disciplines and levels of organization was a major preoccupation for biologists after World War II. Both Bartholomew and Mayr used the term "integration" repeatedly in their two articles discussed here. In her perceptive analysis of postwar organismal biology, Erika Milam emphasized that despite differences, Dobzhansky, Mayr, George Gaylord Simpson, and other contributors to the modern synthesis broadly agreed on the distinction between proximate and ultimate causation as a way to integrate biological levels of organization while at the same time defending organismal biology from the threat of reductionism posed by molecular biology and physics.[66] Equating "ultimate" causation with evolutionary explanations was a not-so-subtle way of privileging the modern synthesis that Mayr and Dobzhansky were so involved with promoting. I argue that Bartholomew was involved with creating a somewhat different synthesis of disciplines and explanations. Although he later recalled reading the canonical texts of the modern synthesis as a graduate student, Bartholomew rarely cited this literature in his own research.[67] Evolutionary theory, broadly construed, was an important component of the study of adaptation that Bartholomew promoted, but genetic mechanisms were not a central part of the explanatory framework that he used even when discussing populations. Adopting Mayr's sharp dichotomies between proximate versus ultimate causation and functional biology versus evolutionary biology was never a critical issue and, indeed, ran counter to Bartholomew's strong interdisciplinary motivations. Bartholomew and his students played an important part in transforming the American Society of Zoologists into the Society for Integrative and Comparative Biology. Although not a complete break with the past, the change in title reflected a decisive shift in emphasis. Bartholomew was lionized for his role in setting the philosophical agenda for a new society based both on the integration of biological specialties and on the organism as the focus of integration for levels of biological organization.[68]

Despite the differences highlighted here, Mayr and Bartholomew shared a deep commitment to using homeostasis as a basis for understanding adaptation. From Bartholomew's philosophical point of view, homeostasis was coextensive with the life of an organism in its environment. Mayr routinely turned to homeostasis, self-regulation, and cybernetic feedback to characterize integration in organisms and populations, and by extension all levels of biological organization. This was part of his argument in "Cause and Effect in Biology," and he greatly expanded his ideas a few years later in *Animal Species and Evolution* (1963). Mayr's discussion of complexity and organization in his magisterial *The Growth of Biological Thought* (1982) leaves little doubt that even at the end of his career, he viewed feedback, cybernetics, and systems thinking as bedrock principles for philosophy of biology.[69] Mayr's use of homeostasis and systems thinking was shared by many biologists who were attracted to cybernetics without necessarily delving deeply into the more technical and mathematical aspects of the new discipline. This informal use of systems thinking provided a way to link levels of biological complexity from cells to ecosystems. It suggested ways to discuss goal-directedness without recourse to discredited ideas of teleology. It also provided a way to discuss highly integrated biological systems that avoided vitalism but also had a more sophisticated perspective than older forms of mechanism. Describing populations, communities, or ecosystems in cybernetic terms of negative feedback control highlighted the similarities with the physiology of organisms without falling back on simplistic analogies that plagued earlier organicist approaches. Thus, the attenuated influence of Cannon's original idea of homeostasis encouraged applying self-regulation broadly in areas of biology far removed from medical physiology, but it also raised thorny problems of combining functional and evolutionary explanations.

2

Bodily Wisdom or Stupidity?

Cannon's *Wisdom of the Body* presented biologists with a flexible concept of self-regulation used in an array of biological disciplines and levels of biological organization. Despite the broad appeal of homeostasis, the concept was also highly problematic, and from the publication of Cannon's book to the present, it has met intense scrutiny. Critics have complained that Cannon overemphasized optimality and normality. This, critics charged, led Cannon to a too-optimistic perspective both in physiology and society. Despite Cannon's claims for a holistic concept of self-regulation involving integrated responses by various organs, critics often dismissed homeostasis as a mechanistic and purely reactionary response to environmental disturbance. Indeed, the complexity of the external environment and the ability of organisms to respond proactively were underdeveloped elements of homeostasis that a succession of critics have been quick to point out. For some, this meant abandoning homeostasis, replacing it with alternative concepts or elaborating it in ways that Cannon had not originally envisioned.

Self-Regulation and Chronic Disease

In his 1944 Harvey Lecture on traumatic shock, the future Nobel laureate Dickinson Woodruff Richards recounted the case of a young man admitted to the emergency room at Bellevue Hospital bleeding profusely from a severe scalp wound.[1] Despite the injury and the fact that he was inebriated, the man was conscious and had a steady pulse. Concerned about the bleeding, an attendant had the man rise to a sitting position, after which the patient became pale, lost consciousness, and suddenly died. That such a seemingly inconsequential change in position could bring about a fatal circulatory collapse highlighted Richards's point that despite recent advances in understanding traumatic shock, the condition remained unpredictable. It also reflected Richards's broader concerns about the tenuousness of physiological self-regulation. Although acknowledging the important influence of Cannon's early studies of traumatic shock conducted during World War I, Richards took a jaundiced view toward the Harvard physiologist's later ideas about homeostasis. The fact that acute alcoholism likely contributed to the fatal event in the Bellevue patient was simply one of several reasons to doubt the wisdom of the body. As Richards quipped a decade later, any good pathologist could write a book on the stupidity of the body.[2] Indeed, Richards claimed that the body often was not just stupid but "egregiously, calamitously stupid."

Richards later recalled that his misgivings about Cannon's ideas arose during his government service with the National Research Council during the Korean War, although the evidence from his Harvey Lecture suggests that the roots of his critical attitude had developed almost a decade earlier.[3] What Richards took as Cannon's cheery attitude toward social harmony seemed at odds with unsettled wartime political conditions, and it conflicted with Richards's conservative nature. In the realm of physiology, Richards believed that Cannon had placed far too little emphasis on disease, particularly the type of chronic, debilitating diseases that were a major cause of death and that had been so central to Richards's own medical experience. For Richards, the reality of this physiological degeneration seemed to contradict homeostatic balance. Just as importantly, Richards's earlier studies of traumatic shock had convinced him of the limitations of physiological self-regulation in many cases of imminent danger. The simple mechanisms that Cannon presented in *The Wisdom of the Body* were at best incomplete explanations for phenomena that affected the entire organism. From Richards's critical perspective, Cannon's "undiluted optimism" was the attenuated expression of an outdated positivism and reflected an uncritical acceptance of a simple, mechanical teleology.

Richards gained fame as one of three Nobel laureates who shared the 1956 prize in physiology or medicine. Together with his younger colleague André Cournand and the Austrian physician Werner Forssmann, the Nobel Foundation honored Richards for research leading to successful cardiac catheterization. Applying this technique revolutionized cardiopulmonary physiology, because for the first time scientists could directly measure pressure, flow, and blood gases in the heart itself. The device also provided a means for performing delicate surgical procedures on the heart without opening the chest cavity. Richards and his colleagues applied the technique to the study of traumatic shock as well as emphysema and other chronic degenerative diseases—areas of interest that Richards and Cournand had worked in for a decade before perfecting the use of catheters.

As a cardiopulmonary physiologist, Richards acknowledged intellectual debts to a lineage of researchers who had perfected techniques for measuring cardiopulmonary function during the late nineteenth and early twentieth centuries.[4] Most importantly, he credited Lawrence J. Henderson for encouraging his earliest research and for elucidating the "essential concept" that the heart, lungs, vessels, and blood formed a coordinated cardiopulmonary system for exchanging gases between the atmosphere and tissues.[5] Despite his deeply acknowledged intellectual debt to Henderson, Richards was skeptical about applying organic analogies to society, as both Henderson and Cannon had done. Behind the elegant mechanisms for gas exchange and transport that Richards helped discover lurked the specters of physiological dysfunction and chronic disease, which preoccupied him throughout his career. From his conservative perspective, analogies to physiological self-regulation provided little useful guidance for understanding or improving social relations. Richards was also a classicist and historian of medicine whose medical idols were Hippocrates and Harvey.[6] Although acknowledging the importance of balance in Hippocratic medicine, Richards was more deeply impressed by the tragically precarious nature of human existence.

Richards was born in Orange, New Jersey, in 1895.[7] His maternal grandfather was a physician, as were three of his uncles. All of these forebears trained or worked at Columbia University and Bellevue Hospital, where Richards spent most of his professional career. He studied humanities at Yale but decided on a career in medicine after serving in the army during World War I. Entering the Columbia School of Medicine, he earned both a medical degree and a master's degree in physiology in 1923. Although he never worked with Henderson, the two men corresponded during this period, and the Harvard physiologist both encouraged and guided Richards's early research on oxyhemoglobin dissociation curves. Richards continued research on anemia and its effects on the circulation of the blood during his internship and residency, before spending a year (1927–1928) in London working in the laboratory of Sir Henry Dale, whom Richards also credited with guiding his career in research. Unlike Henderson, whose critical attitude Richards compared to Socrates, Dale was a skilled experimentalist who impressed Richards with his ability to take an idea and develop it into a fruitful line of research. Richards's work with Dale centered on the study of vasodilation in response to histamine but also led to the discovery of the biological activity of acetylcholine, later recognized as an important neurotransmitter.

After joining the Department of Medicine at Columbia University, Richards pursued a distinguished career in clinical research centered on cardiopulmonary physiology and pathophysiology. His early clinical research focused on oxygen therapy for a variety of chronic diseases such as pulmonary tuberculosis, pulmonary fibrosis (silicosis), and emphysema. Administering oxygen through nose catheters or keeping patients in oxygen tents or rooms with elevated oxygen levels often relieved shortness of breath and other symptoms, but the outcomes were unpredictable even for the same disease. Some patients recovered, but others died. Furthermore, providing oxygen had consequences beyond reversing hypoxia, suggesting that maintaining a balance of blood gases was not a simple mechanical process.[8] For example, in some cases carbon dioxide levels in the blood increased in response to oxygen because the breathing rate decreased. Prolonged oxygen therapy carried its own risks, and aside from a few healthy volunteers, most of the subjects of Richards's studies were gravely ill patients whose cardiopulmonary diseases had not improved with other treatments.

Richards's interest in oxygen therapy went hand in hand with his more basic interest in understanding the ventilation of the lungs and the dynamics of blood flow in the pulmonary circulation. Following a lineage of research dating back to the late nineteenth century, Richards attempted to refine methods of measuring blood flow based on the Fick principle developed by the German physiologist Adolph Eugen Fick. Comparing dissolved oxygen in blood leaving the heart and in the mixed venous blood returning to the heart, the Fick equation provided an idealized model for oxygen transport. Although the equation had been widely accepted since the 1870s as a method for determining cardiac output by comparing dissolved gases in venous and arterial blood, sampling these gases posed daunting challenges, particularly in human subjects. One could safely sample arterial and venous blood, but most physiologists and physicians considered it too dangerous to try to sample mixed venous blood from the right atrium. By 1930 Richards was aware of Werner Forssmann's successful, but highly controversial, attempt to insert a catheter into his own heart in 1929. Richards recognized that this new technique held promise for directly sampling blood in the right heart, yet he referred to it as "a somewhat formidable procedure."[9] Led by André Cournand, a talented experimentalist who had recently emigrated from France to join

Richards's lab, animal experimentation on right heart catheterization proceeded slowly during the 1930s. During this period, Richards and Cournand continued indirect methods of determining venous and arterial gas concentrations using rebreathing devices developed earlier by János Plesch, John Scott Haldane, Gordon Douglas, and Yandell Henderson. Although widely used, these indirect techniques assumed a steady state between inspired oxygen and exhaled carbon dioxide. Patients suffering from cardiopulmonary diseases seldom met this condition, and the assumption often did not hold even in healthy individuals.[10]

Cardiac catheterization proved a revolutionary method for obtaining precise data for venous blood directly from the heart. Although at first there was considerable resistance to using the procedure on humans, Richards and Cournand safely implanted catheters in the right atrium and other locations in the heart for several hours without adverse effects to patients.[11] With these successes the procedure became one of the most common methods for measuring cardiac function.

Richards and Cournand perfected cardiac catheterization just as they embarked on a large study of the causes and consequences of traumatic shock. World War II focused renewed medical interest in this pathological condition, and the emergency service at Bellevue Hospital provided a varied and abundant group of accident victims suffering from skeletal trauma, chest injuries, hemorrhage, abdominal injuries, burns, and exposure. Complicating matters, many of the patients also suffered from chronic alcoholism. Members of the team wryly noted that inebriated patients were not the most cooperative human subjects, but the study did provide an opportunity to gather important information on vasodilation caused by alcohol and its destabilizing consequences.[12] The unpredictable outcomes in these cases also provided Richards with an argument for the inadequacy of simple "mechanical" explanations for shock based on loss of blood. Blood loss certainly was the major contributing factor to most cases of shock, but Richards emphasized that it could not fully explain the long and complicated degenerative consequences of trauma.

The Bellevue team studied sixty-one shock victims and thirty injured patients not exhibiting symptoms of shock during a two-year period from the beginning of 1940 through the end of 1941.[13] Richards's 1944 Harvey Lecture provided an opportunity to discuss this research within a broader historical context.[14] He acknowledged the importance of Cannon's groundbreaking *Traumatic Shock* published two decades earlier, much of which remained valid. Like Cannon's treatise, which was largely the result of medical experiences during World War I, Richards's work was heavily indebted to the ongoing war efforts during the early 1940s supported by the Office of Scientific Research and Development. He was also able to highlight the uses of cardiac catheterization for making precise physiological measurements. Although members of his lab had published a few articles, Richards claimed that the technique was still largely unknown outside of a small group of specialists. From Richards's perspective, the rigorously quantitative data gathered from catheterization experiments put the understanding of acute shock on a firm ground of "fact rather than of hypothesis."[15]

Research by the Bellevue physiologists supported the general causes of shock and the body's responses as outlined by Cannon and later physiologists but suggested a number of new therapeutic procedures that Richards and his colleagues had more recently implemented. Most notably, Richards argued the advantages of using whole blood transfusions instead of plasma in acute cases of shock, a claim that remained open to disagreement during World War II.[16] More broadly, he emphasized the need for a deeper understanding of the chronic effects of shock. Despite the acknowledged debt to his predecessor, Richards expressed dissatisfaction with Cannon's "mechanistic" approach and suggested that a more complete explanation focusing on long-term effects on the whole body was needed.

Cannon had distinguished between primary shock accompanying the initial stunning effect of traumatic injury and secondary shock involving a more protracted condition marked by weakness, pallor, thirst, weak pulse, and progressively decreasing blood pressure.[17] Cannon's claim that loss of blood volume was the fundamental cause of shock remained central to the explanation of the condition. Experiments on animals supported this idea, but Richards noted that until recently there were no precise quantitative studies of venous return or right atrial pressure on humans. In addition, the exact cause of reduced blood volume remained open to question. Some investigators argued that it was entirely due to blood loss from the site of injury, while others argued that blood also leaked from damaged capillaries into the tissues throughout the body. These uncertainties provided the context for a more refined study of traumatic shock using cardiac catheterization: "If new ground were to be broken in developing further the knowledge of shock, particularly clinical shock in human cases of injury, it was clear that new techniques, new methods of study, were needed."[18]

Patients were sedated and a catheter was inserted through the femoral artery to the right side of the heart. The catheter was left in place, and measurements of circulatory function were made throughout treatment, which often lasted from four to eight hours. Despite some differences among various types of shock, reduced venous return appeared to be the primary cause of secondary shock in all cases. Blood volume was often reduced, right atrial pressure dropped, and cardiac output declined. Peripheral resistance to blood flow remained normal, suggesting a compensatory constriction of capillary beds in response to reduced blood volume. Although damage to capillaries might contribute to shock, Richards ruled it out as a primary cause—more likely it was an effect. Similarly, all of the evidence seemed to support blood loss at the source of injury rather than general leakage from capillaries as the major cause of reduced blood volume in victims of skeletal trauma and hemorrhage. Quick replacement of blood was critical to reversing shock, and although transfusions of plasma, saline, albumin, gelatin, or other fluids also rapidly restored blood pressure, the Bellevue team documented the long-term consequences of various treatments. Because none of these alternatives addressed the underlying problem of oxygen transport, the initial restoration of blood pressure provided a false sense of security when patients' conditions often later deteriorated. Even with whole blood transfusions, many patients remained in a precarious state.

Although the data collected by Richards's team supported what he referred to as Cannon's "mechanical concept of shock," Richards cautioned that this was not a complete explanation. In particular, he focused attention on events in the vascular beds. Vasoconstriction to counter the decreased blood volume and maintain blood pressure might be an adaptive, compensatory response, but it had potentially adverse consequences to local tissues. The Bellevue study found a major reduction in blood flow to the kidneys possibly leading to tissue damage that delayed recovery and caused chronic health problems. Deprivation of oxygen, as well as the accumulation of toxic by-products, led to tissue damage. Thus, both clinicians and physiologists needed to think of shock as more than a "mechanical" problem of loss of blood volume solvable by quick replacement therapy. Instead, Richards concluded, "shock must also be considered as the first stage of a profound bodily disturbance whose consequences continue and may progress for a long time."[19]

Hyperexis: The Vaulting Ambition of the Body

If traumatic shock represented a "rude unhinging of the machinery of life," it was but an extreme example of what Richards considered the body's tendency to respond ineffectively or inappropriately to threats. These pathological responses were evidence not of the wisdom of the body, but rather of the body's "vaulting ambition" that overleaped short-term goals of restoring balance in ways that undermined long-term health. Together with both the immediate and long-term effects of shock, Richards's lifelong interest in chronic and degenerative diseases such as emphysema provided a general framework for criticizing Cannon's idea of homeostasis.[20]

Richards presented his most direct and extended critiques of homeostasis in two lectures delivered five years apart to the Practitioners Society in New York, a small group of prominent physicians who met monthly in private homes for dinner and discussion of medical topics. These addresses were later published in the *Scientific Monthly* (1953) and *Perspectives on Biology and Medicine* (1960). He continued to develop his critique during the following decade as he conducted historical studies on Hippocratic medicine and a reconsideration of Lawrence Henderson's ideas on teleological reasoning. This led to the republication of the original essays in Richards's book, *Medical Priesthoods and Other Essays* (1970). The self-published book also included a historical essay that unfavorably contrasted Cannon's simplistic "mechanical" teleology with Henderson's more sophisticated teleological view of fitness as a mutual or reciprocal relationship between organism and environment. Unlike Henderson, Richards complained that Cannon and his followers tended to look for a simple purpose or function in every physiological process and viewed homeostasis as a reaction that inevitably restored balance. This emphasis on normal function and balance led physiologists to ignore pathology or to view it only as a deviation from normality, rather than as a long, degenerative process leading to death.[21] Although Richards never delved deeply into social homeostasis, brief comments in the essays reflect his conservative misgivings about what he considered utopian views of society. In contrast to the Pollyanna attitude of Cannon and his followers that emphasized "life, liberty, and pursuit of happiness" Richards took a grimmer view that the essence of life is struggle and suffering. Failing to face this inconvenient truth resulted in social policies that might seem homeostatic but had unintended and unpredictable consequences that sometimes disrupted the social order.

Writing in *Scientific Monthly*, Richards complained that Cannon had taken an overly optimistic view of physiology, emphasizing normal function, while downplaying degeneration and disease.[22] "More specifically," he wrote, "I argue that the concept of homeostasis, as developed by Walter Cannon, useful and sound though it is, has now too strong a hold upon us, so that it seems almost to have possessed all our physiological thinking. Other balancing concepts are needed." For Richards, the excessive responses of the body typical of chronic, degenerative diseases were examples of homeostatic overreaching that often led to the detriment or death of the organism. Viewed from the perspective of pathology the body behaved in ways that were not wise, but rather "egregiously, calamitously stupid."

Homeostasis ignored the fact that short-term attempts to maintain constancy sometimes resulted in long-term pathological consequences. For example, scar tissue might be homeostatic within the context of wound healing and prevention of infection, but ultimately pathological, resulting in damaged joints in rheumatoid arthritis, glomerular nephritis in the kidney, or cirrhosis of the liver. Such long-term consequences were not simply due to the breakdown of homeostatic mechanisms but represented a more fundamental tendency for the body to overreact to stimuli, create imbalances, and react in inappropriate and maladaptive ways. To characterize this tendency, Richards used the Greek term "hyperexis," meaning "having too much." Hyperexis provided a countervailing concept to overly optimistic notions of homeostasis, while at the same time emphasizing that physiological balance and pathological disturbance often shared common causes. Increased volume in a diseased and weakened heart might be homeostatic for meeting immediate circulatory needs but contributed to the underlying problem and often led to congestive heart failure. Such cases of double-edged responses that were at once homeostatic and destabilizing were a common feature of chronic, degenerative diseases.

According to Richards, accepting such a balanced perspective was a useful corrective for a physiology too focused on the "normal" state but was even more necessary to counter homeostatic analogies applied to human societies. This was not pessimism, Richards claimed, but rather "a vigorous and fearless acceptance of pathology both physiological and social."[23] Thus, although one might think in terms of governmental checks and balances, it was equally necessary to recognize that both bodies and social systems fell victim to disorder—often the result of misguided attempts to regulate normal function.

Richards sharpened and extended his critique of homeostasis in a longer article published in *Perspectives on Biology and Medicine* three years after being awarded the Nobel Prize.[24] The tone of this essay was considerably harsher than his earlier article. Richards dismissed Cannon's ideas as unoriginal and misleading. Homeostasis was little more than a restatement of Claude Bernard's *milieu intérieur*, made worse by the fact that Cannon had not adequately acknowledged his precursor. Homeostasis added nothing useful to the original idea. Indeed, by ignoring pathological and degenerative conditions, Richards accused Cannon of overemphasizing normal states in ways that other followers of Bernard such as Henderson and Barcroft had avoided.

In addition to hyperexis, Richards described three additional un-homeostatic responses characteristic of chronic or debilitating diseases. Often homeostatic mechanisms are deficient and fail to compensate for pathological conditions such as necrosis, atrophy, and congenital defects. Such failures of self-regulation resulting from insufficiency were examples of "ellepsis." In other situations, such as heart failure, the body might react inappropriately, for example by reabsorbing sodium and water, leading to congestion and edema. Thickening of the arteries resulting from hypertension was another example of an inappropriate physiological response that Richards identified as "akairia." Finally, Richards described unbalanced, disordered mental states, or "taraxis," a term for confusion that he borrowed from Hippocratic medicine. "Confusion, taraxis, is omnipresent in greater or lesser degree in all our lives," he wrote; "it constitutes a considerable segment of total experience. It occupies some of the time for all of us; very much of the time, unhappily, of some. It makes up many, if not most, of those troublesome things that will not let us rest, 'that perilous stuff, that weighs upon the heart.'" Struggling against taraxis to restore some psychological balance to life might be an ennobling homeostatic response. For example, Beethoven's composing the Ninth Symphony, which he could never hear, constituted for Richards a compelling example of a reestablishment of order, saneness, and direction in the face of psychological confusion. Such heroic acts might be "the greatest human achievement," yet they were notable largely by their rarity. More common were situations such as that of a polio survivor whom Richards attended. Although she had made a remarkable recovery, she would remain permanently paralyzed and would never walk, even with crutches. "We can call this [recovery] homeostasis if we like," Richards wrote, "but it is not completely so. As a total event, for the girl and her family, it was not homeostasis. It was a thing confused, senseless, cruel beyond understanding, taraxic in the extreme."[25]

Richards turned to the broader philosophical issues of teleology as it related to health and disease in his J. Burns Amberson Lecture to the National Tuberculosis Association in 1966.[26] The intent of the talk was to place his research within the philosophical context of the teleological reasoning that he traced ultimately to Harvey, and more proximally to Henderson. Half a century before Richards's lecture, Henderson had struggled to reconcile the apparent goal-directedness of life processes with an evolutionary worldview that was mechanistic and devoid of design or purpose.[27] For Henderson, biological fitness was a mutual adaptation of the organism and the inorganic environments that surrounded it. The physical and chemical environment was not just fitted to life as we know it, but to any sort of life that might exist. Teleology was a necessary tool for recognizing adaptations and understanding the functional interactions between organic and inorganic systems. Henderson's perspective was evolutionary, but only loosely Darwinian. For Henderson, adaptation was fundamentally a physical and chemical problem, rather than a question of adaptation and reproductive success. He failed to see natural selection as a creative process rather than as simply trial-and-error elimination of the unfit.[28] Based on chemistry and physics, Henderson's understanding of adaptation as the exquisite "fit" between the organism and its environment was embedded in proximate causes, rather than ultimate causes of random mutation and natural selection.

Richards had little interest in evolution, but Henderson's teleology, which emphasized organic complexity, provided him with a philosophical basis for launching an attack on Cannon's homeostasis. By oversimplifying the body's responses to change and overemphasizing the maintenance of stability and normal function, Cannon's teleology provided an inadequate explanatory framework for understanding both short- and long-term changes in the body. For example, Richards pointed to the physiological changes that occur during acclimation to high altitude: increases in blood volume, red blood cells, hemoglobin, and pulmonary arterial pressure. These adaptive changes involved homeostatic responses to hypoxia, but they also paralleled the pathological changes that occur in some chronic cardiopulmonary diseases such as emphysema. Permanent residents of high elevations exhibited these responses in an exaggerated manner, and Richards pointed out that the incidence of heart failure was higher in these populations than those at sea level. The line between adaptive homeostatic response and pathological hyperexis was blurry and indistinct. For Richards, the orderly plan of maintaining constant oxygen levels in the blood could easily spin out of control, leading to acute mountain sickness or chronic hypertension.

In retrospect, Richards's idea of hyperexis might have benefited from a more Darwinian perspective on adaptation and the environment. His criticisms of homeostasis and his desire to broaden the scope of self-regulation notwithstanding, Richards shared Cannon's reduction of the external environment to a set of disturbances to which the body adapts in a purely reactive way. Indeed, like many medical practitioners, Richards viewed adaptation primarily as the body's pathological accommodation to such disturbances. Almost completely absent from this medical perspective was an understanding of organisms as active agents capable of adaptively modifying the environments that surround them. Transferred to a medical context, a broader environmental perspective might have emphasized and explored the social dimension of health to reduce the incidence of chronic disease through the preventive measures of public health. Richards's career preceded the widespread recognition that smoking was a major cause of emphysema and other cardiopulmonary diseases that he studied. Indeed, during the 1950s some prominent physicians dismissed such a link.[29] Nonetheless, working in a large urban hospital certainly exposed Richards to other social factors such as poverty and alcoholism that contributed to chronic disease. His conservative stance toward homeostasis—both physiological and social—evinced an intellectual pessimism that seemed to preclude an effective attack on these broader social ills.[30]

Understood from an evolutionary perspective, the body's short-term homeostatic wisdom would not necessarily be at odds with long-term pathological consequences emphasized by Richards. Selection favors physiological responses or adjustments to environmental stress that contribute to reproductive success even though they may lead to chronic, degenerative conditions later in life. Richards failed to grasp this Darwinian view of adaptation, and he never developed it within his discussion of hyperexis.[31] Postreproductive senescence might be as tragic as Richards described, but it is an unavoidable side effect of natural selection promoting physiological maintenance only to the extent that it contributes to successful reproduction. From an evolutionary perspective, the excessive and apparently maladaptive pathological conditions that Richards devoted his career to studying were, in many cases, the unremarkable result of the degeneration of physiological systems whose "purpose" is not longevity, or even Henderson's "fit" with the environment, but reproduction.

Homeostasis within a Darwinian, Cybernetic Context

Richards's critique of homeostasis, so influenced by his classicism, might seem a curious throwback in the context of post–World War II biology and medicine. Cannon's younger colleague Arturo Rosenblueth dismissed Richards's arguments as a "heteroclite jumble of statements and events."[32] Rosenblueth was equally dismissive of a somewhat different critique of Cannon's concept by David Drabkin, a biochemist in the medical school at the University of Pennsylvania. Drabkin's arguments went to the heart of the cybernetic version of homeostasis that Rosenblueth was pioneering. Drabkin gained recognition for crystallizing hemoglobin and developing an eponymous, and widely used, reagent for measuring hemoglobin levels in blood. Hemoglobin, with its iron-containing heme moiety, provided Drabkin with an example of the precarious and imperfect balance typical of homeostatic regulation. Iron is necessary not only for the oxygen-carrying function of hemoglobin and its close biochemical relative myoglobin but also for the function of other important proteins such as cytochrome c, which plays a critical role in cellular respiration. Despite its centrality, iron is often inadequately supplied in the diet and difficult to absorb from the gut. Curiously, this difficulty is largely the result of the body's own biochemical barriers to iron absorption. From a homeostatic perspective, these barriers seemed perverse by making anemia the most common deficiency disease encountered by physicians.

Although blocking the absorption of iron seemed contrary to the body's best interests, Drabkin pointed out that too much iron is even more dangerous than too little. The difficulty of absorbing iron from the gut matches the difficulty of excreting it by the kidneys. As the lesser of two evils, the body tolerates mild anemia to protect itself from toxic excess. Although impressed by the intricate anabolic and catabolic pathways linking the production and destruction of various iron-containing proteins, Drabkin warned that an optimistic homeostatic perspective should not obscure the "tight squeeze" that cells and organisms encountered between "the evils of impoverishment and overabundance."[33]

Unlike the classical influences on Richards's critique of homeostasis, Drabkin's argument focused specific attention on misleading mechanical analogies linking physiology to cybernetics. Although acknowledging, and even expanding, Cannon's discussion of organisms as self-regulating, open systems, Drabkin was skeptical of close comparisons between organisms and high-fidelity amplifiers or other feedback systems devised by engineers. For Drabkin, the currently popular cybernetic perspective simply reflected an updated mechanistic philosophy that substituted automatic control systems for the steam engines, storage batteries, and other devices that previous generations of physiologists had turned to when explaining organic structure and function. Without denying the usefulness of cybernetics, Drabkin objected to its uncritical application to homeostasis. Superficial analogies with automatic control systems reinforced Cannon's ideas of normal function and the inviolability of the organism while largely ignoring the instability inherent in both organisms and automated devices. Instead, Drabkin warned, "homeostasis only rarely functions perfectly, and breakdowns may occur frequently. A more mature philosophical approach, as a helpful guide to those who would heal the sick, is the recognition that *the body is not always wise and the environment often unfit*."[34]

Drabkin emphasized how deeply dependent organisms were upon the uncertain supply of energy and materials from the external environment. From this critical perspective, constancy was an illusion in open systems that exhibited only constant, often unpredictable, change.[35] The self-regulating mechanisms of homeostasis did not free the organism, but rather enslaved it to the imperfections of design that often failed: "It is remarkable how a mechanism designed for stabilization eventually can doom us to instability."[36] These failures included not only the pathological conditions that Richards referred to as hyperexis, but also a more fundamental inability to balance the inflow and outflow of energy and materials. The resulting "confusion of purpose" left the organism like a tightrope walker precariously teetering between poverty and overabundance.[37] If iron metabolism illustrated how this balance tended toward chronic impoverishment, Drabkin claimed that Cannon's own research on homeostasis of blood glucose highlighted the tendency toward pathological overindulgence.

Cannon had emphasized the broad margins of safety that buffered the effects of fluctuation in blood glucose, but Drabkin pointed to the inadequacy of stored energy held in reserve by cells and the tendency for blood glucose to drop when food was unavailable. Homeostatic regulation of blood glucose in the absence of food intake required "raiding" the surrounding tissues of fat and protein once the scanty supplies of carbohydrates were exhausted.[38] This precarious balance was more problematic because the cells of the brain were totally dependent on glucose as the sole source of energy. Ideally, Drabkin argued, the body would maintain adequate stores of energy to prevent a potentially unstable oscillation, but in reality that was not the case. On both the cellular and organismal level, this instability manifested itself in dangerous tendencies toward extreme deprivation or overindulgence. Although one could speak in the abstract about a "normal" body weight, Drabkin pointed out that humans tended toward the extremes of overly lean "Hollywood" or obese "beerhouse" body forms, each with attendant health problems frequently faced by clinicians. These pathological extremes were manifestations of a deeply irrational, selfish behavior of cells that undercut the idealized feedback controls of homeostasis.

Skeptical of the explanatory adequacy of cybernetics, Drabkin turned to a curious "chemical psychopathology" of cells controlled by drives, compulsions, phobias, and a "will for survival."[39] The metabolic importance of glucose and the limited ability to store this source of energy were responsible for cytoglucopenia or "cellular sugar hunger." This compulsive drive to obtain adequate energy from the surrounding internal environment might be expressed even in situations of overabundance when a fat cell "sees" a deficiency that doesn't actually exist. In such situations, the inevitable will to survive pitted the self-interested fat cell against the good of the whole, endangering the tenuous homeostasis holding the organism together. Taking a swipe at Cannon's social homeostasis as a misleading metaphor, Drabkin concluded, "If anything, the social community of functioning cells behaves compulsively. There is some foundation to encourage speculation about the possibility that the psychosomatic element in disease states may be an expression of the 'unconscious' cellular biochemical phobias."[40]

Writing in a broadly interdisciplinary journal, Drabkin claimed literary license in anthropomorphizing cellular behavior. That apology did not prevent Rosenblueth from aggressively attacking what he took to be a thinly veiled vitalism at the heart of Drabkin's argument. Like Richards's taxonomy of pathological concepts, Rosenblueth dismissed Drabkin's cellular psychopathology as "non-cognitive, operationally meaningless, animistic," "panmentalistically tainted," and "unnecessarily pessimistic."[41] For Rosenblueth, criticizing homeostasis because organisms did not regulate iron or glucose with absolute precision did violence to the original intent of Bernard and Cannon. Rosenblueth also complained that Drabkin's critique flew in the face of evolutionary thinking that demanded only adaptive adequacy rather than perfection.

Trained as a physician in Paris and Berlin, Rosenblueth taught physiology for several years in Mexico City before receiving a Guggenheim Fellowship to join Cannon's lab in 1930. After the fellowship expired, Rosenblueth eked out a precarious existence at Harvard tutoring medical students. During this period, he collaborated with Cannon on experimental studies of the sympathetic nervous system but was also actively engaged in interdisciplinary discussions that led to the development of cybernetics. He organized a seminar on philosophy of science regularly attended by Norbert Wiener. The two men became close friends and collaborators on several articles dealing with both the science and broader philosophical implications of self-regulation.[42] Despite Cannon's active lobbying, Rosenblueth was unable to secure a permanent academic position in the United States. He returned to his native

Mexico to take a faculty position at the National Autonomous University after Cannon retired in 1944.[43] Rosenblueth continued his work in cybernetics and physiology, returning to Cambridge on two extended visits to work with Wiener as a visiting fellow during the late 1940s.

Although he acknowledged that deviations from a homeostatic set point might indicate a pathological condition, Rosenblueth denied the implication that disease, in general, should be understood as a failure of homeostasis. It was equally unreasonable, he argued, to criticize the concept of homeostasis on the basis that self-regulation is often imperfect. Neither Bernard nor Cannon had claimed that organisms could be entirely free from the external environment. From Rosenblueth's evolutionary perspective, understanding homeostasis required only adequacy and specificity of response. Homeostatic mechanisms had evolved not to promote general health or perfect stability, but rather survival in the face of specific, imminent harm. Take, for example, the polycythemia that Richards had used as an example of hyperexis, or the body's tendency to overreact or correct one problem only to cause another. The increase in red blood cells might contribute to congestive heart failure, but Rosenblueth argued that it represented the body's homeostatic response to lack of oxygen, which constituted an immediate threat to survival. Patients with emphysema might eventually die from heart attacks, but they would have died sooner from anoxia if the body had not increased hemoglobin production. According to Rosenblueth, Drabkin's claims about the inadequacy of stored energy and the shortcomings in the self-regulation of blood glucose was an even more egregious example of "misconceptions and errors" to which critics of homeostasis were prone. Citing the law of diminishing returns, Rosenblueth pointed out that size considerations alone explained the limits of energy storage in both cells and organisms, and that increasing storage capacity was unlikely to increase survival. The evidence, Rosenblueth argued, supported the existence of homeostasis and its importance in the evolutionary struggle for existence. "The fact is that when the blood sugar of most mammals rises or falls, mechanisms come into play which correct these deviations from the mean; this fact justifies the conclusion that there is a homeostatic control of the blood-sugar concentration. The fact is also that most mammals do not fall into hypoglycemic or diabetic coma, even under a wide range of changes in the sugar input or in its consumption; this fact justifies the conclusion that the homeostatic control is adequate, i.e., that it is sufficiently accurate for the survival of these individuals and species."[44]

Homeostatic mechanisms were never perfect adaptations but simply good enough to increase chances of survival in a perilous world. Furthermore, Rosenblueth dismissed claims made by Drabkin and Richards that homeostasis rested on a simplistic teleology: "Neither Bernard nor Cannon suggested that the purpose of evolution was the development of a constant internal environment; they merely noted that evolution has often, not always, resulted in increasing stability of this environment."[45]

Rosenblueth combined his broadly Darwinian defense of homeostasis with a more full-throated support for the cybernetic approach to physiology that he was pioneering with Wiener. This explanatory perspective promised multiple benefits. Analogizing homeostasis to automatic control systems encouraged thinking in engineering terms that dissolved artificial barriers among mechanical, biological, and social systems. It provided a new explanatory terminology of closed-loop feedback systems involving receptors, effectors, and stabilizers. In the case of poorly understood physiological processes, thinking in terms of thermostats or other automatic control systems provided an important heuristic tool. This cybernetic approach could help isolate and elucidate particular homeostatic processes but also held out the possibility of a unified theoretical explanation for self-regulation in general. Rosenblueth's combination of functional and evolutionary explanations was deeply attractive to a number of biomedical researchers after World War II, even though a thoroughgoing Darwinian medicine never developed. The challenges of combining homeostasis, cybernetics, and evolution were exemplified by the highly successful career of the Harvard-trained physician and physiologist Arthur C. Guyton.

Autoregulation and Evolutionary Autonomy

In 1956 Guyton published his *Textbook of Medical Physiology*, which quickly became the most popular and enduring textbook in the field.[46] Only thirty-seven years old, Guyton had already chaired the Department of Physiology and Biophysics at the University of Mississippi School of Medicine for eight years. Dissatisfied with available textbooks and traditional pedagogy, Guyton began preparing mimeographed lecture notes that he distributed in class. These notes later formed the nucleus for the textbook that rapidly established itself as an authoritative source of information for medical students and advanced undergraduates. It continued to be widely used more than half a century after it first appeared, and more than a decade after Guyton's death in 2003.

In organizing his textbook, Guyton set an ambitious goal of unifying human physiology on the foundations of evolutionary theory, cell theory, and homeostasis. Evolution was fundamental because the many physiological mechanisms shared by humans and other mammals were the result of common ancestry. This justified generalizing conclusions drawn from dogs and other experimental animals used by Guyton and other laboratory physiologists. More profoundly, humans and other multicellular organisms had evolved from ancient unicellular life forms. Because of this evolutionary heritage, each of the trillions of cells making up the human body continued to exhibit some autonomy, despite the communal structure of tissues, organs, and the organism as a whole. Despite specialization and division of labor, each cell remained a living entity in its own right, surrounded by a fluid environment providing needed electrolytes, nutrients, chemical buffers, and other necessary substances. Both individually and together, cells maintained these chemical constituents of the internal environment at remarkably constant concentrations. Although some chemical substances were required for intracellular activities of individual cells, others such as sodium and potassium ions played larger roles in integration and control of tissues, organs, and the whole body. Thus, the body provided the internal environment for cellular activity, and the cells—individually and collectively—maintained this environment in a relatively constant condition. This two-way causation was central to Guyton's conception of homeostasis, which informed his presentation of human physiology. "This entire text," he wrote, "whether it calls the word by name or not, emphasizes over and over again the principles of homeostasis."[47]

It would be easy to dismiss Guyton's evolutionary thinking as speculative and superficial. Despite his claim that the *Textbook of Medical Physiology* rested on an evolutionary foundation, there was little explicit reference to evolution except in the introductory chapter. Guyton apparently derived his ideas about evolution of multicellularity from reading George Gaylord Simpson's popular book *Major Features of Evolution* and Edward O. Dodson's 1952 *Textbook of Evolution*, both of which were listed as references at the end of his introduction. Later editions of Guyton's textbook omitted even the brief introductory comments about evolutionary biology, focusing more on systems theory and a broad engineering perspective. Using Ernst Mayr's classification of biology into functional and evolutionary fields, Guyton's *Textbook of Medical Physiology* seems to fall squarely in the former camp. Guyton's teaching and research were strongly oriented toward answering "how" questions using proximate causation rather than "why" questions using ultimate causation. Yet Guyton's commitment to autoregulation and his cybernetic account of homeostasis rested heavily upon his belief that the evolution of multicellularity did not erase a fundamental cellular autonomy. Although homeostasis had evolved into a complex whole-body phenomenon, the mechanisms of self-regulation retained a strong foundation of local control by cells and tissues. Individual cells contributed to the body and benefited from this social existence, but they never completely gave up their freedom and independence. In contrast to Drabkin, who emphasized the pathological consequences of cellular autonomy, Guyton championed cellular individuality. Like Cannon, he was willing to generalize his physiological ideas by drawing analogies with the evolution of human societies. In doing so, Guyton argued that social stability rested upon individual initiative and responsibility mediated through decentralized local control.

Guyton's remarkable career in physiological research and teaching came only after a budding career as a surgeon ended abruptly after he contracted poliomyelitis as a young adult. The son of a prosperous ophthalmologist, who also served as dean of the School of Medicine at the University of Mississippi, the younger Guyton completed his undergraduate degree with honors in three years at the university.[48] After graduation, Guyton entered Harvard Medical School at the beginning of World War II. He completed his medical studies with an internship at Massachusetts General Hospital, but wartime service in the navy interrupted his surgical residency. During this brief hiatus, he conducted bacteriological research at Fort Detrick, Maryland, where he developed an electronic method for counting and measuring the sizes of airborne particles. The technique had important defense implications for detecting aerosolized bacteria, viruses, and toxins but could also be used to measure dust and other public health hazards. The work resulted in Guyton's first scientific publication and presaged his lifelong interest in designing scientific instruments.[49]

After the war ended, Guyton returned to Massachusetts General Hospital to complete his surgical residency, but polio dashed his plans. For the rest of his career, Guyton remained partially paralyzed in his right leg and left arm, which required him to use crutches or a wheelchair during much of the day.[50] Turning down academic positions in Boston, he returned to the University of Mississippi in 1948 to become the chair of the Department of Physiology and Biophysics at only twenty-nine. At the time, the medical school had only a two-year program, and none of the faculty members conducted research. As the medical school expanded its curriculum to four years, Guyton successfully gained external funding for his own research and the growing group of students and coworkers who joined his laboratory. When he retired forty years later, he had published more than five hundred articles, monographs, and a variety of highly successful textbooks—most notably his *Textbook of Medical Physiology*.

As a medical student, Guyton was interested in hypertension, which was thought to result from overstimulation by the sympathetic nervous system. This belief in sympathetic control of the heart may have stemmed tangentially from Walter Cannon's influence, although by the time that Guyton arrived at Harvard Medical School, the eminent physiologist was ill and no longer actively engaged in teaching or research. Recalling that Cannon's departure led to a decline in physiology at Harvard, Guyton took no specialized courses in the subject and did no physiological research as part of his education.[51] However, as an intern he trained under the cardiac surgeon Reginald Smithwick, who had perfected a technique for treating hypertension by severing the sympathetic nerves to the heart and peripheral blood vessels—a procedure hearkening back to Cannon's early experiments with cats. The operation resulted in an immediate drop in blood pressure, although hypertension eventually returned in most cases. Guyton later recalled that observing a surgical procedure with ambiguous results taught him the danger of being too wedded to an attractive hypothesis. It also shifted his thinking away from simple, centralized conceptions of self-regulation toward an emphasis on local control and the relative autonomy of various tissues in the body. This decentralized conception of homeostasis shaped his thinking about physiology but also resonated with his general evolutionary perspective and his broader social and political views. Temperamentally and philosophically, Guyton was an individualist but at the same time deeply loyal to his local community.

Guyton's decentralized view of self-regulation rested on ideas of reciprocity, redundancy and automaticity. He emphasized that the human body is composed of 100 trillion cells, each of which benefits from homeostasis and each of which contributes to maintaining a constant internal environment. "Thus," Guyton wrote, "each cell benefits from homeostasis, and in turn each cell contributes its share toward the maintenance of this state. The reciprocal interplay provides continuous automaticity of the body until one or more of the functional systems loses its ability to contribute its share of function. When this happens, all the cells of the body suffer."[52] Cellular reciprocity had evolved with the first appearance of multicellular organisms, and the complex homeostatic control exerted by the vertebrate central nervous and endocrine systems were evolutionary overlays on older and less centralized self-regulatory processes that continued to form a foundation for maintaining the internal environment.[53] The evolution of self-regulation did not involve progressive centralization so much as evolutionary tinkering resulting in the gradual elimination of destabilizing positive feedback loops and accumulation of partially overlapping and redundant control systems.[54] As a result, if one system weakened or failed, others could sometimes compensate for the defect.

Both the automaticity and local control of circulatory function impressed Guyton as he began performing experiments on cardiac output. The traditional view held that the nervous system and the heart itself controlled cardiac output. An alternative perspective, championed by Guyton, viewed the supply of blood returning through the vena cava as the regulator. In Guyton's theory the heart was an automaton or "permissive organ" (at one point he compared it to a sump pump) that pumped as much blood as was available to it.[55] Indeed, the heart was physically capable of pumping over twice as much blood as it normally did. Conversely, cardiac output dropped whenever insufficient blood returned to the heart through the vena cava. Although venous return was a complex phenomenon, ultimately the oxygen needs of the local tissues mediated by the opening and closing of capillary beds regulated cardiac output. From this perspective, Guyton claimed, the circulatory system was the servant of the body's tissues, rather than their master.[56]

Dilation and constriction of blood vessels in capillary beds regulated blood flow in response to changes in oxygen, carbon dioxide, or electrolytes. This idea of autoregulation dated back to the work of the Danish physiologist August Krogh, who won the Nobel Prize in physiology or medicine in 1920 for the discovery of small muscular sphincters that controlled blood flow through capillaries in the muscle tissue of frogs and other animals.[57] What set Guyton apart were the broader implications that he drew from earlier studies of autoregulation. Autoregulation in certain important organs, notably the kidneys and brain, became widely accepted, but Guyton believed that autoregulation was a universal phenomenon that applied to all tissues and organs in the body. Not only did autoregulation occur at all levels from cell to organism, Guyton argued, it involved a temporal sequence of adaptations from short term to long term. For example, capillary beds regulated blood flow through tissues in response to immediate needs for oxygen, removal of carbon dioxide, and maintenance of arterial pressure. These responses occurred rapidly, often in seconds or minutes. More powerful autoregulatory responses involving changes in vascularization occurred over periods of weeks in response to long-term needs of the tissues. In either case the primary drivers of adaptive change were the cells and tissues: "The basis for autoregulation is the fact that each tissue is capable of controlling its own blood supply in proportion to its needs and primarily in proportion to its need for oxygen."[58] Although in some cases this local control could be countered or overcome by actions of hormones or the central nervous system, from Guyton's perspective homeostasis was fundamentally grounded in the ability of individual cells to regulate their immediate environments.

Guyton and his students demonstrated the independence of autoregulation from central nervous control through a series of macabre experiments on decapitated dogs.[59] After anesthetizing the dog, the researchers destroyed the central nervous system by injecting ethanol into the spinal cord, crushing the cervical vertebrae in a vice, and removing the head from the animal. They manipulated blood pressure by transfusing blood from a reservoir or removing it by bleeding. These manipulations resulted in significant changes in cardiac function. However, over the course of several hours, cardiac output, oxygen consumption, and arteriovenous oxygen differences all returned to nearly normal levels. The experiments demonstrated that self-regulation could occur without central nervous control, and that locally mediated control was capable of effectively regulating cardiac function after acute disturbances.

Guyton's broad interpretation of autoregulation fit neatly with his developing interests in systems analysis and computer modeling. Combining boyhood interests in electronics and early professional interest in designing physiological instruments, cybernetics and systems thinking came naturally to Guyton. He was reading deeply in the literature of this subject while writing his *Textbook of Medical Physiology*. He also taught himself Fortran to program the hundreds of equations required for a complete model of circulatory regulation. Supported by his first large NIH grant in 1960, Guyton attracted a cadre of graduate students trained in physics and mathematics to work on computer modeling in his lab.

Even before completing large-scale modeling on digital computers, the group made major conceptual breakthroughs using analog computers. Although less powerful and slower than digital computers, analog computers provided some important advantages for modeling physiological systems. The electrical circuits that the programmer needed to physically construct were a close and instructive analogy to physiological processes such as blood flow. Analog models allowed for increasingly complex analysis by adding components in a modular fashion. Eventually Guyton's systems analysis of circulatory regulation included eighteen major systems composed of 354 modular blocks.[60] The interlocking modularity of the analog models provided a strong and instructive analogy for Guyton's decentralized autoregulation while also presenting both local and whole-body self-regulation as a complex cybernetic system of closed feedback loops.

Modeling was not simply a progression to more-sophisticated representations but a symbiotic interplay between the physiologist's mental constructs and the electronic circuitry of analog computing. The formal logic of programming forced the physiologist to critically analyze assumptions. This was particularly the case when models led to unrealistic predictions. Just as importantly, modeling and experimentation were mutually reinforcing activities. Modeling highlighted unanswered questions and gaps in physiological knowledge and thus suggested where experimentation was necessary.[61] More directly, physiologists used models to investigate processes for which experimental techniques were presently inadequate. Finally, in the case of the total system model, Guyton's group discovered that individual processes often could act abnormally without adversely affecting overall performance because other processes compensated for the abnormality.

The total system model became a powerful metaphor for the body's ability to compensate for dysfunction, and it reinforced Guyton's belief in a decentralized whole-body autoregulation based on multiple, local control systems beginning at the level of individual cells.[62] More-centralized control systems involving hormones and the central nervous system overlay these basic regulatory mechanisms, in a sense acting as "backup control systems."[63] Thus, the modularity of the total system model reflected the evolution of increasingly sophisticated control systems that never completely replaced fundamental cellular and tissue-based regulation. Because negative feedback loops overlapped and partially compensated for one another, the total system was remarkably stable and resistant to perturbations.

In 1974–75 Guyton served as president of the American Physiological Society, which provided an opportunity to discuss the broader implications of evolution and homeostasis in an address to the society when he left office.[64] At the height of his career, Guyton was in an expansive mood. Optimistic about the state of physiology, he was less ebullient about the broader social context within which scientists worked. Environmental degradation, overpopulation, and social inequality were pressing national issues. Adequately rewarding merit while avoiding exploitation also concerned Guyton, as did the growth of centralized government and loss of local control. What insights might physiology provide for these pressing social problems?

In his address Guyton argued that physiologists had a unique understanding of humanity unrivaled by physicians, theologians, philosophers, or others who studied the human condition. Physiological self-regulation provided not only an explanation for how organisms operated but also deep insights into political and social systems: "Civilization is a communal organization of men in the same way that the human body is a communal organization of its organs and the organs in turn communal organizations of cells."[65] Homeostasis must have been a characteristic of even the first organisms, but the advent of multicellularity allowed specialization of cells and division of labor. This communal life also required reciprocity and the need for increasingly elaborate self-regulation. Evolutionary trial and error gradually perfected homeostatic mechanisms and compensatory self-regulation, but it accomplished this through blind mutation and the brutal elimination of the weak and survival of the strong. For the past century, physiologists had come to realize the dynamics of the natural laws governing self-regulation, but according to Guyton, it was only in the post–World War II era that they understood the feedback mechanisms of homeostasis as cybernetic systems.

What insights might this modern physiology provide for understanding social, political, and economic problems? For Guyton, civilization was a communal organization of individuals in the same way that the body was an organized community of cells, tissues, and organs. The difference was that although the human lineage stretched back over billions of years to the first forms of life, civilization was a relatively recent experiment. The progressive elaboration of negative feedback regulation, and elimination of the vicious cycles of positive feedback that characterized human physiology, had only begun to evolve in human societies. Environmental degradation, the human population explosion, and social and economic inequality were examples both of the absence of effective social regulation and of the tendency for greed, hatred, and racism to destroy nascent negative-feedback controls. Human intelligence might partially replace natural selection, but it was no guarantee of social harmony. Pessimistically, Guyton suggested that social regulation might need to evolve through a trial-and-error process taking thousands of years.

Guyton was aware that the ideas of "control" and "negative feedback" might suggest the need for strong centralized government that was anathema to him. Indeed, he expressed concerns about the growth of the federal government, excessive taxation, laws that hindered equality of opportunity, and the threat of mandatory retirement that penalized older but still productive members of society. Here too physiology provided insights into appropriate social homeostasis. Just as the complex homeostatic mechanisms of the human body rested on a foundation of local autoregulation at the tissue level, so a just society needed to have strong local control to counterbalance centralized government. A stable society needed to avoid the extremes of both unregulated, laissez-faire capitalism and over-regulated socialism. Unregulated capitalism led to economic inequalities and exploitation of workers. Although labor unions served as a useful negative feedback mechanism that balanced labor and big business, Guyton worried that an unintended consequence was a decline in the work ethic without any decrease in human greed. Guyton was even more critical of socialism, claiming that three-quarters of people in socialist countries lived under brutal dictatorships. Social democracies might provide a middle way, avoiding the extremes of untrammeled individualism and centralized authority, but Guyton was skeptical of their success. Democratic socialism was a recent development, and it was not clear how successful this experiment in social evolution would be. Guyton also pointed out the potential for positive feedback cycles to destabilize a political system in which the central government provided for most of the needs of people. Such a system would almost inevitably result in corruption and inefficiency. The solution, Guyton thought, was a decentralized meritocracy that rewarded hard work and demanded responsibility from its citizens in a way that mirrored cellular reciprocity in the human body. "The principles of physiology teach us that each cell in the body has its own equal right to opportunity, to exist, and to flourish, but these principles also teach us that the cells have these rights only so long as they perform their individual responsible duties to the body—or otherwise the entire body will vanish."[66]

Guyton's belief that physiology held a privileged perspective for the critique and prescription of public policy was reminiscent of Cannon's social homeostasis a generation earlier. Both scientists were willing to draw extensive analogies between physiology and politics. Both framed their arguments within a broad evolutionary framework that held out hope that human wisdom might replace natural selection, while at the same time cautioning that the evolution of self-regulation had occurred through millions of years of evolutionary tinkering that led to frequent extinction. In other ways the physiological basis for the social homeostasis of Cannon and Guyton differed in subtle but profound ways. Cannon's research focused on hormones and the sympathetic nervous system, the components of which often acted antagonistically to maintain constancy through a system of checks and balances. Although advocating social and political autonomy, Cannon was willing to grant the need for centralized control during the emergencies of the Great Depression and World War II. Guyton was more uncompromising in his commitment to individualism and local control, both of which mirrored his belief that cells maintained autonomy while regulating the overall functioning of the multicellular body. The overlapping nature of this autoregulation also allowed the body to compensate for local defects.

Guyton's emphasis on self-reliance and individualism reflected a life story of perseverance and achievement in the face of great adversity. Deprived by polio of his aspiration to become a surgeon, Guyton plunged ahead as a prolific experimental researcher, computer modeler, successful teacher and mentor, and creative instrument maker. Even during his initial recovery in a polio hospital at Warm Springs, Georgia, Guyton designed both the prototype for an electric wheelchair controlled by a joystick and a mechanical lift for moving patients from bed to wheelchair. With no formal graduate training in physiology, he quickly mastered the discipline to write the most popular medical textbook of the late twentieth century, while building a large, productive laboratory that attracted students and postdoctoral researchers involved with experimentation and computer simulation.

Guyton's localism and wariness of centralized government was more ambiguous. His fond childhood memories of life in Oxford, Mississippi, evinced a strong sense of place, as did later recollections of his frequent chess matches with William Faulkner and selling the famous author a sailboat that Guyton had designed and built during college. So too the realization that the main building of the newly expanded medical school was named for his father was a constant reminder of family ties to the university. On the other hand, Guyton failed to acknowledge the pernicious effects of states' rights and local control when he mentioned the scourges of racism, social inequality, and environmental degradation during his presidential address to the American Physiological Society. Federal agencies, rather than state or local governments, had been responsible for passing and enforcing laws against segregation and discrimination during the two decades preceding Guyton's lecture. The university of which he was a part had been forcibly desegregated by federal authorities in the face of intense opposition from state and local officials. Guyton's own professional success was highly dependent on federal support for research after World War II. Guyton's success in transforming medical education at the University of Mississippi was part of a modernizing trend that was heavily dependent upon external resources, national political forces, and broad social movements, rather than local interests. His own professional success was as much a product of educational opportunities at elite universities as it was of local upbringing.[67] Tellingly, although Guyton and his wife had ten children, all of whom became physicians, none of them remained in Mississippi. Although the eight sons were educated at the University of Mississippi, they all took their medical training at Harvard University. The two Guyton daughters graduated from Harvard College, before they too pursued successful medical careers that took them far away from Mississippi. This remarkable Guyton dynasty gained national renown through articles that appeared in *Reader's Digest*, the *Los Angeles Times*, and other popular periodicals.[68] These stories celebrated individual initiative as well as the importance of a stimulating and supportive home environment. Other important factors far removed from the local environments of Jackson and Oxford, Mississippi, remained unacknowledged.

Guyton's *Textbook of Medical Physiology* highlights the continued importance of homeostasis in biomedical contexts but also, in a less obvious way, the influence of evolutionary thinking in medicine. Despite his early enthusiasm for using evolution as part of a three-legged foundation for medical physiology, Guyton never fully elaborated an evolutionary medicine—certainly not a thoroughly Darwinian perspective on medicine. Yet his presidential address demonstrates his continued interest in evolution, particularly as he applied it to cellular autonomy and autoregulation in multicellular organisms. A more thorough Darwinian perspective would require deeper thinking about adaptation and ecological complexity. It would also require bridging the boundary between functional biology and evolutionary biology by medical researchers engaging questions and problems requiring explanations using both proximate and ultimate causation. Although such thinking might not come naturally in the context of laboratory experiments on dogs or computer simulations of self-regulation, medical researchers sometimes left the confines of hospital-based laboratories to study homeostasis in broader evolutionary and ecological contexts.

3

Free and Independent Life

By his death in 1962, Homer W. Smith was the leading authority on renal physiology, and his work over three decades demonstrated how medically oriented human physiology and a broad comparative physiology could be mutually reinforcing. Although he conducted his research on water and ionic balance in various groups of fish primarily during summer retreats from his medical research, the outpouring of publications suggests that he considered this work much more than an avocation. So did his enthusiasm for using physiological evidence to confront the paleontologists who questioned his claim that ancestral vertebrates evolved in freshwater, rather than in marine environments. Evidence from fossils was equivocal in this case, and Smith believed that comparative physiology of the kidney provided compelling support for the freshwater origin of vertebrates. He was quite successful in making this argument in scholarly articles, popular essays, and his best-selling book *From Fish to Philosopher* (1953). The broad outlines of his phylogenetic analysis of the vertebrate kidney remain largely intact, even though some of the details continue to provoke criticism, inspire revision, and stimulate research.

The intellectual core of Smith's work was Claude Bernard's famous observation that the internal environment provides the basis for free and independent life. Historically, Smith believed, studies of osmotic regulations in freshwater and marine fish starting in the late nineteenth century were a direct outgrowth of Bernard's distinction between the *milieu intérieur* and *milieu extérieur*. In an early paper on the renal physiology of sharks, he wrote, "It was to be expected that biologists, with Bernard's thesis in mind, would turn immediately to the marine and fresh-water fishes where problems concerning the distribution of salt and water between blood and environment were presented in their most obvious form."[1] His own phylogenetic studies were a self-conscious continuation of that intellectual lineage. Smith also used Bernard's idea of free and independent life as the basis for an evolutionary philosophy of science that he applied broadly to unify his physiological research with his interests in consciousness, human creativity, and ethics. Smith also credited Walter Cannon, with whom he worked for two years, for ideas of self-regulation and automaticity as applied to maintaining internal constancy.[2] Reworking the ideas of Bernard and Cannon in an explicitly Darwinian framework, Smith interpreted homeostasis as the dynamic opposition between the organism and the external environment. Life was a constant struggle against forces that tended to destroy it. Maintaining freedom and independence—however tenuous and temporary—provided the only goal in an otherwise purposeless world.

The "Smithian Era" in Renal Physiology

Smith originally trained as a chemist. After completing an undergraduate degree at the University of Denver in 1917, he worked under the direction of Eli Marshall at the Chemical Warfare Station in Washington, DC.[3] The two men became lifelong friends and eventually collaborated on evolutionary studies of the kidney, but during the war Marshall directed military research on the biological effects of nerve gases. After the war Smith continued graduate studies at the Johns Hopkins University School of Hygiene and Public Health, where he earned a doctor of science degree in 1921 with a dissertation on the pharmacology of arsenic. After a short stint in the research laboratory of the Eli Lilly pharmaceutical company, Smith spent two years as a National Research Fellow in Walter Cannon's lab at Harvard. Smith later claimed that working with Cannon was an "important turning point" in his life and that his later professional success rested on this opportunity.[4] In 1925 Cannon helped Smith secure his first academic position, as the chair of the physiology department at the School of Medicine of the University of Virginia. Smith held this position for three years before becoming director of the physiology laboratory at the New York University College of Medicine. He spent the rest of his career there, although he pursued evolutionary research during the summers at the Mount Desert Island Laboratory in Maine, which he helped establish as an important site for marine research.[5]

During his three decades at NYU, Smith established himself as "the uncontested patriarch of modern nephrology."[6] His three textbooks of renal function in health and disease, published at various stages in his career, became medical "classics." The "Smithian Era" of renal physiology, as Robert Pitts dubbed the period, was a particularly exciting time in the field. Disagreements about whether the kidney produced urine through filtration or by secretion of metabolic wastes subsided as a more comprehensive explanation grew around the recognition that filtration, reabsorption, and secretion were all important processes in renal function.[7] The recognition of an energy-dependent, active transport mechanism for moving substances across membranes against a concentration gradient was especially important for this expanded view of kidney function, although little was known about the enzymes involved or how energy was coupled to the transport system.[8] Together with diffusion and osmosis, active transport provided a mechanistic conceptual framework that Smith and his contemporaries celebrated for eliminating lingering remnants of nineteenth-century vitalism.

Important technical innovations also provided the basis for greater precision and quantification of kidney function. Improved micropipettes and the use of mercury or oil to block fluid movement allowed physiologists to sample liquid at various places in the nephron, to precisely quantify chemical changes in the filtrate, and to identify the functions of the various anatomical structures of the nephron.[9] The invention of the flame photometer and automated analytical techniques during World War II allowed rapid and precise measurements of ions such as sodium and potassium in the fluid flowing through the nephron, largely replacing tedious titrations and gravimetric methods of analysis.[10] Smith's own development of the "clearance method" was a major innovation for measuring filtration rates, as well as rates of secretion and reabsorption of various substances during urine formation. The procedure involved injecting inert substances into the bloodstream and later measuring their presence in the urine. During the 1930s Smith experimented with the nonmetabolizable sugar inulin, which because of its physical characteristics became the "gold standard" for measuring glomerular filtration.[11] During the Smithian era, physiologists convincingly explained all of the basic processes involved in urine formation, laying a firm quantitative foundation for later cellular and molecular studies of the nephron.

Philosophy and Physiology in Evolutionary Studies of the Kidney

The degree to which Smith intertwined philosophy and physiology from the outset of his career is evident from his early work on the lungfish (*Protopterus aethiopicus*). With support from a Guggenheim Fellowship, in 1928 Smith traveled to Lake Victoria in eastern Africa to find specimens of the lungfish, which he brought back to the United States for laboratory experiments. His early research on metabolism and water balance in *Protopterus* was among the first physiological studies done on these unusual fish.[12] He accidently discovered that the adult fish depended on aerial breathing when he initially kept the specimens he collected submerged in water. More carefully controlled experiments in the lab confirmed that without access to air the fish quickly died of asphyxiation.[13]

Asphyxia also seemed to be the primary stimulus that brought the fish out of estivation. Smith studied the physiological changes accompanying this state of inactivity that occurred during the dry season when the fish burrowed into the mud. Encased in a cocoon and surrounded by dried mud, the fish had access to air through a small hole leading to the burrow, although the rest of body was effectively sealed from the external environment. Smith found that fish could remain in a state of "profound sleep" for more than a year when kept in the laboratory.[14] Despite the low metabolic rate during dormancy, the fish quickly depleted stored carbohydrates; and although relying heavily upon fat reserves, it catabolized proteins to provide half of its energy needs. Other vertebrates resort to protein catabolism only during starvation, and Smith characterized the estivating lungfish in a similar dire state. Because the kidneys did not form urine and the gills could not excrete ammonia during estivation, nitrogenous waste from protein metabolism was stored in the body as urea. With the exception of sharks and other elasmobranchs, which use stored urea for osmotic regulation, no vertebrates—in sickness or health—had levels of urea as high as the estivating lungfish. Carbon dioxide also increased in body tissues during estivation. At the beginning of the wet season, when their swampy habitats flooded again, the lungfish quickly resumed activity but had to struggle to free itself from the surrounding mud to reach surface air before drowning.

Despite the importance of the lung for vertebrate evolution, Smith considered the lungfish to be an evolutionary blind alley. Natural selection had shaped the fish's unique adaptations for a particular way of life but also trapped the species in a precarious existence that seemed to limit its future possibilities. Indeed, the way that the lungfish spent much of its existence imprisoned in rock-hard mud was a potent metaphor for Smith's worldview that combined evolution with organic self-regulation. The lungfish barely experienced Bernard's free and independent life, but then, the same could be said of many terrestrial vertebrates that had sprung from the ancient lunged ancestors of *Protopterus*. Mammals, with their exquisite regulatory mechanisms, might have greater freedom from the vicissitudes of the external environment, but even humans only imperfectly experienced free and independent life. Later in his career, Smith explored the evolutionary implications of his research for understanding human creativity and consciousness in essays and in his popular book *From Fish to Philosopher*, but he had already outlined the philosophical interplay between evolution and homeostasis when he began building his reputation as a renal physiologist and evolutionary biologist in the early 1930s.

While he was preparing the initial reports of his research on *Protopterus* for scientific journals, Smith wrote a fictionalized account of his experiences in Africa. Eventually *Kamongo* became a popular success, but the publisher, Alfred Knopf, initially rejected the manuscript, because he doubted that there was a market for a book that straddled science and fiction. Smith's stark evolutionary worldview may also have unsettled the publisher, as it later did some reviewers.[15]

Set on a steamer passing through the Suez Canal during a stifling heat wave, *Kamongo* consists of an extended conversation between a biologist named Joel and an Anglican priest, "Padre." At the beginning of the book, Joel relates the challenges of finding specimens of *Protopterus* and the results of experiments done on lungfish estivation. The conversation then predictably turns to evolution and the human condition. Smith's literary approach provided a rich context for exploring issues of science and religion because Padre is no Bible-thumping fundamentalist, but a sophisticated, Oxford-trained theologian with an abiding interest in science. The priest's persistent probing of Joel's "pessimistic" outlook provided Smith with entrées for his stand-in to elaborate the implications of a Darwinian worldview devoid of purpose. At one point early in the conversation, Joel fears that he himself might be guilty of a form of "scientific evangelism" no better than religious varieties.[16] Smith later acknowledged that this young biologist did too much talking in the story, and Padre too little—a criticism also leveled by several reviewers.[17] Those critical observations notwithstanding, the philosophical ideas that Smith expressed in this early literary work continued to form an important intellectual foundation for his evolutionary and physiological research throughout his later career.

Early in their conversation, Padre expresses admiration for evolutionary theory and his belief in the compatibility of science and religion. Both the poetry of the Bible and the scientific theory of organic change speak to the magnificence of life.[18] Joel will have none of this. The "cold, hard fact" is that the lungfish is an evolutionary blind alley destined to extinction despite the apparent progressive innovation of the lung. "This lung of his," Joel argues, "which had come into being, promised to bring him freedom from the old way of living, promised to break the bonds that chained him to a life beneath the water, but it only left him chained alternately beneath the water and the mud. If anything, he was worse than before."[19] Nonetheless, although Smith denied progressive evolution and argued that adaptations were always partial and relative to particular environments, he could not free himself completely from the idea of progress. The lung had opened new opportunities for vertebrate evolution even if the lungfish seemed caught uncomfortably between two habitats. Though it is tightly circumscribed by chance mutations, contingencies of an unpredictable environment, and the constant pressure of natural selection, one might still find in the evolution of freedom and independence a highly attenuated notion of progress. According to Smith's fictionalized stand-in, "The only time when we can properly speak of evolution as being upward is in those instances where the theatre of life's activity has been enlarged, where some new acquisition or way of living permits the animal to move about more freely and independently within the circumscriptions of its environment."[20]

Variations on this statement continued to appear in Smith's later writings, indicating the importance of finding some small degree of "upwardness" in a phylogenetic tree that he preferred to orient sideways. Toward the end of *Kamongo*, Padre challenges Joel's attempt to place life in a completely materialistic framework devoid of purpose; he exclaims, "But life has a purpose, it has power and knowledge—there must be something to it beside a mere spin of energy like a whirling dervish in the wind—."[21] Joel interrupts him in mid-sentence and begins an extended mixing of metaphors comparing life, alternately, to a whirlpool and a gyroscope. Both of these inanimate systems seem purposive in the way they maintain equilibrium against external perturbations, but eventually they come to a halt when they have dissipated the energy that maintains them. "So it is with life," Joel continues, "for life's purpose is to keep on living although at every hand its environment tends to arrest it; its power is the force with which it opposes the destructive forces that tend to put it down; and its knowledge is the sentience by which it selects from the world about it those means and conditions best suited to its ends."[22] Joel's "gyroscopic whirlpool" is purely mechanical and, in fact, very similar to automated machines capable of sensing external stimuli, resisting change, and maintaining internal equilibrium. However, the living "whirlpool" also has the ability to accumulate energy, grow, and give rise to descendants. It thus also perpetuates itself as part of an evolutionary lineage.[23]

Homeostasis and Vertebrate Phylogeny

Although the comparative anatomy and physiology of the vertebrate kidney and the evolutionary conclusions drawn from them became a major focus of Smith's career, they apparently fit somewhat uncomfortably with traditional medical research. Even his friend Eli Marshall, who would soon be drawn into the comparative research, initially expressed incomprehension that Smith took a serious interest in fish. Smith recalled that when Marshall first learned of this new line of research, he remarked, "Fish? My God, man! Don't you know that fish are lower than frogs?"[24] While medical physiologists used dogs as experimental models or stand-ins for humans, Smith considered fish, with their complex systems for maintaining internal stability against a hostile osmotic environment, interesting in their own right. The creative tension in Smith's thinking between the struggle for existence and organic self-regulation provided insights into general physiological principles that probably wouldn't have occurred if he had focused only on mammals. For example, although Smith often compared the mammalian kidney to a "master chemist" playing the key role in maintaining the integrity of the internal environment, he also came to realize that in most vertebrates important elements of this self-regulation are performed by gills, skin, and other organs. At the same time, the mammalian kidney demonstrated the "extravagance" of nature in constructing a jerry-rigged mechanism that no engineer would design. This extravagance made sense only from a historical perspective that considered the evolution of new structures, repurposing of existing structures, and degeneration and loss of structures no longer adaptive in particular environments.[25] Comparative studies of lungfish, sharks, and other vertebrates whose kidneys were structurally quite different from those of humans were just as important as experimental studies on dogs whose kidneys were very similar to humans'. Indeed, the initial development of Smith's clearance method for measuring filtration and other renal functions—his major technical contribution to renal physiology—owed much to early studies by Eli Marshall on how aglomerular fish produce urine without filtration.[26]

Smith's earliest studies of the physiology of lungfish became the starting point for a much broader comparative study of the structure and function of kidneys—or more specifically the nephrons—of various fish in relation to the habitats in which they live (figure 2). This comprehensive study, which was a pioneering effort in using physiological ecology to understand phylogeny,

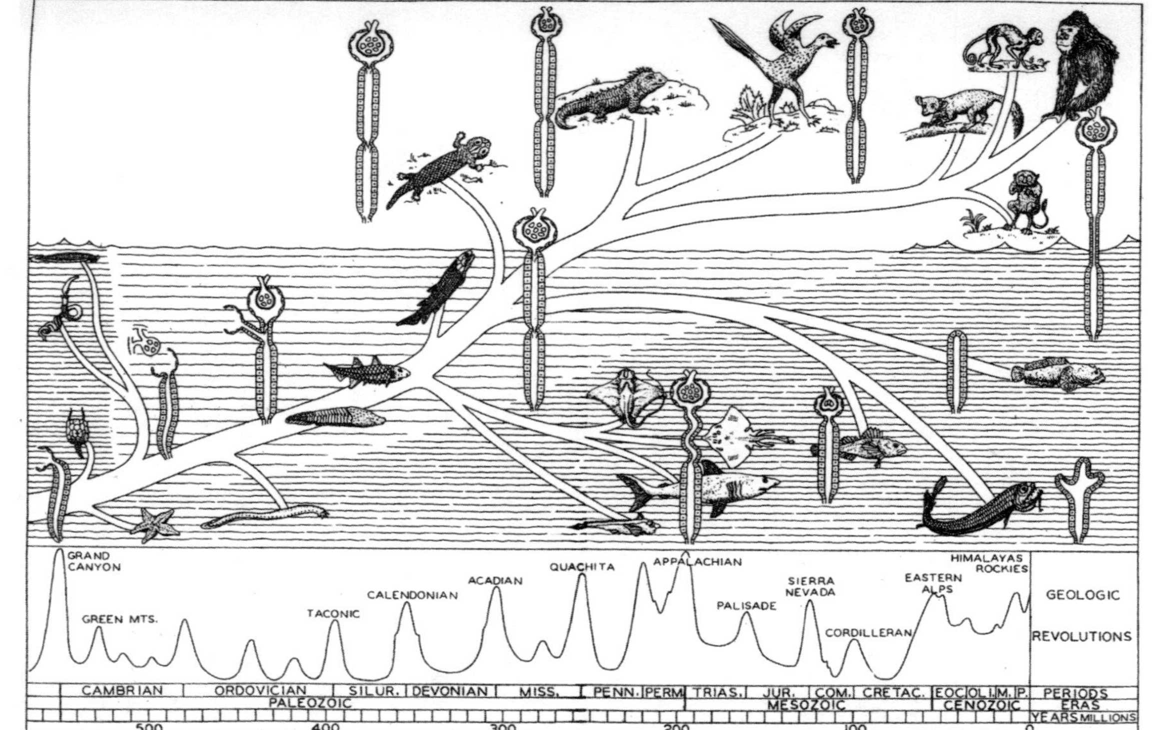

Figure 2. Phylogenetic tree of the major groups of vertebrates based on renal anatomy and physiology. Note the schematic diagrams of nephrons superimposed on the groups. This is the original version of Smith's phylogenetic diagram; in later versions, he corrected the spelling errors ("Calendonian"; "Quachita"). Homer W. Smith, *Lectures on the Kidney*, University Extension Division, University of Kansas, Lawrence, 1943, 8–9.

occupied Smith from the late 1920s until his death in 1962. He worked out the broad outlines of the phylogenetic scheme together with Marshall, who was Smith's neighbor during the summers at Mount Desert Island Laboratory. The two men had a friendly but competitive relationship and aside from one seminal article that they wrote together in 1930, they conducted most of their work independently.[27]

Central to the phylogenetic scheme developed by Smith and Marshall in 1930 was the evolution of the glomerulus.[28] This structure at the head of the nephron is a filtering device consisting of a cuplike tubule (Bowman's capsule) surrounding a tuft of capillaries. During the first step in urine formation, plasma containing ions and small molecules is forced out of the highly permeable capillaries into Bowman's capsule. The filtrate is further processed by selective reabsorption and secretion in the tubular sections of the nephron to form urine. In an extensive histological survey of twenty-five species from twelve different families of fish, Marshall and Smith found wide variation in the development of the glomerulus. Based on these differences they categorized the species into four groups. Freshwater fish had kidneys with numerous, well-developed glomeruli, while marine fish had kidneys with rudimentary glomeruli or lacked them altogether. Marshall and Smith also identified two intermediate groups between these two extremes.

From these comparisons Marshall and Smith argued that the glomerulus evolved as an adaptation to freshwater. Contrary to the widely accepted belief that the vertebrates arose as marine organisms, Marshall and Smith claimed that the protovertebrate was a freshwater organism facing the challenge of removing excess water that continually entered the body by osmosis. Urine production was, first and foremost, a mechanism for water balance and only secondarily a method of removing metabolic waste products. Liquid forced out of the blood during filtration in the glomerulus became a copious, dilute urine, which balanced the excess water flowing into the body. When fish invaded marine habitats, they faced a very different problem of water balance, because water tended to leave the body by osmosis. In this situation the glomerulus and the copious urine that it produced became liabilities. Consequently, mutation and natural selection resulted in an evolutionary reduction in the size and number of glomeruli. In the case of aglomerular marine fish, the structure disappeared completely.

Although it was based on a comparative survey of fish, Marshall and Smith extended their ecological and phylogenetic scheme to include other vertebrate groups. Amphibians, which are primarily freshwater, tended to have well-developed glomeruli. Terrestrial vertebrates face a water balance problem fundamentally similar to that of marine fish; they are constantly in danger of dehydration. As expected, Marshall and Smith found reduced glomerular development in birds and their reptilian relatives. In contrast, mammals retained well-developed glomeruli, but the mammalian nephron evolved into a highly efficient filtration-reabsorption system that could eliminate metabolic waste while conserving water. In particular, the loop of Henle provided a mechanism for concentrating salts and metabolic waste products in the urine, while reabsorbing water. In the original scheme, Marshall and Smith claimed that mammals were unique among vertebrates in their ability to form urine that was osmotically more concentrated than the body fluids—a claim that later required some significant revision.

The broad outlines of the evolution of the vertebrate kidney presented by Marshall and Smith in 1930 provided a durable phylogenetic scheme that is still widely accepted. Yet its simplicity was deceptive. For example, a few aglomerular fish have reinvaded freshwater habitats, and understanding how they maintain water balance remains an area of active research.[29] Birds posed another complication that bedeviled the scheme. Marshall and Smith admitted that the avian nephron has a loop of Henle, though not as highly developed as in mammals. Indeed, the kidneys of birds contained an odd mixture of primitive reptilian nephrons and more advanced "mammalian type" nephrons.[30] Critics of this overtly hierarchical view of birds and mammals have pointed out that both groups have successfully invaded the same broad range of habitats, indicating that their homeostatic mechanisms for water balance are equivalent. Through convergence, birds and mammals have arrived at similar adaptive peaks by slightly different anatomical and physiological routes. By focusing on fish and mammals, the groups they knew best, Marshall and Smith inevitably gave the scheme an appearance of linearity, despite Smith's strident opposition to progressive evolution.[31] Although he drew a branching tree, it is much less bushy than most biologists would posit today. While accepting the broad outline, critics have tended to focus on details of groups such as reptiles and birds, which were never the central focus of Smith's own experimental research.[32] Given the complexities, one can perhaps sympathize with Eli Marshall's decision to abandon phylogeny after the 1930 article. According to Marshall's biographer, the decision "freed him from the intellectual torments that Homer Smith suffered as he followed and classified the twisting trails of renal physiology decade after decade."[33]

The Comparative Physiology of Marine and Freshwater Fish

The anatomical differences in the nephrons of marine and freshwater fish paralleled two broad homeostatic patterns corresponding to problems of water and ionic balance now familiar to every student.[34] Because their body fluids have a lower ionic content than seawater, marine fish continually lose water by osmosis. Although scales provide some protection, water loss is inevitable across exposed membranes of the gills. Thus, a marine fish constantly faces dehydration. Freshwater fish face the opposite problem and have to combat the constant influx of water that dilutes their body fluids.

Smith played a key role in understanding the broad pattern of renal and extrarenal adaptations that marine fish use to regulate water balance. By adding phenol red dye to aquarium water, Smith found that marine teleosts, or bony fish, drank seawater, as demonstrated by the presence of the dye found in the digestive tract. He also discovered that water was absorbed along with some ions through the gut wall.[35] Although drinking replaced body water lost by osmosis, the marine fish faced the problem of excreting the large amounts of sodium, chloride, and magnesium ions contained in seawater. Fishes' kidneys are incapable of producing urine more concentrated than the blood, so this organ could not be primarily responsible for either water or ionic balance—other organs must be involved. Smith found that both sodium and chloride levels decreased as fluid passed through the gut, but magnesium concentrations increased. Smith reasoned that the movement of sodium and chloride across the gut wall was responsible for the osmotic movement of water from the gut into the body fluids. Some nonrenal mechanism eliminated these excess ions, and Smith proposed that the gill membranes were most likely pumping sodium and chloride into the surrounding seawater against a concentration gradient. Although the exact mechanism of this movement was unknown, it clearly required an expenditure of energy. Conversely, Smith reasoned that magnesium and other divalent ions remaining in the gut exited through the anus. Gills, rather than the kidney, also proved to be important for removing the nitrogenous wastes from protein metabolism. By placing fish in a divided chamber with a rubber seal separating the head (including the gills) from the rest of the body, Smith measured increases in ammonia and urea in water in the front section of the chamber but not in the back.[36] Smith concluded that these diffusible substances were moving passively from higher concentration in the blood to lower concentration in the surrounding water across the gill membrane. Thus, fish used discrete anatomical structures in different parts of the body to accomplish important excretory functions combined in the mammalian kidney.

Important exceptions to this basic pattern were the elasmobranchs, which include sharks, rays, skates, and related cartilaginous fish. Smith was particularly interested in the unusual mechanism that this group had evolved for maintaining water and ionic balance.[37] Elasmobranchs maintain water balance by retaining urea in the body. Indeed, in most cases the levels of urea are sufficiently high to make these marine fish slightly hypertonic, reversing the flow of water from outward to inward. Although relatively nontoxic compared to other organic forms of nitrogen, urea in elasmobranchs was at a much higher concentration than in any vertebrate other than the estivating lungfish that Smith was also studying at the time. This "uremia" was higher than in pathological cases of human renal failure, and this led some zoologists, notably Joseph Needham, to characterize the elasmobranchs as a primitive or degenerate group. Smith came to see the group in quite a different light. As he pointed out, even in renal disease uremia was a symptom, not a cause, and high levels of urea were usually harmless. Furthermore, there was no evidence that sharks and their relatives were at a competitive disadvantage to other marine fish. Indeed, because they needed to excrete the excess water that entered their bodies, elasmobranchs had retained well-developed glomeruli. This evolutionary retention had allowed some sharks to reinvade brackish or freshwater environments. From Smith's perspective, the retention of the glomerulus made the entire group preadapted for this type of evolutionary radiation back into the ancestral vertebrate habitat.[38] Thus, elasmobranchs had escaped the blind alley that marine teleosts entered when they lost their glomeruli. According to Smith, "the accumulation of urea automatically liberates the organism from osmotic enslavement to a marine environment, and relieves it of the necessity of continuously doing osmotic work at the gills by hypertonic salt excretion, as is the case in the teleost."[39] Both physiologically and evolutionarily, the elasmobranchs had attained a high level of freedom and independence from the external environment.

Smith was less directly involved with studies on the physiology of freshwater fish, but the broad pattern elucidated by the Danish physiologist August Krogh and others dovetailed neatly with the mechanisms of water balance in marine teleosts.[40] Faced with the problems of constant influx of water by osmosis, freshwater fish do not drink, and they excrete copious dilute urine. Inevitably, some necessary ions are lost in the process. The fish compensates this loss by actively transporting sodium, chloride, and other ions from the dilute watery environment into the body across the gills. Both freshwater and marine teleosts use gill-mediated mechanisms for ionic regulation, although the direction of movement is opposite in the two groups. In both cases the movement of ions against a concentration gradient is an energy-requiring process. Working in Krogh's lab at the University of Copenhagen, the polymath Ancel Keys measured the rates of ionic transport in eels using a complex heart-gill perfusion and identified the cells likely involved with the transport.[41] This was an early demonstration of the process of active transport, and Smith approvingly noted that Keys had shown that the transport of ions by gills of fish was comparable to the salt-concentrating work in the kidney of mammals.[42] Although no fish or other nonmammalian vertebrate could produce hypertonic urine, Smith pointed out that the gills were a nonrenal mechanism for accomplishing the equivalent physiological function.

Based on the survey of renal anatomy in various fish, the comparative physiology of water balance in freshwater and marine fish, his understanding of renal physiology in humans and other mammals, and a broad review of the existing literature on other vertebrate groups, Smith articulated three tightly linked arguments for phylogenetic reconstruction. First, the glomerulus was a unique vertebrate structure that had evolved to remove excess water from the body. Second, the later evolution of the glomerulus strongly correlated with habitat: freshwater vertebrates retained well-developed glomeruli, while in marine vertebrates these were lost or reduced. To prevent water loss, the nephrons of terrestrial vertebrates other than mammals also were evolutionarily reduced. Third, among vertebrates, only mammals (and perhaps some birds) were able to form urine more concentrated than the body fluids. In the mammalian lineage, the tubules leading from the glomerulus had become specialized to reabsorb needed ions and other solutes, while actively secreting others. In particular, the hairpin loop of Henle played a key role in this process, although the cellular and molecular mechanisms involved were only starting to be unraveled during Smith's lifetime. Enough was known, however, to recognize that the mammalian kidney had become a "master chemist" capable of regulating the internal environment to maintain water and ionic balance, pH, and blood pressure, as well as eliminating the nitrogenous waste products of metabolism.[43] The multitasking nephrons in the mammalian kidney allowed unprecedented freedom and independence from the vicissitudes of the external environment. With tongue in cheek, Smith noted in a lecture on the evolution of the kidney, "Superficially, it might be said that the function of the kidneys is to make urine; but in a more considered view one can say that the kidneys make the stuff of philosophy itself."[44]

From Fish to Philosopher

From Fish to Philosopher was the epitome of Smith's career as a comparative physiologist and evolutionary thinker. Written in an engaging style and aimed at a general audience, it was a commercial success reprinted several times. Like many popular scientific books, however, *From Fish to Philosopher* was motivated by more than profit or public education. Smith also provided a succinct and synoptic argument for his phylogenetic claims about vertebrate evolution and the evolutionary development of the kidney combined with a concise critique of the opposing view that vertebrates had originated as marine organisms. The forty pages of detailed endnotes suggest that he directed this argument toward professional biologists, as well as general readers. Perhaps just as importantly, Smith juxtaposed his broad evolutionary philosophy against those of other prominent evolutionary biologists with whom he profoundly disagreed. This required a nuanced approach, because despite serious philosophical disagreements, Smith was intellectually indebted to evolutionary biologists, notably George Gaylord Simpson and Alfred S. Romer. Because Smith relied on both of these scientists to support his claims about vertebrate evolution, he muted his criticism of their broader evolutionary views. Nonetheless, Smith borrowed from them selectively, and he could not completely disguise the great intellectual gulf that separated his philosophical views from Simpson's and Romer's, particularly on human nature and evolutionary progress.

Despite his attempts to distance himself from the notion that evolution is directional, the title of his book retains a whiff of progress. It would have been easy for general readers to find linearity in Smith's vivid presentation of the evolutionary development from the primitive kidney that allowed the first vertebrates to survive in freshwater, to the multitasking mammalian kidney that liberated the human organism from the external environment to such an extent that consciousness and creativity were possible. Indeed, the science editor for the *New York Times* described Smith's evolutionary account as a "wonderful pageant" culminating with "man as nature's masterpiece."[45]

This apparent progress had occurred not due to any necessity or purpose, Smith continued to argue, but to random mutations sifted by natural selection. "Progress can take a variety of forms," Smith wrote, "but it is only in the development of increased physiological independence of environment, which for mobile forms, involves increased awareness and perception of the environment and increased ability to react accordingly, that we can refer to evolution as upward rather than just sideways."[46] Although the phylogenetic tree that Smith drew was indeed largely horizontal, mammals are perched on the limb branching at the upper right corner of the diagram—and the marine teleosts with their degenerate nephrons on a slightly descending set of branches beneath. This phylogenetic account highlighted the uneasy tension in Smith's combination of Bernard's free and independent life with Darwinian evolution. Evolution inevitably led to extinction, and Smith continued to emphasize how both specialization and degeneration of structures often led to the type of evolutionary "blind alleys" that he had described in the lungfish and marine teleosts. By giving up the glomerulus to meet the osmotic challenge of living in salt water, the marine teleosts had entered a kind of Faustian bargain that "enslaved" them to the marine environment and seemingly prevented any future possibility of reentering freshwater. The evolution of the lung from a primitive swim bladder preadapted some vertebrate groups for the conquest of the land but condemned the hapless lungfish to a muddy "imprisonment" for much of its life. In comparison, Smith seemed to suggest, the freedom and independence from the external environment afforded by the mammalian kidney allowed for a progressive evolutionary enlargement of the brain. But although he was deeply interested in consciousness and creativity, Smith drew a gloomy conclusion from this evolutionary relationship between kidney and brain. "The kidney is required to operate only upon our internal environment whereas the brain is free to operate upon as much of the cosmos as it can apprehend. It is perhaps this imbalance between demand and operational capacity that accounts for the fact that so many more hospital beds are filled by subjects of mental disease than with renal disease."[47]

In the end, a big brain and efficient kidneys did little to change the existential dilemma that linked humans with all other living creatures. Returning to an analogy that he had first developed in *Kamongo*, Smith wrote that "the momentum that carries the simplest organism on and on in its never-ending battle with its environment, actively opposing every circumstance that tend[s] to slow it down. Won't die!—Must die! . . . Won't die!—Must die! . . . I conceive of life as some whirlpoollike deflection of energy which has beat its way up the long road of evolution, into every organismal reaction, into every homeostatic state, in every adaptation [ellipses in original]."[48] For Smith, life was "perpetually in search of the free and independent life, perpetually failing for one reason or another to find it."[49]

Smith's emphasis on the incessant struggle for existence, the inevitability of extinction, and the fact that once lost a character is unlikely to evolve again regardless of need, placed him well within orthodox Darwinian thought. Nonetheless, his emphasis on the dire consequences of overspecialization and his descriptions of specialized species being in a phylogenetic "blind alley" or "enslaved" or "imprisoned" by their environments evinced an evolutionary fatalism not necessarily shared by other evolutionary biologists.[50] For example, recent critics have pointed out that the reduction or loss of the glomerulus in marine fish, birds, and reptiles did not limit the evolutionary potential of these successful vertebrate groups, because other renal processes were amplified to accomplish homeostatic functions typically associated with glomerular filtration.[51] Within his own historical context, Smith's evolutionary philosophy was also more pessimistic than alternatives proposed by prominent evolutionary biologists such as Simpson. From reading Simpson and other contributors to the modern synthesis, Smith's ideas on evolutionary innovation shifted in subtle ways, but in the end he remained committed to a reductionist and mechanistic materialism that went well beyond what Simpson endorsed.

Simpson's most popular book, *The Meaning of Evolution* (1949), particularly influenced Smith. He referred to it repeatedly in *From Fish to Philosopher*, although Smith was highly selective in the passages that he borrowed. *The Meaning of Evolution* originated in the Terry Lectures that Simpson had delivered at Yale University in 1948. The lecture series was devoted to discussions of science and religion, and although an atheist, Simpson presented views on progressive evolution and human nature that might have been acceptable to some liberal theologians. In doing so Simpson's evolutionary philosophy was in strong opposition to some of the views that Smith had articulated repeatedly, beginning in the 1930s with *Kamongo*. Like other leaders of the modern synthesis, Simpson advocated a form of evolutionary humanism that struck a balance between mechanistic materialism and purposive progressionism.[52] He was a strong selectionist and adaptationist but also a committed humanist. Simpson criticized what he considered extreme views on these issues, and he formulated his own delicate balance by viewing humans as a product of evolution, but in their social and cultural development "an entirely new kind of animal."[53]

Throughout his career Simpson criticized paleontologists who misused assumptions of progress and linear trends to interpret the fossil record. He also distanced himself from the idea of overall evolutionary progress with humans as the endpoint, so forcefully argued by Julian Huxley.[54] By contrast,

Simpson held a more pluralistic view of progress.[55] He denied progress had any meaning in biology without specifying criteria for measuring it a priori. Progress was always relative to some point of reference, and even then it was prone to misuse. Simpson balanced this criticism of simple-minded anthropocentrism with equal disdain for biologists who rejected anthropocentrism out of hand. Anthropocentrism could be justified in a limited and highly constrained way because humans are naturally inclined toward a human-centered understanding of the world.

In the chapter "Man's Place in Nature," Simpson made a stronger argument that humans are unique but also that they are the highest form of animal life. Humans are so different from all other animals, he argued, that if a fish could think, it would be amazed that any human would question the proposition that *Homo sapiens* is the highest form of life. Of course, he added, if the fish could think in such an abstract way, it would, in fact, have become a human. Humans' unique intellect allowed them to know that they are products of evolution but also to free themselves from many evolutionary constraints, radically modify environments, and perhaps direct the future course of evolution. This power—not always used for good—also entailed a unique sense of responsibility, ethics, and values. "It is important to realize that man is an animal," Simpson wrote, "but it is even more important to realize that the essence of his unique nature lies precisely in those characteristics that are not shared with any other animal. His place in nature and its supreme significance to man are not defined by his animality but by his humanity."[56] Simpson was sharply critical of unnamed biologists and philosophers who argued that humans are "nothing but animals." Such thinking was not only false; it was viciously misleading, because it distorted human nature and undermined human values and ethics.

Smith could easily have been one of those unnamed biologists targeted by Simpson. Contrary to Simpson's claim for the uniqueness of human beings, including their highly developed sociality and unparalleled freedom from the environment, Smith viewed the human condition in terms of individual struggle with the environment no different from that faced by all other living beings. Stating a belief in the "essential indignity of man," Smith concluded, "He is part of the natural universe, and his problem like that of every other natural organism, is to seek the maximization of his individual freedom against storms and droughts, depressions, parasites, plagues, and, not least, other men."[57] When he wrote *From Fish to Philosopher*, Smith referred approvingly to Simpson's rejection of purpose and overall direction in evolution, without mentioning the great paleontologist's guarded defense of anthropocentrism in biology and the idea of a limited progressive evolution.

Although much less accommodating of these ideas, Smith accepted an even more restricted evolutionary progressivism and tied it to homeostasis. In *From Fish to Philosopher,* he repeated the claim that he had periodically made since *Kamongo* that progress could be identified with those adaptations that had allowed certain lineages greater freedom and independence from the environment. To strengthen this position, he enthusiastically adopted Simpson's view of natural selection as a creative and constructive process.[58] After reading Simpson, he had come to see natural selection not simply as a winnowing process but also as a mechanism for assembling adaptive combinations of random mutations. Viewed in this way, natural selection could repurpose old, and sometimes not very useful, structures in new adaptive ways. This emphasis on creativity partly counterbalanced the inherent pessimism in Smith's earlier thinking. It provided a way for Smith to discuss innovation and apparent progressive trends in evolution without abandoning his commitment to a reductive, mechanistic materialism. It also harmonized with his view of homeostasis as a dynamic opposition between organism and environment. For Smith, evolutionary homeostasis also came to mean a balance of maintaining adaptive structures while turning other structures to new purposes. In successful lineages this also meant avoiding overspecialization that led to evolutionary blind alleys.

Smith highlighted this evolutionary creativity and homeostasis in his discussion of the unique adaptations of elasmobranchs, which he contrasted with the more specialized physiology of marine teleosts. By using elevated levels of urea to maintain water balance, sharks and their relatives had maintained well-developed nephrons. From Smith's perspective the retention of the glomerulus allowed great evolutionary flexibility unmatched by any marine teleost. Although the high urea content prevented most sharks from surviving a rapid transfer from salt water to freshwater, over the course of evolutionary time several species had reinvaded brackish and even freshwater environments.

The retention of urea in elasmobranchs was a unique adaptation that allowed an unusually high level of freedom and independence from the external environment. Using a metabolic waste product to maintain osmotic balance was a prime example of the "putting of old things to new uses," and doing so had spared the nephron from the degeneration that had occurred in other marine fish.[59] One supposes that Smith took some pleasure in recasting elasmobranchs from the role of pathological misfits to evolutionary heroes. The only time that Smith explicitly referred to homeostasis in *From Fish to Philosopher* came when he described the unique adaptations for water balance found in the group, writing, "It is one of the most strikingly simple means known for automatically maintaining an important homeostatic state."[60]

Smith also applied natural selection as a creative process to the mammalian kidney, although he walked a fine line to avoid presenting this organ as the terminus of a progressive evolutionary development. As the leading authority on the kidney, Smith could marvel that it had become a "master chemist" capable of maintaining the constancy of the internal environment to an unprecedented degree. At the same time, he described the kidney as a kind of Rube Goldberg device that natural selection had cobbled together from existing structures, some repurposed to improve the reabsorption of water and necessary solutes while excreting metabolic wastes. The result was an organ unique in its ability to produce a highly concentrated urine, but by an improbable process that no engineer would design.[61] As in other vertebrates, the excretory function was secondary to maintaining water and ionic balance, but the mammalian kidney had an unparalleled ability to keep the internal environment in an "ideal, balanced state."[62] Not only had the mammalian kidney provided the necessary milieu for the evolution of thermoregulation and a brain capable of philosophizing, but the very integrity of the body depended upon the multiple functions of the organ. According to Smith, "Bones can break, muscles can atrophy, glands can loaf, even the brain can go to sleep, and endanger our survival; but should the kidneys fail in their task neither bone, muscle, gland nor brain could carry on."[63] A single organ could do what required the combined efforts of the kidney and nonrenal tissues such as gills and skin in other vertebrate groups—and it could do it more efficiently.

The philosopher's kidney as the apparent terminus of renal evolution certainly gave Smith's account the appearance of progress, and this sense reinforced his rather curious treatment of birds. During the 1950s the avian kidney was less understood than its mammalian counterpart, although it was generally believed that birds were incapable of concentrating urine to the same extent that mammals can. Some of the nephrons in the bird kidney contain loops of Henle, but others do not. Even when present the loops are shorter than in most mammals. But as Smith was also aware, some marine birds have well-developed nasal glands that excrete excess sodium chloride. In addition, birds and reptiles had evolved a unique method for excreting nitrogenous waste not in the urine but in a concentrated paste eliminated along with the feces. Because uric acid contains more nitrogen than urea and because it is relatively insoluble, this mechanism efficiently removed metabolic wastes from the body while conserving water. In short, the avian kidney and digestive tract work together to accomplish what the mammalian kidney does by itself.

Comparing the birds and mammals might have provided an opportunity for Smith to elaborate on how natural selection follows alternative pathways to create equivalent adaptations for salt and water balance. However, from the beginning of his phylogenetic research, Smith had difficulty fitting birds into his scheme. He always presented the group to be degenerate in comparison to mammals. In taking to the air, birds had literally flown into an evolutionary blind alley.[64] Indeed, in *From Fish to Philosopher* Smith argued that wings were a major cause of the problem. These structures were a specialization that constrained future evolutionary potential by limiting manual dexterity. Furthermore, he argued that wings had allowed birds the luxury of migrating from hostile environments, while mammals had to tough it out against inclement environmental challenges. These were quite unusual claims, given the ecological and evolutionary success of birds, including the ability of desert species to maintain water balance in extremely hot, dry conditions. It was a curious and quite unsatisfactory account that highlighted the difficulty that Smith faced in freeing his thinking from a lingering evolutionary fatalism and a bias toward mammals, perhaps inherited from his work in medical physiology. Taking a more nuanced perspective on adaptation and convergent evolution might have allowed a less tortured account of birds and marine teleosts, for which he was later criticized.

Later criticisms notwithstanding, Smith highlighted the importance of understanding physiological function within a broad ecological and evolutionary context. His early publication of research on the lungfish in *Ecology* detailed the organism's unusual adaptations to a fluctuating environment. His later phylogenetic survey of vertebrates explicitly linked physiology and anatomy to environmental constraints. His evolutionary perspective on the relationship between organism and environment deepened and broadened the early insights of Bernard and Cannon. To a greater extent than most medically oriented physiologists, Smith was able to move beyond a narrow dichotomy between normal and abnormal function. Understanding homeostasis in evolutionary and ecological contexts focused Smith's attention on the whole organism as an adapted and adaptable agent. His forays into renal physiology beyond the confines of the biomedical laboratory contributed importantly to the development of physiological ecology after World War II. As we shall see, Smith used his professional stature to encourage younger physiologists entering the new field and to actively promote their work.

4

Living Water

Near the end of *From Fish to Philosopher*, Homer Smith included a chapter entitled "Animals That Live without Water," devoted to life in the desert. It was the most ecological chapter of Smith's book, detailing the diversity of habitats categorized as deserts, their physical characteristics, and the physiological demands placed on both plant and animal life in arid environments. Central to the discussion were mammals, which face the conflicting demands of self-regulating temperature and water. Cooling the body either by sweating or panting requires evaporative water loss, thus working at cross-purposes with the animal's need to prevent dehydration. Balancing these demands requires an integrated set of physiological and behavioral adaptations. Remarkably, Smith reported, some small desert mammals never drank liquid water but apparently subsisted entirely on metabolic water derived from the breakdown of proteins, fats, and carbohydrates.

Despite the fact that he was reporting secondhand the research of younger physiological ecologists, Smith's account of the integrated homeostatic processes allowing small rodents to survive, and even thrive, in the desert revealed an infectious enthusiasm. Kangaroo rats and other diminutive desert mammals became exemplars of homeostatic regulation and adaptation to a hostile physical environment. At least partly because of the writings of Smith and other popularizers, studies of kangaroo rats became early classics of physiological ecology (a topic that will be discussed in the next chapter). Accounts of the animal's ecology, physiology, and behavior are still used in biology textbooks more than half a century after the original research was completed.

Somewhat buried in his discussion of nonhuman desert dwellers was Smith's acknowledgment that the physiological ecology of adaptations to the desert owed a great deal to military research conducted during World War

II, particularly by Edward F. Adolph, a physiologist in the medical school at the University of Rochester. Reporting his studies conducted primarily on soldiers, Adolph's *Physiology of Man in the Desert,* published shortly after the war, was immediately recognized as a landmark scientific study.[1] Smith and other physiologists applauded the book for basing the challenges of life in the desert on experimental evidence, rather than anecdotes. Adolph's interest in the desert predated his important wartime research. Previously, he had been a member of a team of researchers hired to study the debilitating effects of heat and thirst on construction workers at the Boulder Dam (later renamed the Hoover Dam). The research convinced Adolph of the precariousness of human life in the desert—even as the new dam provided water and electricity that propelled explosive growth of cities and farms in the desert Southwest. His research on both humans and nonhuman animals also reinforced Adolph's conviction that homeostasis was more complicated than the simple examples provided by Cannon's *Wisdom of the Body.*

A Critique of Homeostasis

In 1961 Adolph wrote a history of self-regulation—an area of research in which he claimed some expertise.[2] Motivating the review was Adolph's irritation that physiologists assumed that Walter Cannon originated the modern field of study with his idea of homeostasis. But Adolph claimed that Cannon's homeostasis was merely an "invention" for ideas that he traced back to Hippocrates. Each generation of scientists had fashioned its own notions of stability, balance, and equilibrium. These ideas had not so much developed over time as "crystalized" within particular historical and cultural contexts.[3]

According to Adolph, the current enthusiasm for homeostasis was only partly due to Cannon's explanations of neural and endocrine control of bodily functions. Less propitiously, it also rested on a facile fascination with self-regulating machines that seemed to mimic biological systems. Just as an earlier generation of physiologists invoked vital forces to explain physiological phenomena, Adolph claimed that late twentieth-century physiologists uncritically turned to automatic control systems as analogies for biological self-regulation.[4] Adolph's impatience with contemporary notions of homeostasis was rooted in his deep mistrust of analogizing physiological self-regulation to thermostats, circuit breakers, or other single-purpose control mechanisms. In part, this critical view stemmed from Adolph's pioneering studies of life in the desert. His research on temperature regulation and water balance dramatically demonstrated that two regulatory functions sometimes acted at cross-purposes. Sweating to stay cool inevitably resulted in water loss. Alluding to the popular metaphor of the thermostat, Adolph insisted that understanding temperature regulation required considering the entire house, not just the furnace and its thermostatic control mechanism.[5] According to Adolph, Cannon and his followers ignored this important point by focusing on isolated mechanisms, rather than on the organism as an integrated whole.[6] In contrast, Adolph tied his own important insights about the physiology of desert animals to an older holistic perspective on the fit between organism and environment. He made no secret about his intellectual debt to his mentors Lawrence J. Henderson and John Scott Haldane, whose views he contrasted with Cannon's overly mechanical concept of homeostasis.[7]

Adolph's repeated critiques of homeostasis are intriguing, if enigmatic. In an unpublished fragment, Adolph contrasted his intellectual debts to Henderson and Cannon, highlighting the differences in intellect and temperament of the two older scientists. Although expressing admiration for Cannon's *Wisdom of the Body*, Adolph claimed that Henderson was the more creative thinker and perhaps had instigated Cannon's initial thinking about homeostasis. From Adolph's perspective Henderson's holistic account of the fit between organism and environment was more compelling than Cannon's mechanistic and reductionist account of homeostasis. Adolph wrote the brief sketch as part of a historical article on ideas of self-regulation, but he removed his frank reminiscences before submitting the paper for publication.[8] Intellectual differences aside, Adolph may have felt some defensiveness toward homeostasis because his own book, *Physiological Regulations*, published in 1943, was less successful than *The Wisdom of the Body*. Physiologists routinely cited Adolph's work, but the book failed to match the broad popularity and wider influence of Cannon's account of homeostasis.

It is ironic that Adolph's critique of homeostasis and his attempts to erect an alternative philosophical understanding of self-regulation had a limited impact, while his research on desert physiology exerted an important influence on physiological ecology after World War II. The two sides of his career would seem closely related, because life in the desert requires exquisite regulation of body temperature in the face of excessive heat, and water balance in the face of scarcity. How organisms meet this dual challenge poses questions of perennial interest to physiologists, ecologists, and evolutionary biologists—not to mention the military. During World War II, Adolph led a group of researchers investigating the ability of humans—specifically soldiers—to survive and work in the desert. In the preface of *Physiology of Man in the Desert*, Adolph described the work as a contribution to a newly emerging field of environmental physiology.[9] This work served as an important point of departure for postwar studies of how mammals and other animals adapt to life in the desert. Despite his criticism of homeostasis, Adolph's wartime research also contributed more broadly to the development of a physiological ecology focused on adaptation and self-regulation in a wide range of hostile environments.

Early Research on "Living Water"

After earning a BA from Harvard University in 1916, Adolph began graduate studies in chemical physiology and zoology, only to have his education interrupted by World War I. During the war he worked in hospital laboratories of the US Army Medical Services. Afterward he returned to Harvard to complete a PhD under the direction of Henderson. He then held a Sheldon Traveling Fellowship at Oxford University in 1920–21, where he worked in Haldane's laboratory. Adolph later acknowledged the important intellectual influence of these prominent mentors, noting that they "perhaps unwittingly implanted a curiosity about the regulative aspects of physiology."[10]

During his year at Oxford, Adolph conducted a series of experiments on excretion with Haldane's associate John Gillies Priestley.[11] These experiments were an outgrowth of water-drinking studies published by Haldane and Priestley five years earlier. Using themselves as experimental subjects, Priestley and Adolph measured the chemical composition and quantity of urine produced after drinking a variety of isotonic, hypertonic, and hypotonic solutions. In one test Priestley was able to drink an impressive 5.5 liters of water in three hours, which far exceeded his maximum urine production of 800 ml per hour. Urination remained elevated for several hours after he finished drinking, although the chloride concentration of the urine sharply declined as the experiment progressed. By contrast, the concentration of hemoglobin in Priestley's blood remained constant throughout the experiment, suggesting that little of the excess water was stored in the circulatory system.

Adolph was not able to ingest such prodigious quantities of water, but the results were qualitatively similar: he excreted all of the ingested water over a period of several hours. Any tendency for the body to lose electrolytes was compensated by a reduced excretion of these dissolved substances as the experiment continued. When the researchers consumed similar amounts of an isotonic salt solution, the results were quite different, because the body required nearly twenty-four hours to excrete the excess water and salts. Consequently, Adolph and Priestley measured a slight dilution of hemoglobin in the blood and an increase in body weight. Ingesting strongly hypertonic salt water resulted in decreased production of concentrated urine, requiring in some cases more than a day to rid the body of excess salt.

The results of these experiments cast doubts on earlier claims that the body could store excess water and hold it in reserve much the way fat is stored, an important point that would be made more forcefully two decades later during Adolph's studies on soldiers in the desert. For Adolph, water balance seemed to be an exquisite example of the body's ability to regulate its overall composition and function. Adolph aptly captured his lifelong fascination with water balance in the title of an article later published in the *Quarterly Review of Biology*, "Living Water."[12] Water was not simply a fluid, solvent, or chemical substance. It was an integral part of the complex living substance of the organism, which was precisely regulated in the face of environmental fluctuations.

Returning to the United States, Adolph worked as an instructor in the Department of Zoology at the University of Pittsburgh for three years and taught summer courses at the Marine Biology Laboratory in Woods Hole, Massachusetts. During these early years, he continued the self-experimentation that he had begun at Oxford to understand how the body maintained water balance while excreting metabolic waste products and excess ions.[13] Using laboratory facilities provided by the US Bureau of Mines in Pittsburgh, Adolph completed a series of over fifty daily recordings of his own water intake and urine production. These experiments set a pattern that his research would follow throughout his career. He made careful quantitative measurements of the amount of urine and also the concentrations of urea, chloride, bicarbonate, and other excreted substances. He directed this input-output approach at the level of the whole organism, with little emphasis on the functioning of individual tissues or organs. Adolph presented the results in graphical form that highlighted the correlations that he found between water and various other constituents of the urine.

After leaving Pittsburgh, Adolph held a one-year National Research Council fellowship at Johns Hopkins University before moving to the physiology department at the University of Rochester School of Medicine, where he spent the rest of his career. The brief stint at Johns Hopkins was particularly important for broadening Adolph's interests in comparative physiology. Encouraged by Herbert Spencer Jennings and the cellular physiologist S. O. Mast, Adolph began experiments on water balance using a wide variety of organisms, including protozoans, earthworms, daphnia, and frogs.[14] What united this eclectic set of studies was Adolph's opposition to the idea that osmosis or any other purely physical process could completely explain water balance in organisms.

A series of quantitative studies of the contractile vacuole in the protozoan *Amoeba proteus* illustrated the difficulty in defending this claim. Using hanging drop preparations, Adolph timed the formation, growth, and expulsion of vacuoles in amoebae placed in media of different salt concentrations. He measured the diameters of vacuoles using a micrometer and attempted to estimate the total volume and surface areas of the protozoans. Based on these experiments, he criticized earlier claims by the British zoologist Marcus Manuel Hartog that contractile vacuoles served to remove excess water that entered the organism by osmosis. According to Adolph, Hartog's conclusion rested on only "a modicum of experimental evidence," but primarily on comparisons with parasitic and marine amoebae that lacked contractile vacuoles. The lack of vacuoles in protozoans living in the isotonic environment of a host or the hypertonic marine environment was consistent with the osmotic hypothesis, but Adolph remained skeptical. Although admitting that his own measurements were subject to error, Adolph claimed that there was no consistent effect of salt concentration on the growth of contractile vacuoles. Rather, the growth of the vacuoles appeared more closely correlated to the volume of the amoeba than to its surface area. This suggested to Adolph that the vacuole was removing metabolic wastes rather than eliminating excess water. Although he cast some doubt on the earlier claims of Hartog about osmosis, Adolph was forced to admit that the results of his own experiments could not conclusively explain the activity of vacuoles in microscopic protozoans.

The results of Adolph's experiments with frogs were equally ambiguous. He attempted to demonstrate the importance of the skin as well as the kidneys in water regulation.[15] He determined the water content of his experimental subjects by measuring changes in body weight and volume of urine produced in frogs placed in solutions of different salt concentrations. He also measured water movement across isolated pieces of skin stretched over thistle tubes. He skinned frogs in an attempt to determine the external surface areas of his experimental animals, although measuring the irregular surface proved frustratingly difficult. Adolph's results supported the hypothesis that water balance involved more than kidney function. Urine production did not increase in frogs placed in various hypotonic solutions. Surgically removing a kidney did not result in weight increase, even though the remaining kidney produced only half as much urine as the original pair. When he tied off the cloaca to prevent urination, his experimental frogs were still able to maintain water balance even in hypotonic solutions. All of these results pointed to the important regulatory role of the skin in maintaining water balance in amphibians, although Adolph strongly resisted any simple, mechanistic explanation for these phenomena. Summarizing five years of research in his article "Living Water," Adolph concluded, "The skin is not a mere 'sieve,' or 'semi-permeable,' or possessed of 'one-way permeability'; instead it is the seat of certain forces which modify the exchanges of water in compliance with conditions."[16]

Although he admitted that the mysterious "forces" were presently unknown, he suggested that further studies of electrical conductivity, aerobic metabolism, and temperature effects might shed light on the organism's ability to regulate water movement across the skin. Central to his perspective was Adolph's belief that organisms actively regulated the internal environment in a holistic way that involved the coordinated activity of multiple organ systems irreducible to simple, passive mechanisms such as diffusion or osmosis, or to the activities of single organs such as the kidney.

Adolph's *Physiological Regulations* expanded this holistic approach to understanding water balance but also highlighted difficulties in applying the perspective in a concise and convincing way. When Adolph approached Charles Thomas about the book in 1940, the publisher expressed interest, although even before seeing the manuscript, Thomas had reservations about the title.[17] *Physiological Regulations* struck the publisher as rather dull—"Let's try to brighten up that title," Thomas suggested. None of the alternatives that Adolph suggested pleased Thomas, and after reviewing the manuscript, the publisher rejected it. Skeptical that the book would appeal to even a modest scientific audience, Thomas advised the liberal application of a "blue pencil" to the five-hundred-page manuscript to make the argument more concise and tightly focused. The manuscript met a similar fate when Adolph sent it to two other publishers, both of whom doubted that the book would be commercially successful. Adolph finally convinced Jacques Cattell to publish *Physiological Regulations* three years later, but only after the University of Rochester agreed to pay half of the costs for a limited run of eight hundred copies.[18] When the ichthyologist Carl Hubbs wrote a review for the *American Naturalist*, he expressed disappointment that despite its title Adolph's book was not a general treatise on self-regulation, but a rather amorphous collection of data without an adequate explanatory framework or connections with other physiological research on osmoregulation.[19]

Unfortunately, Adolph wrote *Physiological Regulations* at the same time that the Nobel laureate August Krogh was preparing a much more influential book on water balance. Despite the growth of the field, Krogh noted that at the end of the 1930s much remained obscure about osmoregulation. The time was ripe for a careful stocktaking, and in his *Osmotic Regulation in Aquatic Animals*, Krogh provided a concise but probing analysis of studies on a wide range of organisms from protozoa to mammals.[20] The book set out to describe the osmotic steady states maintained by organisms and to explain the various mechanisms responsible for these dynamic relationships.[21] From Krogh's perspective these steady states were fundamentally cellular phenomena. Cells, on a "crude first approximation," were droplets of a solution separated by a semipermeable membrane from another solution—usually with a very different ionic concentration. For protozoans and other unicellular organisms, this semipermeable membrane was the boundary between internal and external environments. Although in less-intimate contact with the external environment, the same cellular principles applied within multicellular organisms. The regulation of water and ions always involved semipermeable membranes that allowed certain substances to pass more or less freely while restricting the movement of others. These movements included the physical processes of diffusion of ions and osmotic flow of water from higher to lower concentrations, and probably an energy-dependent active transport of substances against concentration gradients. This was a bold thesis, because biologists knew little about any of these processes at a cellular level. The existence of active transport, in particular, rested on logical necessity more than on experimental evidence.

Krogh combined a rare talent for designing careful and precise quantitative experiments, often using instruments of his own design, with an ability to criticize the weak points of competing arguments.[22] As a young scientist, he had performed a series of experiments that undermined the idea that the lungs can secrete oxygen into the blood against a concentration gradient. Krogh's teacher Christian Bohr championed this idea, as did Haldane, who played such a formative role in Edward Adolph's early career.[23] Krogh's experiments caused a rift between him and Bohr but also propelled the younger physiologist to a brilliant independent career. Although best known for his work in respiratory physiology, particularly the function of capillaries in supplying oxygen to muscle during work, Krogh devoted considerable attention to questions of water and ionic balance in the cells of various aquatic organisms from unicellular organisms to vertebrates. Krogh's young protégé Ancel Keys presented the first tentative demonstration of active transport with experiments on chloride ion transport across the gills of eels during the early

1930s. How this energy-requiring process worked on a molecular level was unknown. Nonetheless, together with osmosis and diffusion, active transport provided a simple conceptual explanation for transport of substances across membranes. These three cellular processes formed the intellectual framework for Krogh's *Osmotic Regulation in Aquatic Animals,* which quickly gained recognition as a critical assessment of the state of the field from the perspective of one of the leading practitioners.

Krogh was highly critical of Adolph's experimental techniques and theoretical interpretations, particularly in his early studies with amoebae. He was scathing in his criticism of the experimental evidence that Adolph had presented in opposition to the idea that the contractile vacuole was primarily an osmotic regulator. Rather than using modified Ringer's solution, Adolph had employed crude salt solutions that were harmful to protozoans. His experiments were too short to demonstrate changes in cellular volumes in hypotonic and hypertonic solutions. Krogh also claimed that by focusing only on the contractile vacuole, Adolph failed to measure changes in volume of the whole amoeba. The criticism was not quite fair, because Adolph had attempted to estimate cell volumes, but it highlighted Krogh's contention that Adolph's methods were imprecise and his conclusions ambiguous. Careful measurements in balanced ionic solutions by other scientists demonstrated to Krogh's satisfaction that filling and expulsion of the contents from contractile vacuoles was more rapid in hypotonic solutions than in hypertonic conditions. Other evidence also supported the osmotic regulatory function of the contractile vacuole. For example, temporarily damaging the vacuole by ultraviolet radiation caused the amoeba to swell until the vacuole began to function once again. Poisoning with cyanide also reduced the function of the contractile vacuole, resulting in swelling of the microorganism. Transferring the protozoan to uncontaminated water gradually reversed this effect.

Krogh published his classic study of osmotic regulation while Adolph was writing his *Physiological Regulations,* and one might have expected a rebuttal to the Danish physiologist's harsh criticism. Although Adolph listed *Osmotic Regulation in Aquatic Organisms* in his bibliography, he barely mentioned Krogh in the text. Adolph's curious skepticism about osmosis and semipermeable membranes, combined with his rather amorphous discussion of "forces" responsible for maintaining water balance, compared unfavorably with Krogh's concise set of cellular mechanisms.

Krogh's criticism notwithstanding, Adolph's antireductionism and his holistic emphasis on the organism interacting with its environment later proved more successful for studying human adaptation to desert environments, where water balance and temperature regulation were equally important challenges. It allowed Adolph to answer practical questions of considerable strategic interest to military planners. For example, is it possible to become acclimated to dehydration through continuous or repeated water loss? Can drugs counter the unpleasant effects of dehydration? What about simpler remedies such as sucking on a pebble or chewing a piece of rubber? Faced with inevitable dehydration on a forced march, is it better to drink available water all at once, or to ration it over the duration? How does clothing affect water loss and temperature regulation? Through a series of "sweat of the brow" experiments conducted during World War II on soldiers, university students, and scientists themselves, Adolph's team made a number of important discoveries, some of which were counterintuitive. Combining human experimentation with comparative studies on dogs and laboratory rats also raised new questions about how organisms, more generally, adapt to extreme environments. This early work provided a springboard for a nascent physiological ecology that rapidly expanded after the war.

Human Performance in the Desert

Adolph briefly considered the physiological effects of elevated temperatures in humans early in his career at the University of Pittsburgh.[24] Supported by the Bureau of Mines, the US Public Health Service, and the American Society of Heating and Ventilating Engineers, Adolph conducted experiments at high temperatures and humidity under standardized laboratory conditions meant to simulate the environment encountered in deep mines. Adolph closely patterned the study on earlier research dealing with temperature regulation and the causes of heat stroke conducted by Haldane and other physiologists. Experimenting on himself, as well as other human subjects, Adolph recorded the physiological effects of high temperatures (40°C) and humidity. All of the subjects reported discomfort, including muscle cramping and sensations of tingling in the skin, within a few minutes after entering the experimental chamber. Despite individual variation, most subjects found it impossible to endure these extreme conditions for more than an hour even sitting quietly. Standing or exercising greatly reduced the time most subjects could tolerate these conditions, although a few hardy individuals pedaled a bicycle ergometer for forty minutes in the moist heat. After sitting in the hot room for fifty minutes, Adolph's pulse rate nearly doubled, and his core body temperature steadily rose to 39°C during this experiment. Despite the high humidity, he lost nearly a kilogram of weight, primarily through sweating.

Adolph's brief foray into temperature regulation in humans resumed a decade later when David Bruce Dill invited him to join a research expedition to Boulder City, Nevada.[25] Heat-related deaths and disabilities among workers constructing the nearby Boulder Dam, later renamed Hoover Dam, were of such concern that the Bureau of Reclamation recruited Dill to conduct field research on water balance and temperature regulation using humans, dogs, and donkeys. Serving as director of the Human Fatigue Laboratory at Harvard University, Dill had already conducted physiological research in extreme environments, notably the high Chilean Andes.[26] Adolph had recently spent a sabbatical at the fatigue lab, but although he was part of the 1937 Harvard Desert Expedition to Bolder City and collaborated with the team, he published several articles independently. Both the field research in Nevada and related laboratory experiments at the University of Rochester built on his earliest water-drinking experiments done in Haldane's laboratory a quarter of a century earlier, as well as the Haldane-inspired experiments on temperature regulation conducted during Adolph's brief tenure at the University of Pittsburgh.

Unlike the earlier experiments, desert research brought Adolph face-to-face with the problem of adaptation to a complex and fluctuating natural environment, rather than a controlled laboratory setting. Because he and the other members of the expedition were experimental subjects as well as researchers, the close interaction between organism and environment was impossible to escape in the searing desert heat.[27] Strenuous exertion involved genteel activities such as hiking or playing tennis, rather than heavy, daylong construction work, but the physiological measurements left little doubt about the stress imposed on laborers at the Boulder Dam. Facing daytime temperatures that often exceeded 38°C in the shade, the researchers experienced problems of temperature regulation, water loss, and ionic balance that were both visceral and intellectually challenging. Indeed, personally experiencing the symptoms of these complex physiological processes was an important part of the research. Tables of data identified the researchers by initials, highlighting individual differences in physiological responses but also suggesting that an informal competition coincided with the objectivity of the measurements. The way that Adolph and Dill reported personal experiences with water loss and elevated body temperature reinforced this resemblance to a scorecard. For example, describing the differences, Adolph self-reported, "The greatest accumulations [of heat] were in subject A who perspired with difficulty and suffered discomfort." By contrast, he wrote, "Subjects who did not accumulate much heat were more comfortable, felt more fit, and were able to continue exercise for longer periods. Even these individuals, however, were limited in the rates with which they could do work for any considerable time."[28] From this highly subjective perspective, Adolph's holistic understanding of regulatory physiology expressed itself in a more compelling way than his earlier discussions of frog skin or the contractile vacuoles of protozoans. This narrative account also meshed well with the simple experimental procedures conducted in a makeshift laboratory in the Boulder City Municipal Building. The researchers monitored body temperature using rectal thermometers and evaporative water loss by changes in body weight. Despite limited laboratory equipment, the team also conducted chemical analyses of chloride levels in the blood. This was particularly important in documenting the loss of salt from perspiration and its deleterious consequences for construction workers.

Of particular interest to Adolph during the Boulder City study was the relationship between sweating and body temperature. Conceptually, one could construct a simple balance sheet of water intake via food and drink versus loss in sweat and urine, all of which the researchers accurately measured. Although human subjects tended to quickly acclimate to working in the hot desert climate, this came at the high cost of water loss. Even with decreased urine production, individuals consumed large amounts of water to compensate for the water lost through sweating. The amount consumed depended upon the environmental temperature and amount of activity, but could be as much as eleven liters of water per day. Copious sweating sometimes actually resulted in a drop in the surface temperature of the body during exercise, although Adolph reported an elevated core temperature reaching nearly 40°C while playing tennis in the sun. Although it might be common knowledge that heat stress is greater in sun than shade, the stress of exercising brought home the fact that body temperature was a complex result of metabolic heat production and solar radiation, as well as convection and conduction from hot surfaces.

Experimenting upon themselves, members of the team found considerable variation in sweat production. Adolph sweated less than others even after acclimating to the heat, and his body temperature continued to be elevated compared to his companions. Thus, although training led to an increased ability to work in the desert heat, considerable individual variation persisted. It also became clear that acclimation to heat did not decrease the need for water intake. Indeed, this need might actually increase. Working to near exhaustion,

Adolph's sweat production during his first week in Boulder City increased 10% over the baseline set at Harvard, but it nearly doubled a month later. The findings had practical significance both for newcomers to the desert and experienced laborers working on the dam throughout the year. Dill noted, "We have here an explanation of the tolerance for high temperature acquired as summer advances and of the disastrous effects of early heat waves on men doing hard work."[29] Equally important, the team found that during acclimation, although sweat production increased, it became more dilute as the body retained chloride ions. "If this observation is confirmed by further study it is quite significant," Dill concluded, "for it implies that the sweat glands respond protectively to excessive salt loss, even though the response is passive."[30] Despite this finding, the scientists recommended that workers at the dam site take salt tablets to prevent muscle cramps resulting from the inevitable loss of electrolytes.[31]

The human body appeared remarkably adaptable to the high temperatures encountered in the Nevada desert, at least so long as abundant water and, perhaps, salt tablets were available. However, an unexpected and surprising discovery complicated a neat balance sheet of body temperature and water. Adolph and Dill found that during heavy exertion, subjects often had little interest in drinking, and this resistance to restoring lost water persisted even after exercise. Indeed, subjects often reported being satisfied when they had consumed only enough water to replace half of what had been lost due to sweating. Furthermore, even restoring this fraction of lost fluid often took half an hour, and many subjects felt satisfied after taking a few large mouthfuls of liquid. To complicate matters, most drinking occurred during meals or in the evening before sleep, rather than immediately following exercise. As a result, there was a tendency for humans in the desert to remain in a continual state of dehydration, even when water was readily available.

Adolph later referred to this phenomenon as "voluntary dehydration," a physiologically maladaptive behavior apparently not found in other species. An experiment conducted with a donkey at the station highlighted this distinction between humans and desert animals. Five men took turns walking with the donkey for two hours each. After ten hours of exercise throughout the heat of the day, the donkey lost about 7.5 kilograms of water weight, which was similar to the rate of dehydration in the men. However, when offered water at the end of the experiment, the donkey consumed a prodigious twelve liters of water—most of which was quickly consumed during the first few minutes. Adolph and Dill concluded, "The human type of ingestion is unsuited to precarious water supplies. No other animal than man has been studied in which the amount of water taken at one time is insufficient to restore permanently the water balance."[32]

This seemingly casual comparison between humans and donkeys formed an important basis for categorizing animals in an informal taxonomy that both Adolph and Dill emphasized, and that later physiological ecologists adopted. The donkey was a true desert animal adapted both by its ability to withstand dehydration that would be fatal to other animals and by its ability to quickly replenish water when it was available. The relatively large body size of the donkey provided a kind of thermal inertia that enhanced these adaptations to heat and water loss. Because of its mass, the donkey's core temperature rose more slowly during the heat of the day but also cooled more slowly during the cold desert nights. Smaller animals were much more at the mercy of the environment, in their relative inability to resist both temperature changes and water loss. The only adaptive recourse available to small desert animals constrained by the physics of volume and surface area was to find shelter from the sun. In Dill and Adolph's classification scheme, humans formed a unique category of animals poorly adapted to desert life but able to survive—and even thrive—because of technological and cultural innovations.

If Boulder Dam sparked Adolph's interest in desert physiology, World War II both expanded and focused his research. Through Dill's influence, the Office of Scientific Research and Development awarded Adolph a government contract in 1942 to study human performance at a desert training center established by General George S. Patton on the Blythe Army Air Base near Indio, California.[33] Together with a team of researchers from the University of Rochester, Adolph spent the rest of the war conducting a wide range of field and laboratory studies using soldiers, students, and the researchers themselves as subjects.

The strategic importance of northern Africa and the ongoing battles in the Sahara Desert during the early years of the war focused considerable attention on human survival and performance in hot, dry environments. Heat-related deaths among recruits during training was also a pressing military concern.[34] Seemingly healthy recruits would occasionally collapse and die from heatstroke when doing the same work as others who were unaffected by the heat. The research conducted by Adolph and Dill at Boulder City, together with the scattered literature on the subject, provided some basis for understanding human adaptability to desert conditions, but a number of important issues remained unresolved. Military officers often believed that their soldiers could become "hardened" to thirst, or from a physiological perspective that they could acclimate to dehydration. Other practical questions about the advisability of rationing water, methods for alleviating thirst, and designing clothing to protect the body from solar radiation had not been systematically studied by military planners. In contrast to the somewhat anecdotal reporting of results gathered from the small group of researchers at the Boulder City site, the army provided Adolph with a much larger pool of subjects for experimentation in various field studies conducted during the summer heat, when temperatures sometimes reached 45°C (figure 3). Dressed in various uniforms, soldiers would hike or work for several hours, often with limited amounts of water. Adolph elaborated on these field studies during the academic year using student volunteers who exercised in a controlled hot room constructed at the University of Rochester. Adolph also used dogs as experimental models to explore the lethal limits of heat and dehydration, for which

Figure 3. Soldiers carrying five-gallon cans filled with various amounts of water on GI packboards. The water presumably was for weight rather than drinking. In some experiments, subjects hiked for up to eight miles with only a one-quart canteen of water to drink. Edward F. Adolph Papers, Edward G. Miner Library, University of Rochester Medical Center, Rochester, New York.

there was only fragmentary and anecdotal information gleaned from cases of humans stranded in the desert or adrift at sea.

The studies of soldiers training in the desert reinforced the early findings of Adolph and Dill. Although extended exposure to desert heat acclimated individuals to tolerate work at elevated temperatures, Adolph's team found no evidence that the human body could acclimate to dehydration. Hardened soldiers who had experienced life in the desert for several weeks required just as much water as novices. Nor did it matter whether they consumed water in one dose or gradually sipped. What mattered was the overall water content of the body, and dehydration led progressively to a syndrome of debilitating physiological symptoms. Human volunteers provided insights to the earliest stages of this "vicious cycle," but experiments with dogs documented an "explosive heat rise" culminating in heart failure resulting from dehydration and the accompanying increase in the viscosity of blood.[35] As with the earlier studies at Boulder City, Adolph documented that dehydrated soldiers consistently failed to drink enough water, even when it was freely available. Over time this "voluntary dehydration" had exactly the same deleterious consequences as "involuntary dehydration" due to water deprivation. Indeed, voluntary dehydration was insidious because soldiers unwittingly became incapacitated even when they had access to water.[36] To maintain peak physiological condition, soldiers needed to drink more water than they desired.

Adolph's team also carefully documented other counterintuitive strategies for desert survival. Although there was a tendency to take off clothing during the heat of the day, soldiers who were fully clothed lost significantly less water than those who were nearly naked. Surprisingly, the color and texture of the clothing seemed to have little effect on water loss and body temperature, so long as it allowed air circulation. Adolph acknowledged that this might have been predicted, because Berbers and other indigenous desert dwellers in northern Africa wore relatively heavy, but loose-fitting, wool clothing. Not only did this provide protection from the sun and an insulating temperature gradient between the skin and the hot external environment during the day, but it also provided warmth during the cold desert nights.

Voluntary dehydration and the conflicting demands of heat dissipation and water balance provided Adolph with examples of what he referred to as "nonhomeostatic" regulation. The human body was an impressive yet imperfect temperature regulator. Especially in dry climates, sweating provided an excellent method of heat dissipation, but at the high cost of water loss. Under extreme conditions of heat and exercise, the body could lose as much as three liters of water per hour. Even with this expensive evaporative cooling, core body temperature did not remain constant but often increased, especially under physical exertion. The failure of most humans to drink enough water to alleviate dehydration was also nonhomeostatic. Living in the desert seemed to entail a chronic state of mild dehydration that could easily become perilous. There was no evidence that acclimation to dehydration occurred. Even repeated experience with dehydration did nothing to lessen the danger of life in hot, dry environments.

Humans became desert dwellers only through a combination of adequate water supply, technological innovation, and changes in behavior.[37] Humans needed to drink more than they desired. They also needed to wear protective clothing and avoid exertion during the day, although this might run counter to military strategy and the exigencies of battle.[38] In conclusion Adolph wrote, "The desertworthy man is a person who is good at avoiding dehydration and at finding water; but his body, thick-skinned as it appears to be, sweats just as fast and compensates for water shortage just as poorly as the tenderfoot's."[39]

Toward the end of *Physiology of Man in the Desert*, Adolph noted that understanding the limitations of human adaptation to the desert might benefit from a broader comparative study of desert animals. Adolph emphasized that this environmental physiology was in its infancy, noting that most information about desert animals was anecdotal rather the experimental. Although the informal experiments that he and Dill had conducted on a donkey at Boulder City suggested that size and rapid rehydration might be special adaptations that some larger animals evolved for living in hot, dry environments, physiological limitations impressed Adolph more than unique adaptations. Adolph claimed that even camels, known to be able to exist for days in the sun without water, probably had few special adaptations for water balance or temperature regulation. Repeating a common belief, Adolph claimed that camels likely were able to store some water in the rumen of the stomach. However, once this water was exhausted, the camel faced the same physiological constraints as humans, dogs, or white rats. Indeed, from Adolph's perspective a camel was no better adapted to life in the desert than a human carrying a canteen.[40] This speculation turned out to be unwarranted. When physiologists more attuned to ecology took a closer look at camels after World War II, they found no evidence for water storage, but rather a robust array of physiological and behavioral adaptations that allowed the large ungulates to thrive in the most inhospitable deserts on earth.

Although Adolph's desert studies during the war pointed toward an expanding field of physiological ecology, his own research remained largely within the orbit of traditional medical physiology. He used dogs and rats as stand-ins for humans when performing unpleasant or lethal procedures. In other cases he was interested in answering general questions about the physiology of water balance by relying on standard laboratory animals. For example, he designed a series of experiments using white rats to determine whether they would drink salt water and, if so, whether they could excrete the excess salt without losing precious body water in the process. The focus of the study was the general function of mammalian kidneys, rather than adaptations to particular environments. Although rats had more efficient kidneys than humans, Adolph found that the experimental rats avoided drinking salt water and failed to thrive when even small amounts of excess salt were added to their drinking water. At the end of the article, Adolph compared these results with studies in other animals, but not within the context of adaptation and homeostasis. More ecologically oriented physiologists later turned to Adolph's early experiments on laboratory rats as a basis for contrast with small rodents whose ancestors evolved in desert environments devoid of drinking water. Adolph's earlier research served as both a model and foil for what became classic studies in desert adaptations of small mammals. Although later physiological ecologists credited Adolph's wartime research as a pioneering contribution to desert biology, in some cases his early conclusions about desert adaptations were dramatically overturned.

5

Physiological Ecology from an Engineering Perspective

Edward Adolph recognized that his later work fell within an emerging area of research that he referred to as environmental physiology and that others later called physiological ecology. The insight was prescient, even though historically Adolph's research appears somewhat anomalous. His focus on humans, particularly soldiers in wartime, might seem artificial to those interested in how desert species adapt to a complex environment that involves more than heat and lack of water. As Adolph was quick to point out, the desert was not a natural environment for humans. A reliable supply of water, not to mention air conditioning, was required for modern, urban civilization to flourish in the southwestern United States. Anthropocentrism aside, it is striking how little Adolph's work reflected the evolutionary idea of relative adaptation shaped by immediate demands of fluctuating local environments but also by phylogenetic constraints and historical contingencies. Adolph's rather cavalier statement that a camel is no different than a soldier with a canteen highlights the point and stands in stark contrast to the physiological ecology that developed rapidly after World War II.

The contrast between Adolph and those who followed him to the desert is illustrated in a striking way by a pair of intriguing laboratory studies with nearly identical titles published in the *American Journal of Physiology* seven years apart.[1] During the war Adolph performed experiments to explore the ability of white rats to survive drinking salt water, and he reported the results in an article that asked a rhetorical question: Do rats thrive when drinking seawater? The experiments had implications for medicine and human physiology, because wartime research had confirmed that humans stranded on the ocean cannot survive by drinking seawater. Most physiologists believed that mammalian kidneys varied only modestly in their ability to conserve water and concentrate salt in urine. Adolph chose white rats because they were mainstays of laboratory physiology but also because they produced a more concentrated urine than humans and other mammals commonly used in physiological research. Rats avoided drinking salt water unless they were dehydrated, but some of Adolph's rats could survive drinking water half as salty as seawater. Rats drinking seawater survived longer than rats completely deprived of water, although both groups constantly lost weight through a combination of evaporation, excretory water loss, and diminished food intake. None of Adolph's rats could live indefinitely on seawater, although a few hardy animals managed to survive for three weeks. Adolph also found no evidence for acclimation when he gradually increased the amount of salt in the water fed to his experimental rats. To answer the question posed in the title of his article, Adolph concluded that rats, and presumably other mammals, most certainly did not thrive on seawater.

Half a decade later Bodil and Knut Schmidt-Nielsen, Scandinavian scientists who had recently immigrated to the United States, published a counterarticle, "Do Kangaroo Rats Thrive when Drinking Seawater?" The Schmidt-Nielsens had spent two years studying kangaroo rats and other desert rodents in Arizona, both under natural conditions and in the laboratory. They found that their caged kangaroo rats never drank water, even when it was freely available. All of the animals' water needs seemed to be met by the food they ate, even when that consisted only of dry seeds and oatmeal. To explore the limits of the animal's ability to regulate water, the Schmidt-Nielsens fed some kangaroo rats a high-protein diet of soybeans. Excreting the excess nitrogen from protein metabolism required substantial amounts of water, and under these unfavorable conditions, kangaroo rats could be induced to drink water. However, it made no difference whether this was freshwater or seawater. In both cases, the small rodents maintained body weight and seemed to thrive for weeks on end.

Adolph's earlier study on white rats provided an apt foil for highlighting the outsized ability of the kangaroo rat's kidneys to concentrate urine. Seawater was not a normal constituent of the external environment for either animal, but, living in an environment where free water was scarce or entirely absent for much of the year, kangaroo rats had evolved highly efficient kidneys that eliminated waste products with a minimum of water loss. Although evolutionary theory was not an explicit part of the Schmidt-Nielsens' argument, adaptation as a fit between organism and environment was an important intellectual foundation of the physiological ecology that they effectively popularized during the post–World War II era. In contrast to the controlled laboratory environment used in Adolph's experiments on white rats, the natural environment of the kangaroo rat was complex and multifaceted, although later critics would complain that even the Schmidt-Nielsens glossed over the environmental variation.[2]

In other ways, too, Adolph provided important context for the Schmidt-Nielsens' research. The Schmidt-Nielsens routinely cited Adolph's earlier studies as pioneering contributions to a rigorous scientific understanding of the demands of life in the desert. In particular they used his categorization of large and small mammals as a primary scheme for distinguishing the different challenges that physical size posed for self-regulating water and body temperature. For Knut, in particular, Adolph's publications on quantitative methods in biology formed an important foundation for mathematical approaches to analyzing the effects of scaling in animals of very different sizes.

A Classic Case of Adaptation and Homeostasis

Although Knut Schmidt-Nielsen began studies in an engineering program, he soon returned to an earlier interest to zoology.[3] Both his father and grandfather were avid, though amateur, naturalists. Together with his father, who was a professor of chemistry at the University of Oslo, Knut published a three-year study of feeding and growth of lake trout. After completing his degree, he applied to study under August Krogh at the University of Copenhagen at the beginning of World War II. Following Krogh's interest in osmoregulation, Schmidt-Nielsen began working on marine invertebrates, as well as an inconclusive study of salt excretion in marine birds, although his dissertation involved a study of how the small intestine digests and absorbs fats. Completed at the end of the war, Schmidt-Nielsen's dissertation provided evidence that fats are absorbed as phospholipids through the wall of the gut.

During his doctoral studies, Schmidt-Nielsen married Krogh's daughter, Bodil, who was studying dentistry and teaching at the university. A fortuitous visit to Krogh's laboratory by the American physiologist Laurence Irving shortly after the war led to the Schmidt-Nielsens' immigration to the United States and initiated their ecophysiological research on kangaroo rats. Krogh and Irving had developed a friendship before the war when the Danish physiologist had spent a term in the Biology Department at Swarthmore College, where Irving was chair. Together with another of Krogh's students, Per Scholander, Irving had conducted wartime research in the arctic for the United States Army Air Forces.[4] After the war the two scientists were touring

Europe to investigate the possibility of establishing a physiological institute for Scholander, but during the visit Irving offered the Schmidt-Nielsens temporary positions as research associates. The decision to study desert mammals was apparently only one of several possibilities that Irving considered, but it turned out to be a fortunate choice.[5] As a result of the Schmidt-Nielsens' research, the kangaroo rat quickly became an iconic, textbook example of homeostasis and adaptation.

The Schmidt-Nielsens initiated research on kangaroo rats during the summers of 1947 and 1948 at the US Forest Service research station at Santa Rita near Tucson, Arizona (figure 4). When Knut contracted pneumonia during the first summer, Bodil was forced to shoulder much of the experimental research, and she continued to take the lead on studying the renal physiology of the kangaroo rat.[6] The study quickly produced unexpected results that ran contrary to a number of generalizations about desert habitats and mammalian physiology. Although excited by the rapid progress during the first year, Irving was sufficiently concerned about the data collected by his young protégés that he required them to send an informal report to a number of prominent physiologists (including Adolph) before submitting articles for publication.[7] The couple continued to make refinements, but by 1949 the Schmidt-Neilsens outlined what would become the classic account of the adaptations of small mammals to desert environments.

Although the Schmidt-Nielsens originally studied a number of desert rodents, they focused on two species of kangaroo rats in the genus *Dipodomys*: the bannertailed kangaroo rat (*D. spectabilis*) and Merriam's kangaroo rat (*D. merriami*). The overall plan of the research was conceptually quite simple, involving a balance sheet of the water metabolism of small desert mammals. A balance of water intake and loss applied equally to all mammals but highlighted particular problems for small rodents exposed to high temperatures without obvious sources of drinking water. The evaporative cooling on which all mammals depend works at cross-purposes with water conservation. This is accentuated in small animals with comparatively large surface areas for absorbing heat and losing water. Earlier studies by mammalogists and ecologists had raised more questions than answers about the adaptive strategies used by kangaroo rats and their relatives.[8] Unlike wood rats, which survived by eating cactus and other succulent plant material, the kangaroo rats apparently lived completely on a diet of dry seeds. The Schmidt-Nielsens combined a series of laboratory and field experiments that provided a compelling account of the kangaroo rat's tight water budget. They also effectively

Figure 4. *Left to right:* Bodil Schmidt-Nielsen, August Krogh, and Knut Schmidt-Nielsen in Santa Rita, Arizona. Laurence Irving Papers, University of Alaska Archives, Elmer E. Rasmussen Library, University of Alaska, Fairbanks.

popularized their results to form an iconic account of desert adaptation that focused on the daunting physical constraint of body size and the anatomical, physiological, and behavioral innovations that had evolved to meet these challenges. As Knut later noted, kangaroo rats and other desert animals did not simply "eke out a marginal existence" but thrived in a seemingly hostile environment.[9] Coming to an environment quite different from his Scandinavian homeland, Schmidt-Nielsen recalled being surprised by the rich fauna of the desert Southwest. The realization that the animals they studied were so well adapted to living in extreme heat with little water was an important insight that guided the Schmidt-Nielsens' research, as well as that of other physiological ecologists.

Because water is periodically available during the rainy season, the Schmidt-Nielsens needed to rule out the unlikely possibility that the small mammals somehow stored water in their bodies, which they later used during the dry season. Kangaroo rats fed dry diets in the laboratory maintained their body weights for several weeks without any signs of dehydration. Indeed, when fed dried rolled oats or barley, the small rodents showed no inclination to drink water, even when it was available. Of course, even seeds and other dry foods contain some water and the amount varies depending upon humidity. At the low desert humidity, this might amount to 5% of the weight of the grain. Metabolism was another theoretical source of water. The oxidation of food molecules yields water as a by-product. Whether an animal could rely on this source of water was open to question, because the amount of metabolic water might be equaled or even exceeded by evaporative water loss in air exhaled from the very respiration that was generating the metabolic water.

Even the most optimistic assumptions about water input from metabolism of fats, proteins, and carbohydrates left little room for loss in the water economy of the kangaroo rat. Measurements showed that compared to white rats and other rodents, kangaroo rats produced much drier feces. More startling was the discovery that the kangaroo rat produced more highly concentrated urine than other mammals. Indeed, urine produced by the kidney of the kangaroo rat was saltier than seawater. Bodil Schmidt-Nielsen's initial measurements of chloride concentrations in the urine of kangaroo rats suggested that the animals might actually be capable of drinking seawater without harm. Testing the idea was difficult because kangaroo rats kept in the laboratory never drank water. By feeding animals a high-protein diet of soybeans that required excretion of large amounts of urea, the Schmidt-Nielsens pushed the kangaroo rats' kidneys to the breaking point. Indeed, no rodent could survive for long on this metabolically challenging diet when deprived of water. White rats, wood rats, and kangaroo rats all consistently lost weight as their kidneys struggled to rid the body of urea. The resulting dehydration quickly led to death, although kangaroo rats survived significantly longer than wood rats, which perished after only a few days on the foreign diet.[10] Kangaroo rats provided with either freshwater or seawater eventually learned to drink, and both groups maintained body weight over the course of a three-week experiment. That kangaroo rats could "thrive" on seawater was a remarkable finding. The experimental results were a clever demonstration of the water-conserving ability of the kangaroo rat's kidney that contrasted with Adolph's earlier negative results with laboratory rats. A greatly elongated loop of Henle provided the renal mechanism for concentrating urine and conserving body water for the kangaroo rat by acting as a countercurrent multiplier that "pumped" excess sodium and urea into the urine. Although the details of this mechanism remained conjectural during the early research, Bodil's later comparative studies helped confirm the importance of this anatomical structure. It appeared to be an adaptive feature found among many small desert rodents regardless of phylogenetic relationship. A few species even exceeded the kangaroo rat's dramatic water-conserving ability.[11]

Evaporative water loss was perhaps the most difficult problem facing the Schmidt-Nielsens' research. Unlike humans and some other large mammals, which use the evaporative cooling of sweat to maintain body temperature under hot conditions, kangaroo rats lack sweat glands. Under heat stress the small rodents sometimes moistened the fur under their chins by copious salivation, which the Schmidt-Nielsens interpreted as an emergency measure for evaporative cooling. This was a temporary strategy, because it led to rapid dehydration, and kangaroo rats showed little ability to tolerate increases in body temperature. Experimental animals maintained in conditions warmer than body temperature rapidly lost weight and died. Although kangaroo rats varied in their ability to resist heat stress, none of the animals tested could survive the extreme daytime temperatures found in the desert for more than ninety minutes—and most subjects died much more quickly. This finding had a profound effect on the Schmidt-Nielsens' interpretation of the physiological ecology of the kangaroo rats. Lack of free water in the environment combined with an inability to tolerate increased body temperature or dehydration meant that the kangaroo rat had to conserve water in every way possible. A potential adaptive solution to this problem might be to escape from the desert heat and desiccation in protected underground burrows. The Schmidt-Nielsens became convinced that this behavioral trait was a key to survival for the small nocturnal mammals.

How much protection the burrows provided was not completely clear. Ecologists had recorded temperatures within kangaroo burrows, but there was no information about underground humidity. Bodil's father had devised an ingenious microclimate recorder with a hair hygrometer enclosed in a small pocket watchcase. Although Krogh's device could record moisture for a period of several hours, it was not very accurate at low humidity. The Schmidt-Nielsens were initially skeptical that the device would work in the dry desert heat, but preliminary experimental measurements indicated that air was often much more humid underground than at the surface.[12] To measure the humidity in burrows, the scientists tied microclimate recorders to the tails of kangaroo rats, allowing the animals to drag the devices deep into the burrows. The scientists later retrieved the recorder (sometimes with the rat still attached) by pulling it out by a long wire connected to the instrument. Although temperature and humidity varied greatly in different parts of the burrow, some of the chambers had significantly higher humidity and lower temperatures than the outside environment. Remaining in the relatively cool and humid burrow during the heat of the day provided crucial relief from heat and was a critical strategy for reducing water loss.

During a relatively short period—just two years—the Schmidt-Nielsens outlined what became the classic explanation of self-regulation and adaptation of small desert mammals. The research was primarily physiological but also included important elements of the natural history, ecology, and behavior of kangaroo rats. The physiological ecology of kangaroo rats continued to hold a fascination for the two scientists, even as they moved on to other research questions. For example, Knut later described a countercurrent condenser in the nose of the animal that reduces water loss. Breathing in dry desert air causes evaporative cooling of the mucosa, and warm, moist, exhaled air condenses on the cooled surface rather than being lost to the environment.[13] Bodil's early interest in renal physiology of kangaroo rats expanded to include comparisons with a broad array of other mammals. The significance of the Schmidt-Nielsens' early desert research is attested both by the continued use of the kangaroo rat as a textbook example of homeostasis and adaptation as the starting point for later research that has revised and refined this "classic" case study.[14]

Popularizing the Kangaroo Rat

The fact that the Schmidt-Nielsens chose to summarize their early field and laboratory studies in the popular and widely circulated *Scientific Monthly* as well as more specialized publications suggests that they realized the potential broad appeal of the kangaroo rat.[15] Promotion of their desert research by the Schmidt-Nielsens and others made the diminutive rodent an iconic example of self-regulation and adaptation in harsh environments. Homer Smith devoted five pages to summarizing the Schmidt-Nielsens' research in *From Fish to Philosopher*. He skillfully wove the proximate causes of temperature and water regulation in small mammals into his broad evolutionary account of the vertebrate phylogeny. The kangaroo rat provided Smith with two important lessons for renal evolution.[16] First, comparing the water-conserving strategies of marine fish and small desert rodents emphasized the importance of convergent evolution. Marine fish prevented dehydration by the evolutionary reduction of the glomerulus, whereas the kangaroo rat performed this same function by an elaboration of the nephron. Smith drove this point home by pointing out that the elongation of the loop of Henle to increase water reabsorption evolved not only in kangaroo rats but also in other small desert rodents belonging to several different families. A second lesson—both practical and evolutionary—involved comparing the Schmidt-Nielsens' research to Adolph's wartime experiments on humans. Humans, Smith noted, were quite unremarkable in their ability to cope with desert heat and lack of water because the loop of Henle in human kidneys was relatively short compared to those of small desert rodents. Writing to a postwar audience, Smith emphasized the practical limitations placed on humans living in the desert by the evolutionary legacy of their unspecialized mammalian kidneys. Mammalian kidneys might be master chemists, but few could accomplish the marvels performed by the nephrons of small desert rodents.

Correspondence reveals the deep respect that the Schmidt-Nielsens had for Smith. When wrestling to understand how the kidney of the kangaroo rat functioned to conserve water, the Schmidt-Nielsens would ask each other, "What would Homer say about this?"[17] In return, Smith's intense interest in the younger physiologists' desert research led to an invitation to work at Smith's summer laboratory at the Mount Desert Island Biological Laboratory. There, Smith learned firsthand about the kangaroo rat's renal physiology when Bodil joined him for a summer of research.[18]

Smith's mentoring and his promotion of the couple's work was important for popularizing the kangaroo rat. Smith's *From Fish to Philosopher* brought the Schmidt-Nielsens' early research to a broad audience when the younger scientists were just beginning to establish their professional identities. However, the iconic status of the kangaroo rat as the quintessential small desert mammal was largely due to Knut, whose lively accounts written in his adopted language attracted a large following of students and general readers. Although self-promotion may have been a motivation, at least initially he was reticent about popular writing. When Dennis Flanagan, the editor of *Scientific American*, approached him to write a short article on the desert rat, Knut demurred, citing the unreasonably short timeline for writing the article (four months), his unwillingness to allow editors to alter his writing, and the small size of the stipend ($200).[19] Further negotiations led to a slight increase in the payment and an extended deadline, and Knut agreed to coauthor an article with Bodil. This early popular publication set a pattern for Knut's later research. He popularized all of his major discoveries in *Scientific American* at the same time that he submitted technical reports to specialized journals. Together with his textbooks and other semipopular books, the *Scientific American* articles did much to make him the best-known comparative physiologist of the late twentieth century.

From Kangaroo Rats to Camels

In his autobiography Knut Schmidt-Nielsen gave scant credit to Adolph, despite the fact that the older scientist had contributed to the Schmidt-Nielsens' intellectual development in critical ways. The prominent citations of *Physiology of Man in the Desert* indicate that Adolph's military research provided an important context for the Schmidt-Nielsens' early articles on kangaroo rats. The Schmidt-Nielsens considered Adolph's work to be a pioneering effort to bring experimental methods to a field dominated by anecdotes from natural history and the travel memoirs of explorers. Even though they used Adolph's research on white rats as a point of contrast for the renal physiology of kangaroo rats, the Schmidt-Nielsens acknowledged the importance of the question posed by the older physiologist. Adolph's emphasis on how body size dictates the adaptive strategies of desert animals was even more critical for the Schmidt-Nielsens' work. Adolph and David Bruce Dill distinguished the different adaptive challenges faced by large and small desert animals, although their experimental data were limited to a donkey, which they compared with non-desert-dwelling dogs and humans. The Schmidt-Nielsens greatly extended this comparative approach by including both the smallest and largest mammals that inhabited desert environments. Adolph had also explored mathematical methods for quantifying correlations in comparative physiology, which the Schmidt-Nielsens found particularly useful for explaining the relationship between body size and evaporative water loss. At the beginning of his desert research, Knut wrote an appreciative letter to Adolph about a recent article on quantitative methods that the older physiologist had recently published in *Science*. Repeating this compliment to Adolph in 1951 when he and his wife were completing work on the adaptations of kangaroo rats, Knut acknowledged his deep intellectual debt to Adolph: "Since your article on 'Quantitative Relations in the Physiological Constitutions of Mammals' appeared in 'Science' two years ago, I have used this paper more than any other reprint I have. I use it for a steady reference as soon as a question of estimating quantitative aspects of a physiological mechanism comes up."[20] That the two letters constituted more than a superficial compliment from a younger scientist to an established authority is demonstrated by the way the Schmidt-Nielsens pivoted from studying kangaroo rats to exploring the physiological ecology of camels, the largest mammals that inhabited hot, dry environments. Of course, Adolph's claim that camels likely store water in their stomachs, much like a soldier's canteen, also provided an enticing question for experimental study.

To realize the promise of Adolph's quantitative approach and to more fully explore the distinctive adaptations of small and large desert mammals, the Schmidt-Nielsens began to lay plans for an in-depth study of the dromedary camel in its natural habitat. Not only was the camel the largest desert mammal, but its physiology was almost completely unstudied and shrouded in dubious anecdotal claims. During a year as visiting scholars at Stanford University, the Schmidt-Nielsens scoured the library for information and queried other scientists, as well as students from the Middle East, to learn about the camel. In a report submitted to Irving, who was still supervising their research, the Schmidt-Nielsens wrote that "the information is so variable and contradictory that the literature does not give sufficient evidence for significant conclusions as to the physiology of the camel under stress (heat and lack of water)."[21] A case in point was the hypothetical possibility that camels stored water, highlighted by Adolph's offhand claim about the camel's stomach. Despite conflicting anecdotal reports, the Schmidt-Nielsens concluded that there was "no experimental evidence whatsoever" to decide the issue of water storage.

The Schmidt-Nielsens' decision to study the camel was strategic on a number of levels. Unlike the kangaroo rat, which survived by hiding from daytime heat in underground burrows, camels could not escape problems of temperature regulation in the desert sun. Indeed, because they were widely used for transportation, camels were often active throughout the heat of the day. Understanding how the animals coped with heat and lack of water thus not only was of theoretical interest but might provide economically important information about survival and the limitations of performance in the desert. The fact that so little was known about the physiology of the camel provided a professional opportunity for the Schmidt-Nielsens to make a name for themselves. Indeed, they were quite ambitious to become authorities on desert biology in general. Even at this early stage in their broader program, Irving described the Schmidt-Nielsens' desert research as "leading toward a comprehensive and penetrating view of the subject."[22] From Irving's approving perspective, the Schmidt-Nielsens' plans were part of an even broader "expeditionary physiology" that would involve sending scientific teams to study biology in remote, and often inhospitable, geographic areas. Irving's initiative had geopolitical as well as scientific implications, and the Schmidt Nielsens effectively tied their plans to international science. Allying their research with the recently established UNESCO Arid Zone Research Programme provided a practical justification for pursuing what might otherwise have seemed an exotic but pointless adventure.[23] It also provided opportunities for the Schmidt-Nielsens to present their desert research on both the kangaroo rat and camel to broader international audiences.[24]

UNESCO provided a small grant to purchase camels once they reached Algeria, but finding funding for the trip itself proved to be a frustrating challenge. The Schmidt-Nielsens were unsuccessful in securing grants from either American oil companies or the National Science Foundation. Long after completing the research, Knut continued to harbor a grudge against the NSF for snubbing his initial request.[25] By the time their research plans took final shape, Knut had accepted a faculty position in the Zoology Department at Duke University. A Guggenheim Fellowship paid his faculty salary and expenses during a year in Algeria, as well as a salary for Bodil. Contracts from the Office of Naval Research paid for two research assistants and supplies.[26] The US Army donated a surplus ambulance to transport food and laboratory equipment to the remote research site.

The Schmidt-Nielsens conducted their yearlong research project at the remote oasis of Béni Abbès, 1,200 kilometers from Algiers. The community included a small detachment of French soldiers, a handful of other Europeans, and several hundred ethnic Arabs, Berbers, and Sudanese. Knut transported all of the scientific equipment in the ambulance, which required a five-day drive from the capital over rugged, high-desert terrain. The facilities at Béni Abbès were even more primitive than the research station in Arizona where the Schmidt-Nielsens conducted their earlier research on kangaroo rats. The "laboratory" consisted of a vacant mud brick building with an entrance tall enough for camels to enter. The scientists measured water loss or gain by weighing the animals using a sling balance built specifically for the research project. The research team also performed chemical analyses of blood and urine, and measured respiration, rectal temperatures, and blood volume in their makeshift laboratory (figure 5). Despite the rudimentary conditions and the short duration of the research, the Schmidt-Nielsens completed the most comprehensive study of the physiology of the camel to date. The results were disseminated in technical reports aimed at a growing international audience interested in desert biology and agriculture but were also popularized in the United States by an article Knut wrote for *Scientific American.*[27]

Camels provided a model for adaptations to desert life strikingly different

Figure 5. Knut Schmidt-Nielsen measuring respiration of a camel at the laboratory in Béni Abbès, Algeria. An illustration of the camel with the respirometer was later used as the cover art for the December 1959 issue of *Scientific American* that included Knut Schmidt-Nielsen's popularized account of the research. Photograph provided by Astrid Schmidt-Nielsen.

from the kangaroo rat.[28] Analysis of stomach contents and anatomy discredited the idea that camels could store water in the rumen or other chambers of the stomach. Water vapor lost in breathing exceeded the metabolic water produced from oxidizing fat stored in the hump. On the other hand, camels could tolerate extreme dehydration and elevated body temperature. Unlike humans and dogs, which suffer "explosive heat rise" resulting in circulatory failure when dehydration reduces the volume of plasma and increases the viscosity of the blood, dehydration in the camel occurred primarily in interstitial fluids while sparing the blood. As a result, camels can lose almost one-third of their weight in water without ill effect and can rehydrate quickly when water becomes available. In their ability to lose and gain water, camels far exceeded even the donkey that Adolph and Dill had studied less formally a decade earlier.

The large size of the camel also provided "thermal inertia," allowing the body temperature to change much more slowly than in smaller animals. Body temperature for the camel followed a daily cycle, gradually rising during the heat of the day and falling during the cold desert night. The Schmidt-Nielsens found that unlike the small kangaroo rats they had studied, camels were able to thrive over a broad range of body temperatures. By lowering the gradient between body and environment, the elevated body temperature reduced the need for evaporative cooling. That this was primarily a mechanism for conserving water was suggested by the fact that it operated only when the camel was dehydrated. When water was available, the amplitude of the body temperature cycle was reduced from 6°C to 1°C.[29]

Although camels sweat, their thick fur provides effective insulation from the desert heat and consequent water loss. After shearing one of their experimental camels, the Schmidt-Nielsens found that the animal produced 60% more sweat than before. Deprived of its protective fur, a "naked" camel was at the same disadvantage as a human without clothing. This point of comparison was driven home by the Schmidt-Nielsens' observation that native Algerians relied on a similar form of adaptive insulation by wearing several loose layers of wool as protection against desert heat (and cold) and dehydration.

Behavior played a critical role in the adaptation of kangaroo rats to life in the desert. Even with their powerful kidneys, the diminutive rodents had to remain underground during the heat of the day. From the Schmidt-Nielsens' perspective, this was an inevitable consequence of small size. By contrast, the large camels could not hide from the sun. Although large size and physiology provided the foundation for the camel's adaptation, it also used a variety of behaviors to minimize heat absorption. Camels would seek shade whenever available, although this was often not possible. When standing they oriented their bodies lengthwise to the sun's rays to minimize exposure. When sitting they minimized exposed surface area by folding their legs under their bodies, and they remained in one spot to avoid contact with the hot surface of the ground. Although unable to quantify the water savings due to these behaviors, the Schmidt-Nielsens considered them to be important, if subsidiary, factors contributing to the camel's ability to thrive in the desert heat.[30]

The Engineering Perspective as a Research Strategy

After completing their study of dromedary camels in Algeria, both Schmidt--Nielsens continued research on desert mammals, although they increasingly worked independently. Bodil conducted research in the Zoology Department at Duke University, although she did not have a formal faculty position. Supported by grants from the US Public Health Service and the American Heart Association, she expanded her comparative research of kidney structure and function with Roberta O'Dell, a research associate with the North Carolina

Heart Association. The researchers studied the anatomy and physiology of nephrons in the kidneys of beavers, rabbits, and gerbils. Beavers and gerbils were chosen because they lived in contrasting environments.[31] As inhabitants of Old World deserts, gerbils faced problems of dehydration similar to those of kangaroo rats. Although gerbils drank water and fed on succulent vegetation, their plant diet included halophytic species containing large amounts of salt. Bodil had begun studying these small rodents in Algeria, and together with O'Dell she continued the work in the United States with animals shipped from Egypt. In contrast to kangaroo rats and gerbils, beavers have no shortage of water, both in the food they eat and in their habit of drinking freely. In fact, their tendency to drink and eat at the same time made it difficult to measure how much water they consumed on a normal basis. Even when deprived of drinking water, beavers' normal food of leaves and twigs provided ample water. Thus, beavers would seem to have no need for the elaborate self-regulatory mechanisms for water conservation found in the kangaroo rat or camel. Bodil conducted the studies on beavers during a summer of research in Smith's laboratory at the Mount Desert Island Biological Laboratory in Maine.

Based on the structure of nephron, particularly the loop of Henle, Schmidt-Nielsen and O'Dell identified three distinct anatomical categories. The kidneys of beavers contained only nephrons with short loops of Henle located completely within the outer cortex of the kidney. The inner medulla of kidney was relatively undeveloped compared with those of the other mammals studied. By contrast, the kidneys of gerbils contained nephrons with long loops of Henle that extended deep into the medulla. The kidneys of rabbits contained a mixture of long and short-looped nephrons. The two researchers measured urea and sodium concentrations in different parts of loop of Henle as well as in the urine of each group of mammals. The ability to concentrate these substances was strongly correlated with the relative length of the loop of Henle, particularly with the medullary portion of loop. Lacking elongated loops of Henle extending into the medulla, beavers produced urine with only about 10% of the salt content of urine produced by gerbils. The salt and urea concentration of rabbit urine was intermediate between the two extremes.

At the time, the histology and function of different portions of the loop of Henle were still poorly understood, but the comparative studies of species living in contrasting environments cast some light on the mechanism for salt excretion and water retention. Schmidt-Nielsen and O'Dell described the loop of Henle as a countercurrent multiplier. Fluid flowed in opposite directions in the descending and ascending portions of the loop. In some way the loop produced an osmotic concentration gradient by pumping salt out of the structure, which caused the passive reabsorption of water in the ascending limb. The details of how this process worked on a cellular level remained hypothetical, although different parts of the loop of Henle had thinner or thicker walls, which suggested that they might be more or less permeable to water. Significantly, the ability to concentrate urine was strongly correlated to the relative length of the loops of Henle.

At the same time, Knut was investigating a somewhat similar countercurrent mechanism in the noses of kangaroo rats (figure 6) and, later, camels. In retrospect, he recalled that during the early studies in Arizona he had noticed that the tip of the kangaroo rat's nose seemed cooler than the body, an informal observation confirmed by temperature measurements using miniaturized thermocouples placed in various parts of the nasal passage.[32] Inhaling dry desert air caused evaporative cooling of the nasal tissue. When the animal exhaled warm, moist air, some of the water vapor condensed on the cooler tissue near the tip of the nose. Thus, what might have been a loss of water from a kangaroo rat's meager supply was conserved by a simple mechanism based on dry and moist air moving in different directions. The ability to retain water was accentuated by the convoluted nasal passages, which were relatively long and narrow, providing a large surface area for evaporation and condensation. Unlike the countercurrent multiplier in the loop of Henle that expended energy to form a concentration gradient of ions, the countercurrent exchange mechanism in the nasal passages relied on the passive transfer of heat. Schmidt-Nielsen reasoned that this mechanism worked on the same basic principles as the countercurrent heat exchange that had been earlier described in the flukes of whales and other marine mammals that swam in frigid water.[33] Here, body heat was conserved when cooler venous blood returning from the fluke moved in the opposite direction past warmer arterial blood on the way to the fluke. The blood returning to the body in veins was warmed by a passive transfer of heat from arterial blood. In the case of the kangaroo rat's nose, the movement of air occurred in a single passage, rather than two; but like the countercurrent heat exchange, it involved a passive movement of heat from warmer to cooler areas separated in time rather than space. This was a countercurrent heat exchanger, but one that had evolved primarily for self-regulating water rather than body temperature. Schmidt-Nielsen later described a similar countercurrent exchange system in the nose of the camel. In this case the water-conserving ability was enhanced by hygroscopic mucus

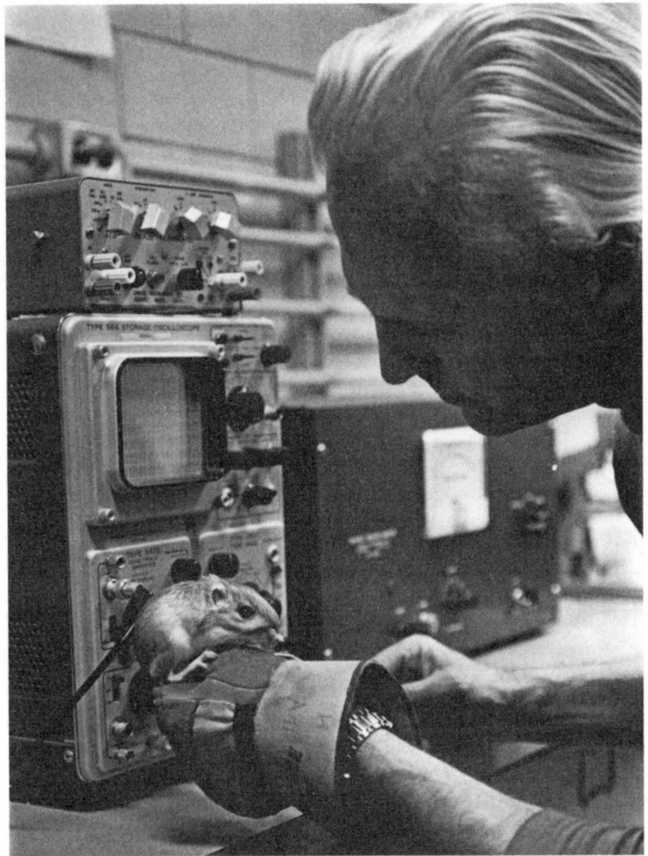

Figure 6. Knut Schmidt-Nielsen with a kangaroo rat in his laboratory at Duke University. The photograph is from the late 1960s, when Schmidt-Nielsen was conducting research on countercurrent heat exchange. University Archives Photograph Collection, Arthur Rubenstein Library, Duke University, Durham, North Carolina.

secreted by the nasal tissues that readily absorbed the condensing moisture from exhaled air, particularly when the camel was dehydrated.

Discussing the anatomy and physiology of the kidney and nasal passages in terms of countercurrent exchange was an example of biological engineering that was fundamental to the Schmidt-Nielsens' approach to physiological ecology. Comparing biological and mechanical structures was more than a loose analogy, because in both cases the same physical principles of diffusion operated. It wasn't that the kangaroo rat's nose was like a heat exchanger designed by engineers. Rather, the two mechanisms were fundamentally the same.[34] This bioengineering approach would seem to be the epitome of functional biology, employing both intrinsic and extrinsic proximate causation. Understanding adaptation to the desert required a thorough understanding of mechanisms for conserving water and dissipating heat. The specific mechanisms were dictated by the physical constraints of body size and physical principles of heat transfer. This intrinsic, proximate causation was complemented by the Schmidt-Nielsens' growing familiarity with the desert environment, as well as the ecology and natural history of the species they studied. Understanding adaptation to the desert required studying native animals within their natural environments, rather than white rats or other model organisms in the laboratory.

What seemed to be missing from the Schmidt-Nielsens' approach to studying kangaroo rats and camels was a deep appreciation for evolutionary accounts of adaptation. They rarely, if ever, mentioned natural selection or other evolutionary mechanisms. Nor were they deeply interested in the evolutionary significance of variation within populations, although they were well aware of individual differences among animals used in their experiments. They accepted that adaptations had evolved to cope with specific demands of the environment, but the explanation of these characteristics was based on physics and chemistry, rather than Darwinian theory. For the Schmidt-Nielsens, the physics of heat gain and loss on small and large objects dictated the alternative adaptive strategies of kangaroo rats and camels. Kangaroo rats might exhibit exquisite adaptations for coping with desert heat, but ultimately their small body size required that they spend much of their lives in underground burrows. Summarizing the importance of physical constraints in his book *Desert Animals*, Knut concluded that "the small animal has no way of escaping the simple laws of physics and its large relative surface—what it can escape by going underground is the parching heat."[35] The physics of heat transfer combined with scaling effects of body mass and surface area set constraints on desert animals that dictated adaptive strategies used to cope with high temperatures.

One might conclude that the Schmidt-Nielsens' apparent lack of interest in evolution is precisely what would be expected of functional biologists who focus on "how" questions, answered in terms of proximate causation. There is a danger, however, in generalizing too broadly on the basis of a sharp distinction between functional versus evolutionary biology. Despite their historical significance and the continuing relevance of their studies, the Schmidt-Nielsens cannot be considered typical of physiological ecologists in general. Physiological ecologists differed in the ways they combined insights from physiology, ecology, evolutionary biology, and behavior. In contrast to the Schmidt-Nielsens, several leaders in the emerging area of physiological ecology used approaches that combined proximate and ultimate causation in ways that defy a simple dichotomy between functional biology and evolutionary biology.

6

An Experimental Naturalist in the Laboratory and Field

Knut Schmidt-Nielsen's engineering approach to understanding how animals work provided one important perspective on physiological ecology of desert mammals. The same year that Schmidt-Nielsen's influential *Desert Animals* (1964) appeared, a strikingly different perspective was put forward by George Bartholomew in the review article "The Roles of Physiology and Behaviour in the Maintenance of Homeostasis in the Desert Environment."[1] Acknowledging Schmidt-Nielsen's broader review of the subject, Bartholomew focused primarily on the field and laboratory studies that he and his students were conducting on a variety of birds, mammals, and reptiles in the California desert. He also used the article to articulate a philosophical perspective for the nascent field of physiological ecology. He continued to elaborate this point of view throughout his career. Unlike Schmidt-Nielsen, who rarely, if ever, referred to homeostasis directly, Bartholomew seemed captivated by the concept and used it in a broad and flexible way to explain adaptations to complex, highly variable environments. While Schmidt-Nielsen formulated broad generalizations based on contrasting types of animals, such as large and small, Bartholomew was more interested in individual variation within and between species. His studies also had a strong biogeographical basis, and he emphasized the variability of the desert environment. Schmidt-Nielsen remained a laboratory physiologist, at heart, who took his tools into the field, but Bartholomew was imbued with natural history, despite the fact that he often used the controlled environment of the laboratory to explore questions of variation and adaptation in desert animals from explicitly ecological and evolutionary perspectives. To a greater extent than Schmidt-Nielsen, Bartholomew combined proximate and ultimate causation in all of his multifaceted studies of adaptation. He considered himself an ecological physiologist, but also an experimental naturalist.[2]

Bartholomew had little interest in the formalities of taxonomy, although his approach to being an experimental naturalist was strongly shaped by early experiences in natural history museums. After graduation from the University of California at Berkeley, Bartholomew earned a master's degree under the direction of Alden Miller in the university's Museum of Vertebrate Zoology.[3] This early research resulted in three articles that Bartholomew published in *Condor* describing the behavior of cormorants in San Francisco Bay. Bartholomew then started a PhD at Harvard's Museum of Comparative Zoology under the mammologist Glover Allen. His training was interrupted by three years of wartime service in a physics laboratory at the Bureau of Naval Ordnance devoted to mine warfare. This practical introduction to electronics would later be important when he needed to design his own physiological instruments. During the war Glover died, and after returning to Harvard, Bartholomew shifted his focus from the museum to the laboratory.[4] In 1949 he completed his dissertation on the effects of light intensity and photoperiod on the development of reproductive organs of house sparrows under the direction of the marine ecologist George L. Clarke. The dissertation set a pattern for much of his later work. Although directed toward the ecological significance of photoperiod on reproduction, Bartholomew conducted all of the experiments in the controlled environment of the Harvard Biological Laboratories.

His later research in the Department of Zoology at the University of California Los Angeles also centered in the laboratory, although Bartholomew also conducted important field studies and maintained a strong intellectual orientation toward traditional natural history.[5] Natural history not only provided an evolutionary context for posing questions about physiological adaptations but also served to integrate biological disciplines around the interaction of organisms in complex environments. From the perspective of an experimental naturalist, Bartholomew emphasized that the organism was not a stuffed skin or pickled specimen in a museum, or even the living creature caged in the laboratory or observed in the field. Instead, the organism "is a complex interaction between a self-sustaining physicochemical system and the environment. An obvious corollary is that to know the organism it is necessary to know its environment."[6] Truly knowing this interaction required a deep understanding and appreciation of variability, both environmental and organismal.

Homeostasis from a Natural History Perspective

Being an experimental naturalist required a much more nuanced interpretation of homeostasis and the interaction of internal and external environments than Cannon had developed in *The Wisdom of the Body*. Organism and environment were inseparable, an insight that Bartholomew credited to Claude Bernard, but understanding the multifaceted interactions required insights from a Darwinian evolutionary perspective. As a naturalist Bartholomew was deeply impressed by environmental complexity.[7] The desert was not a homogeneous hot and dry environment but a jigsaw of microenvironments existing at various scales among which organisms could choose. This biogeographic emphasis was important in a number of ways. Particularly when studying small mammals, it was important to constantly keep in mind that the environment within which the animal lived was not the same as that experienced by humans. According to Bartholomew, "Most vertebrates are much less than a hundredth of the size of man and his domestic animals, and the universe of these small creatures is one of cracks and crevices, holes in logs, dense underbrush, tunnels and nests—a world where distances are measured in yards rather than miles and where the difference between sunshine and shadow may be the difference between life and death."[8] Maintaining homeostasis often involved actively choosing where to be at any given time. Thus, one could differentiate between the microclimate within which the animal might exist and the "ecoclimate" where the animal preferred to live.[9] Even for larger animals, the geographical range of the species often included strikingly different environmental conditions to which separate populations became adapted. This environmental heterogeneity often resulted in a variety of homeostatic mechanisms, even within the same species.

The naturalist's appreciation of environmental complexity also involved understanding the environment as experienced by the animals studied, and not by the scientists who did the studies. This was particularly important in the desert or other environments that seemed hostile to humans. To be successful the physiological ecologist needed to be on guard against anthropomorphic and anthropocentric interpretations.[10] Arid environments might seem inhospitable, but for the inhabitants, the desert is not a hostile environment, but rather just *the environment*, made up of a mosaic of habitat patches within which animals construct their lives.

In contrast to the laboratory physiologist, the naturalist also appreciated that the external environment included not just physical factors such as temperature and water but also the biotic environment of competition, predation, and other forms of symbiosis. Although Bartholomew's physiological ecology focused primarily on the physical environment, he recognized that the homeostatic responses available to a particular species might be shaped by competition with similar species in the same area. Furthermore, the environment was not simply a set of challenges or constraints, whether biotic or abiotic, but also a set of resources that could be manipulated by animals to increase fitness. From this perspective, organisms were not just adapted but also adaptable. Fitness meant more than a static fit with the environment or increased reproductive success; it included the ability to fit into heterogeneous environments through behavioral flexibility.

The experimental naturalist also recognized the importance of phylogenetic history in constraining the adaptations of a species. According to Bartholomew, species were always caught in a "phylogenetic trap" that limited the available adaptive responses to the environment. Anatomy and physiology evolved slowly, but species could often escape from the trap through behavioral responses, which Bartholomew described as being "drastic, rapid, precise, and of exquisite flexibility."[11] The Schmidt-Nielsens had also emphasized behavior in their studies of kangaroo rats and camels, but in a less nuanced way. For the Schmidt-Nielsens, kangaroo rats were forced to spend the day in underground burrows because of the physical constraint of a small body with a relatively large surface area. For Bartholomew, small body size posed an important constraint on life in the desert, but one that could be overcome by behavioral flexibility. Small desert animals were not necessarily nocturnal, and the behavioral responses that allowed them to be active during the day took a number of different forms.

Both the "phylogenetic trap" and the homeostatic behavioral adjustment were well illustrated by the antelope ground squirrel (*Ammospermophilus leucurus*) and its relatives that Bartholomew and his students studied during the late 1950s.[12] Slightly larger than kangaroo rats, the chipmunk-sized ground squirrel still faced the problems of rapid dehydration and hyperthermia when exposed to the desert sun. The ground squirrels conserved water by producing highly concentrated urine, comparable to that of kangaroo rats studied by the Schmidt-Nielsens. They also had an omnivorous diet that included succulent plants, insects, and carrion—all of which provided badly needed water. However, because they are diurnal, the ground squirrels faced problems of evaporative water loss much greater than those of the nocturnal kangaroo rat.

Bartholomew's student Jack Hudson estimated that to maintain a constant body temperature of 38°C, the squirrel would need to evaporate 13% of its body weight every hour—an impossible physiological feat for a small mammal with limited water supply. Maintaining homeostasis while being active during the heat of the day was accomplished through physiological adaptations that combined features reminiscent of both the camel and kangaroo rat, but with unique behavioral responses unlike those described by the Schmidt-Nielsens. In contrast to the kangaroo rat, the antelope ground squirrel tolerated increases in body temperatures as high as 43°C, which allowed it to radiate body heat even at high environmental temperatures. Like the dromedary camel studied by the Schmidt-Nielsens, the ground squirrel exhibited a cyclic pattern of body temperature, although the cycles often lasted only minutes rather than hours. Both the similarities and differences were highlighted in a graph that combined the temperature responses of the two mammals (figure 7). According to Bartholomew, the antelope ground squirrel actively foraged during the day until its body temperature rose to near lethal levels and then darted into its burrow, where it would lie prostrate on the cool floor of the tunnel. Because of its small size, the animal could quickly unload body heat, dropping several degrees in only a matter of minutes. After cooling, the animal once again left the burrow for another cycle of activity. Thus, unlike the camel that had a single temperature cycle during a twenty-four-hour period, the antelope ground squirrel went through repeated short cycles during a day and also allowed its body temperature to drop at night.

This combination of behavioral and physiological adaptations to desert heat illustrated two important points about homeostasis. First, in the antelope ground squirrel it made no sense to speak of "normal" body temperature, because wide fluctuations occurred without any indication of debility. This was also true of many of the birds that Bartholomew and his students studied in hot environments. Rather than maintaining constancy, these cases illustrated an alternative adaptive solution: the "relaxing of physiological homeostasis" by tolerating elevated body temperatures and regulating fluctuations within a rather broad range.[13] Second, homeostasis—whether physiological or behavioral—was always relative to a particular ecological niche. To a much greater extent than the Schmidt-Nielsens, Bartholomew emphasized the ways that natural selection shaped adaptations in closely related populations and species: "natural selection demands only adequacy, elegance of design is not relevant: any combination of behavioural adjustments, physiological regulation, or anatomical accommodation that allows survival and reproduction may be favoured by selection."[14] Unlike the broad comparisons that the Schmidt-Nielsens made between contrasting types of animals, Bartholomew was interested in subtle differences in homeostasis exhibited by closely related and sympatric species facing similar environmental stresses.

Living in the same location with the antelope ground squirrel was the closely related, but much less common, Mohave ground squirrel (*Xerospermophilus*

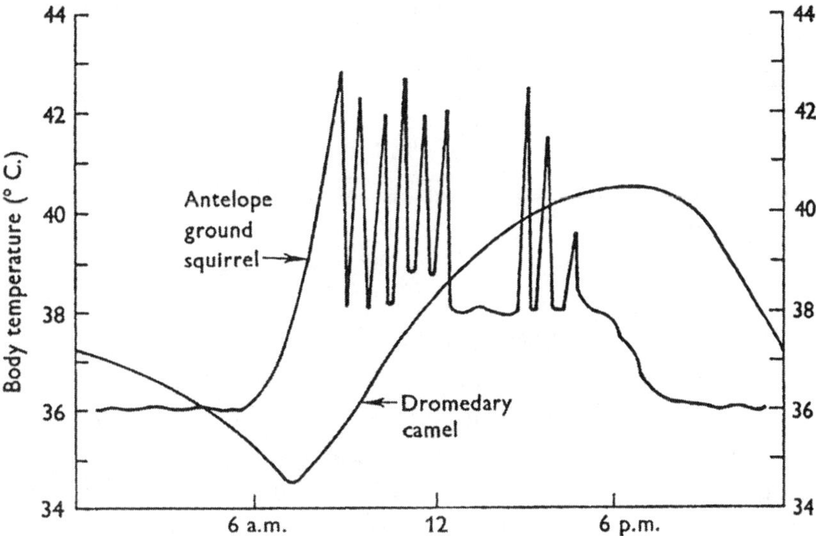

Figure 7. Comparison of temperature cycles in the dromedary camel and the antelope ground squirrel. George A. Bartholomew, "The Roles of Physiology and Behaviour in the Maintenance of Homeostasis in the Desert Environment," *Symposia of the Society for Experimental Biology* 18: 7–29.

mohavensis). Because the two species were sympatric, Bartholomew reasoned that they must have slightly different niches. The fact that the Mohave ground squirrel was so rare also suggested that competition might play an important role in shaping these differences, which were likely both behavioral and physiological. Less tolerant of hyperthermia than its much more active relative, the placid Mohave ground squirrel rarely strayed far from its burrow and spent much of the time in the shade. Unlike the antelope ground squirrel, which was active year-round, the Mohave ground squirrel exhibited extended periods of inactivity, both hibernation and estivation. These periods of torpor did not result directly from changes in ambient temperature but were primarily the result of food deprivation. Regardless of temperature, the animals became torpid whenever food was unavailable. During the spring, when food was sufficiently abundant to support both species, the less competitive Mohave ground squirrel actively foraged, building up large stores of body fat. From August to March, when food was scarce, it escaped competition with the more aggressive antelope ground squirrel by remaining in its burrow. Because this extended period included both hot summer and cold winter temperatures, the inactivity involved torpor related to heat (estivation) and cold (hibernation). From Bartholomew's perspective, these were simply different aspects of a common physiological strategy for minimizing energy consumption when food was scarce.

Exploring the Ecoclimate in a Darwinian Context

As Bartholomew was completing his comparative studies of ground squirrels with his students, he took part on a voyage to the Galapagos Islands organized by scientists at the University of California to assess the state of scientific knowledge about the archipelago and prospects for future study. The eleven-week Galapagos International Scientific Project included a number of prominent international scientists and resulted in a collection of short essays dealing with a wide variety of topics in biology and geology. For Bartholomew, it provided an opportunity to further expand his philosophical point of view on the goals and methods of physiological ecology within the context of a unique ecosystem. Oceanic islands provided "natural laboratories" for studying evolutionary changes of interest to systematists but also to students of physiology and behavior.[15] The simplicity of the islands, with their limited biota, provided particularly promising opportunities for exploring how animals faced the physical challenges of desert heat and lack of freshwater. Here one could explore the distinctions among macroclimate, microclimate, and ecoclimate using poorly studied animal species that were nonetheless widely familiar as a result of the writings of Charles Darwin and other naturalists. Bartholomew used the opportunity to conduct experiments and observational studies on marine iguanas (*Amblyrhynchus cristatus*) to better understand homeostasis in a reptile facing unique problems of temperature regulation. He placed this research within a context of proximate and ultimate causation based on observations made by Darwin in 1835.

In *The Voyage of the Beagle*, Darwin described the unique diet of seaweed that required iguanas to swim in often chilly water, and also the reptile's reluctance to enter the water despite its skill in diving and swimming.[16] He attributed the "apparent stupidity" of the iguana's aversion to entering the water as a response to predation. In contrast to the dangers posed by sharks, the islands were devoid of large predators. Darwin surmised that the fear of oceanic predators had become a fixed hereditary instinct that led iguanas to remain on the safety of land except when they needed to feed. Bartholomew quickly confirmed that iguanas avoided entering the water, even when he provoked them. Yet, once in the water, they also avoided coming on shore when he was present. This suggested that perhaps the iguana's behavior had less to do with predators than with the dangers of hypothermia. He noted that little was known about the physiology and behavior of *A. cristatus*, and he spent two months on Isla Fernandina observing the reptiles and measuring changes in body temperature under different conditions.

Marine iguanas are poikilotherms, which are unable to maintain a constant body temperature physiologically (unlike birds and mammals). However, as with some other reptiles studied by herpetologists, Bartholomew described how the iguanas used behavior to optimize their body temperature by alternately moving from sun to shade, or orienting their bodies to absorb more or less heat from the rocks on which they rested during the day.[17] On sunny days the rocks often exceeded 50°C, but marine iguanas could maintain themselves at a much cooler 35°C, which seemed to be their preferred body temperature. Even without shade the iguanas could maintain body temperatures below 40°C by seeking cool breezes and orienting their bodies to minimize exposure to the sun. However, when tethered the reptiles quickly overheated and died.

When swimming, body temperature decreased to approximate the water temperature, which was often as low as 25°C. How quickly the cooling occurred depended upon body size, with small iguanas cooling much more rapidly than larger animals. In all cases the cooling occurred more slowly than predicted by size alone. This suggested that iguanas might manipulate blood flow under the skin to minimize heat loss. Despite lowered body temperature, the reptiles remained active when swimming, and, indeed, Bartholomew found that it was extremely difficult to catch iguanas when they emerged from the water. Thus, iguanas could function normally over a 15°C range of body temperatures, even though they seemed to prefer a body temperature close to that found in many birds and mammals. Maintaining this optimum temperature might contribute to rapid growth, which Bartholomew suggested was important in a highly social species for which dominance and territorial defense depended largely on body size.

Despite its brevity, Bartholomew's study of iguanas demonstrated the important interaction of behavior and physiology in maintaining homeostasis. It also highlighted Bartholomew's claim that homeostasis meant more than maintaining constancy, and that it included any adaptive response that enhanced reproductive success by improving the fit between organism and environment. Finally, it illustrated the heterogeneity of the environment and the ability of animals to choose the conditions of existence. One could distinguish the equatorial macroclimate of Isla Fernandina from the microclimates of sunny and shady areas on the rocky beach, and also the preferred ecoclimate that iguanas actively pursued. Although able to function normally even in the cold water where they needed to feed, the reptiles avoided the ocean for most of the day and night.

The Engineer and the Naturalist

Despite important differences in perspective, the careers of Bartholomew and Knut Schmidt-Nielsen paralleled one another in important ways and intersected at a crucial period in the development of physiological ecology. Together with their earlier studies of temperature regulation and water balance in small desert mammals, salt regulation in birds highlighted the strikingly different perspectives that the two scientists took on closely related problems. The studies drew a sharp contrast between the engineer's emphasis on elegant design and the naturalist's emphasis on adequacy and relative adaptation to local environmental conditions.

Schmidt-Nielsen had been interested in osmoregulation as a graduate student in August Krogh's lab. He did preliminary studies on how marine birds eliminate the excess salt in their seafood diets. The study did not produce concrete results, and he abandoned the question when he chose a dissertation topic. With an established reputation and stable faculty position at Duke, he returned to the problem with much greater success during the late 1950s and early 1960s.[18] At the time, the Schmidt-Nielsens' marriage was dissolving, and Bodil was not involved with this new line of research. Even before his first publication on the topic, Knut was actively corresponding with Bartholomew, who was beginning his own comparative studies of salt regulation in populations of passerine birds. In particular, he focused on subspecies of savannah sparrows, some of which inhabited salt marshes, while other migratory populations visited freshwater marshes.[19] Bartholomew's preliminary findings suggested that contrary to the widely accepted view, sparrows that inhabited salt marshes were capable of eliminating excess salt by producing highly concentrated urine. Both the design and interpretations of the results of the two sets of studies highlight important differences in the way Bartholomew and Schmidt-Nielsen approached questions of adaptation and self-regulation.

Because seawater contains about three times as much salt as body fluids, marine animals face problems of dehydration somewhat similar to those of desert animals. They must conserve water, while eliminating excess sodium, chloride, and other ions. Mammals have evolved the loop of Henle in the kidney, which acts as a countercurrent multiplier to reabsorb water and to concentrate salt in the urine. The structure is not so efficient as to allow most mammals to drink salt water, but in the case of the kangaroo rat an elongated loop of Henle provides a mechanism for accomplishing this osmoregulatory feat. Because such a well-developed loop of Henle does not exist in birds or reptiles, Schmidt-Nielsen was anxious to discover an alternative mechanism for osmoregulation in marine birds that habitually eat food with high concentrations of salt. A broad comparative approach was particularly appealing because of the differences in the salt content of diets of various marine birds and reptiles.[20]

Knowing that marine birds have nasal glands that are more highly developed than those of terrestrial species suggested a possible mechanism for excreting salt. Schmidt-Nielsen's measurements showed that the glandular secretions were saltier than seawater and that there was a correlation between the size of the gland and the amount of salt in the diet. Schmidt-Nielsen found that the salt concentration of fluids from the nasal glands of petrels, which eat marine invertebrates, was nearly twice that of fish-eating cormorants. The nasal gland was a single-purpose organ for removing sodium chloride, and it did not carry out the many other functions associated with kidney. It functioned only when sodium and chloride ions were elevated. Despite the fact that blood flows in opposite directions in the arteries and veins of the gland, Schmidt-Nielsen concluded that it did not act as a countercurrent multiplier like the one that Bodil described in the loop of Henle.[21]

Schmidt-Nielsen's studies were evolutionary in the sense that he used a broad phylogenetic approach to categorizing comparisons of different animals. His studies of marine birds led to comparisons with salt glands in various marine reptiles, including sea turtles, sea snakes, and marine iguanas. Schmidt-Nielsen concluded that despite differences in efficiency, the salt gland was a common feature of marine birds and reptiles. "The kidney of birds and reptiles cannot produce a urine with high salt concentrations," he wrote, "and one can assume that the extrarenal mechanism for salt excretion is a necessity for successful adaptation to a marine habitat in all sauropsidians."[22]

Bartholomew challenged the widely accepted generalization that only the mammalian kidney is capable of producing a concentrated urine. In his early studies of birds, Bartholomew did not detect salt secretion from nasal glands, but some birds seemed to eliminate excess salt with their kidneys. The results were unexpected and intriguing, but Schmidt-Nielsen was skeptical. In a letter to Bartholomew, he wrote, "I was completely thrilled by what you told me about the savannah sparrow. The more I think about it the more difficult it seems to be to understand. The nasal secretion in marine birds is always conspicuous and easy to observe, and it seems impossible that it could be overlooked even in a small bird. The other possibility, that the kidney does the concentration work, seems even more remote for, I believe, no bird or reptile can produce a urine more than about twice as concentrated as the plasma. I am quite excited about this and look forward to hear more. I do hope that you will have a head or two that you can send me for dissection."[23]

The savannah sparrow (*Passerculus sandwichensis*) was particularly well suited for examining the question of avian kidney function.[24] Two subspecies of savannah sparrow (*P. sandwichensis beldingi* and *rostratus*) are restricted to salt marshes that contain a higher salt concentration than typical body fluids. Several other subspecies spend most of their time in freshwater marshes and only occasionally encounter salt water. Comparing subspecies living in contrasting environments provided a model system for investigating homeostasis and adaptation from an explicitly ecological and evolutionary perspective. The studies highlighted both the importance of individual variation and the role of natural selection in shaping intraspecific differences among populations.

As with kangaroo rats and desert ground squirrels, evaporative water loss posed a challenge for savannah sparrows. Many birds survived for only a few days without water. However, there was considerable individual variation, and half of the savannah sparrows were able to survive for ten days without drinking. One saltmarsh bird lived for several weeks without drinking. Birds also varied in their tolerance for salt water, both as individuals and populations. Saltmarsh birds tended to decrease their intake of salt water that was more concentrated than the urine that their kidneys produced. Surprisingly, many of these birds could live on 75% seawater, and one bird lived for several weeks on undiluted seawater. Sparrows from freshwater marshes increased drinking rates when provided with salt water, even though this led to dehydration.

The ability to tolerate salt water existed to a limited degree in all savannah sparrows, but natural selection enhanced the trait in saltmarsh populations. "From the standpoint of salt intake," Bartholomew concluded, "the [saltmarsh] population can be considered an aggregation of individuals selected from one extreme of the range of salt tolerance shown by the non-saltmarsh races."[25] The experiments did not conclusively demonstrate that the kidney was solely responsible for eliminating salt and conserving water, but they demonstrated that some populations of sparrows closely regulated water and salt even without well-developed nasal glands.[26]

Promises and Problems of Integration and Synthesis

Contrasts between Schmidt-Nielsen's engineering approach and Bartholomew's experimental natural history perspective reflected broader disciplinary divides in organismal biology. In the United States, comparative physiology was institutionalized primarily in two quite different organizations. Knut Schmidt-Nielsen had organized a small, informal comparative physiology section within the American Physiological Society in 1950.[27] At the same time, comparative physiologists, including both Schmidt-Nielsen and Bartholomew, were active members of the Division of Comparative Physiology and Biochemistry in the American Society of Zoologists.

The different orientations of the two societies were highlighted when Schmidt-Nielsen nominated Bartholomew for membership in the APS.[28] Although stressing Bartholomew's physiological qualifications in his letter of nomination, Schmidt-Nielsen privately urged Bartholomew to solicit a second letter of nomination from a medically oriented colleague. From the perspective of many members of the APS, an experimental naturalist might not fit in an organization strongly oriented toward human physiology. Despite their broad interests in physiological ecology, both Schmidt-Nielsens published in the APS flagship publication *American Journal of Physiology*, and Bodil became the first woman elected president of the APS, in 1975. Bartholomew published in a wide variety of journals, but never in the *American Journal of Physiology*. Many of his most influential articles appeared in zoological journals such as *Condor, Auk, Journal of Experimental Zoology*, and *Journal of Mammalogy*.[29] His choice of audiences reflected Bartholomew's strong commitment to a physiological ecology with a broad intellectual foundation in natural history.

Throughout his career Bartholomew was sensitive to problems caused by scientific specialization and the difficulty of integrating perspectives from different disciplines. The problem was particularly acute for physiological ecologists, who of necessity employed interdisciplinary approaches that required moving among various levels of biological integration from cells to organisms to populations and ecosystems. To some extent Bartholomew believed that the problem was unsolvable, because the interactions of an organism with its environment were so complex that any understanding was necessarily partial and incomplete. Combining the traditional physiological focus on organisms as self-regulating "physicochemical systems," the ecologist's appreciation for the complexity of the external environment, and the evolutionist's concerns for adaptation within the context of evolving populations and species, was a challenge that required the integrating perspective of natural history, in both practice and theory. According to Bartholomew, "natural history will always be the touchstone for synthesis (biological significance), analysis (biological mechanism), developing incisive field experiments, and creating theories and models close enough to reality to be taken seriously."[30] Explanations based on proximate and ultimate causation were intimately intertwined in Bartholomew's work, and he seemed capable of effortlessly straddling the divide between functional and evolutionary biology.

In 1996 the American Society of Zoologists was transformed into the Society for Integrative and Comparative Biology, and Bartholomew was hailed for his role in shaping the broad interdisciplinary perspective of the new organization. Bartholomew's legacy was attributable to a number of important contributions: intellectual, academic, and institutional. His holistic perspective on biology encouraged combining approaches from a variety of levels of organization from molecules to ecosystems. But although he considered both organisms and their environments to be "physico-chemical systems," Bartholomew was more inclined than Schmidt-Nielsen to emphasize the unique aspects of biology that demanded more than an understanding of physics, chemistry, and mathematics. Not only were biological systems "staggeringly complex," but they also had a history. Recognizing the role of natural selection in shaping organisms and populations required an emphasis on variability and diversity, as well as the recognition that natural selection resulted in ad hoc adaptive solutions that were "blind to future consequences." Bartholomew inculcated this broadly synthetic and integrative perspective in his students, several of whom took leadership roles in the new society. Faced with increasing specialization in science, Bartholomew served as a "voice of reason" against excessive disciplinary fractionalization for a younger generation who shaped the diverse, but intellectually fragmented, American Society of Zoologists into a more integrated, broadly biological organization.[31]

Bartholomew was certainly not unique in calling for a more integrated and synthetic biology during the latter half of the twentieth century. Nor was he the only biologist during this time to attempt combining homeostasis and evolutionary adaptation to explain the fit between organism and environment. Yet his attitude toward integration was marked by an absence of the defensiveness and antagonism toward molecular biology and the physical sciences found in the writings of some of the prominent evolutionary biologists associated with the modern synthesis. As Betty Smocovitis, Erika

Milam, and other historians have carefully documented, Ernst Mayr, in particular, viewed organismal biology as a bulwark against the encroachment of molecular biology on one hand and a philosophy of science oriented toward physics on the other.[32] In his more conciliatory moments, Mayr might have agreed with Bartholomew's characterization of the organism as a "complex interaction between a self-sustaining physicochemical system and the environment," yet he never embraced physics and chemistry with quite the same enthusiasm that Bartholomew did. Bartholomew could acknowledge the engineering approach of Schmidt-Nielsen, with its mechanistic analogies and explicit reliance on physics, while still championing the role of natural selection and deep insights into adaptation provided by natural history.

Although in a retrospective account of his intellectual development, Bartholomew acknowledged the importance of reading the books of Mayr, Dobzhansky, Simpson, and particularly Huxley as a graduate student at Harvard, he seldom referred to this literature in his professional articles.[33] Conversely, Mayr rarely cited Bartholomew's work, despite the overlap in interests and the shared focus on birds. Indeed, Mayr didn't seem to know what to make of physiological ecology, which he almost completely ignored in his monumental *The Growth of Biological Thought*. In that idiosyncratic history, Mayr briefly mentioned Knut Schmidt-Nielsen, whose work he linked to the tradition of ecological biogeography.[34] This was a curious claim to make about a physiologist who rarely considered problems of geographical distribution and intraspecific variation. For Schmidt-Nielsen, deserts were not so much geographical areas inhabited by populations as they were sets of physical challenges and constraints that determined the limits of tolerance for species and called forth intricate physiological adaptations. In short, for Schmidt-Nielsen, adaptation meant "how animals work" under often stressful conditions.[35] Evolution formed only a general backdrop to his mechanistic studies of functional adaptation.

The connection with geographical distribution would have been more appropriate for Bartholomew, who was genuinely interested in applying physiological ecology to zoogeography.[36] However, Bartholomew did not fit neatly into the categories of functional biology and evolutionary biology that Mayr constructed to organize his historical narrative. Bartholomew was a museum-trained field naturalist but also an avid experimentalist, who viewed controlled laboratory experiments not as a methodology distinct from the comparative method, but as its epitome. Mayr's claim that functional biology (including physiology) was restricted to studying proximate causes, while a largely separate evolutionary biology considered ultimate causes of adaptation and speciation, also ran counter to Bartholomew's interest in integrating biology both on a personal and an interdisciplinary level. As Erika Milam has cogently argued, Mayr emphasized the deep divisions within the biological sciences and presented organismal biology as a distinct community of researchers with a philosophical orientation fundamentally different from that of functional biologists, and more especially physicists and chemists.[37] By contrast, Bartholomew celebrated the unique contributions that naturalists, physiological ecologists, and other organismal biologists made to the study of life, while at the same time arguing for a unified biology. "Biology is indivisible," he wrote; "biologists should be undivided."[38] During the 1960s Bartholomew presented homeostasis and adaptation through natural selection as complementary concepts for bringing about his unification. As we shall see in the next two chapters, Mayr shared this interest in combining homeostatic self-regulation and evolutionary adaptation, but although he sometimes made conciliatory statements about unifying biological specialties, his general perspective remained divisive and often combative.

7

Complexities of Thermoregulation

Most students learn about homeostasis through the exquisite regulation of body temperature in mammals and the mechanical analogy with the thermostatic temperature control in a modern building. Both the example and analogy were crucial for Walter Cannon's *Wisdom of the Body*. Despite the existence of a diurnal cycle of about a degree, the Harvard physiologist pointed to the remarkable stability of human body temperature, noting, "The constancy is so reliable that the thermometer makers can stamp '98.6' on the Fahrenheit scale with assurance that it will mark closely the mean temperature of the healthy person everywhere."[1] If thermoregulation was an example of homeostasis familiar to his readers, it also provided a way of understanding the evolution of self-regulation that Cannon's audience could easily grasp. Emphasizing the popular distinction between cold-blooded and warm-blooded animals, Cannon presented the dichotomy as an illustration of how the evolution of homeostasis led to increasing freedom from the external environment. In contrast to what Cannon considered to be "lower animals" such as amphibians and reptiles that spent much of their lives in a sluggish state dictated by daily and seasonal cold, mammals remained active due to their remarkable ability to maintain a high and constant body temperature by generating, storing, and dissipating heat.[2]

Comparative physiologists and physiological ecologists challenged this simple dichotomy. Studying the natural history, ecology, and physiology of arctic organisms ranging from lichens to caribou, Laurence Irving and Per Scholander noted the active lives of organisms living at the coldest temperatures. They pointed out that arctic fish were just as active as their temperate and tropical relatives. In the spring, even with snow on the ground, insect life was abundant and active, at least in sunny areas. Such phenomena suggested a "homeostatic tendency" even in cold-blooded species.[3] At cold temperatures organisms did not necessarily exist in the sluggish state that Cannon had supposed. The comparative physiologist C. Ladd Prosser was impressed by the ability of fish and other poikilotherms to acclimate to seasonal change. Although the details were poorly understood, he surmised that gradual cooling called forth the production of different suites of enzymes better equipped to function at low temperatures. As a result, the activity of cold-acclimated organisms sometimes approximated that of those at warmer temperatures.[4] This type of compensation by temperature "conformers" was just as homeostatic as the self-regulation by birds and mammals, which were temperature "regulators." Indeed, conformers and regulators were not distinct categories, but rather endpoints on a continuum of adaptive responses.

The linear progressivism of Cannon's evolutionary thinking stands in sharp contrast to a more branching Darwinian scheme, but evolutionary naturalists sometimes emphasized linearity in space, if not in time. Indeed, a major focus of evolutionary studies in the century after Darwin was progressive variation along biogeographic gradients. For example, ecologists, biogeographers, and evolutionary biologists often cited Bergmann's rule that within a taxonomic group, body size tended to increase as one moved from the equator to the poles. In formulating the generalization, Carl Bergmann looked to late nineteenth-century metabolic studies that emphasized the relationship between surface area (through which heat was lost) and volume (by which metabolic heat was produced).[5] The rule found theoretical justification in the fact that as animals get bigger, volume increases more rapidly than surface area. Although widely cited, the thermoregulatory explanation for the rule was problematic, not only because of numerous exceptions but also because poikilotherms also sometimes exhibited the biogeographic trend. Some physiological ecologists raised doubts that small differences in body size studied by Bergmann and his followers were adaptive, emphasizing instead the phenotypic plasticity of organisms and their ability to acclimate to a range of environmental conditions. They also called into question the sharp distinction between cold- and warm-blooded animals, by pointing to numerous cases of an alternative adaptive strategy of heterothermy—the ability to switch between regulating body temperature and allowing the body temperature to conform to that of the surrounding environment. The complexity of homeostatic temperature regulation, particularly in cold environments, involved cases in which the "how" questions of functional biology and the "why" questions of evolutionary biology were deeply intertwined. From the perspective of physiological ecology, the two needed to be studied in tandem.

Heterothermy as an Adaptive Strategy for Small Animals

Among the many birds, reptiles, and mammals that George Bartholomew studied early in his career was the little pocket mouse (*Perognathus longimembris*). Little pocket mice were ideal organisms for studying the physical constraints posed by small size, and the physiological responses to temperature and food supply. Because pocket mice are among the smallest desert mammals, they are particularly vulnerable to temperature changes, especially seasonal and nightly cold. Anecdotal evidence from naturalists suggested that pocket mice underwent hibernation during the winter, although nobody had documented these claims by actually measuring body temperatures of torpid individuals. These small rodents also appeared to be good candidates for physiological experimentation. They were common and easy to trap in the deserts of Southern California. Once captured, they proved quite adaptable to life in the laboratory. They seemed indifferent to having small temperature sensors embedded in their bodies, even when the copper leads protruded for several centimeters from their backs. Despite being connected by wires to recording equipment for several days at a time, the small rodents exhibited a variety of seemingly natural behaviors in the small glass terrariums that were their laboratory homes.

Working with Tom J. Cade, a graduate student who also collaborated on his water balance studies with savannah sparrows, Bartholomew designed experiments with little pocket mice that were informed as much by natural history as traditional laboratory physiology. Circumstantial evidence from field surveys suggested that *P. longimembris* and its close relatives became torpid under a variety of conditions. When recovered from live traps on cold nights, the mice were often inactive, although they quickly recovered when warmed. If, indeed, the mice were true hibernators, as naturalists had suggested, they would be the smallest mammals other than bats to exhibit this physiological state. During preliminary observations in the laboratory, Bartholomew and Cade found that the mice often slept or became dormant even at room temperature.[6]

Controlled experiments manipulating both temperature and food confirmed these informal observations. At room temperatures pocket mice exhibited intermittent activity punctuated by brief periods of sleep, which rarely lasted more than an hour. Body temperature fluctuated by a degree or two centigrade during these irregular patterns of activity and inactivity, although there was no evidence of a diurnal temperature cycle. The mice were able to maintain body temperature within this narrow range at warmer temperatures without evidence of panting or drooling. However, when the environmental temperature exceeded body temperature, the mice could not prevent hyperthermia. Under these conditions, longer periods of torpidity followed bursts of digging, feeding, and grooming. During the inactive estivation, thermoregulation was relaxed, allowing body temperature to rise several degrees. The pocket mice were also unable to maintain constant body temperature even in moderately cool environments. When kept at 20°C, the small rodents intermittently entered short periods of torpor, allowing body temperature to drop to the surrounding temperature. When the mice were aroused and became active, body temperature quickly increased again. At colder environmental temperatures, the mice hibernated, exhibiting slow metabolism and breathing, accompanied by a decrease in body temperature.

Because the pocket mice remained active over a broad range of core body temperatures, from 32–39°C, and because the body temperature could fluctuate rapidly depending upon environmental conditions, Bartholomew claimed that it made no sense to speak of a "normal" or "typical" temperature. This was not an artifact of laboratory life, because in their natural desert environments, the mice experienced large temperature fluctuations both daily and seasonally. Heterothermy was an adaption to short-term changes in temperature in the daily lives of the small rodents. Estivation and hibernation were more profound responses to extreme temperature fluctuations that might occur either daily or seasonally. Meeting these environmental challenges required flexible homeostatic responses that bore little resemblance to thermostatic controlled constancy.

Surprisingly, Bartholomew and Cade found that they could induce the torpor associated with hibernation and estivation even at room temperature simply by removing food from the terrarium. All small mammals have high metabolic rates, but the diminutive size of the little pocket mouse accentuated the energetic requirements that required frequent feeding. When food was unavailable, the tiny rodents resorted to reducing metabolism and relaxing thermoregulation to tide them over. In nature, behavior and physiology worked together to protect against starvation. Behaviorally, the pocket mice were food hoarders, maintaining supplies of seeds in their underground burrows, which they aggressively defended from intruders. Food storage was the "first line of defense" for survival, but torpor associated with both estivation and hibernation was a physiological strategy for extending survival when food was scarce regardless of the surrounding temperature.

The recognition that some small mammals resorted to heterothermy when resting, particularly during cold weather, was not new. Studies on bats dating back to the nineteenth century had documented the inability of the winged mammals to maintain body temperature at rest. Carefully controlled metabolic studies on small mammals after World War II confirmed the generality of this phenomenon. What set Bartholomew's work apart was the way he generalized heterothermy by emphasizing the lability of body temperature, presenting it as a normal part of the everyday lives of small mammals, and underlining the energetic basis of fluctuating temperature by equating hibernation and estivation. If one focused only on temperature, the two processes seemed strikingly different. Estivating animals became torpid under hot conditions, allowing body temperature to rise. Hibernating animals became torpid under cold conditions, allowing body temperature to drop. In both cases, however, deregulation of body temperature conserved energy. Bartholomew and Cade demonstrated this equivalence experimentally by manipulating the ambient temperature of little pocket mice in the laboratory. In the absence of food, the experimenters could shift them from "deep hibernation" to estivation without arousal simply by raising the surrounding temperature.

Recognizing that heterothermy was a broad adaptive strategy that entailed more than a passive response to temperature was an important intellectual shift. It allowed breaking free from "anthropocentric arrogance" and the "tyranny of standard points of view" that emphasized the maintenance of constant body temperature as the paradigm of homeostasis.[7] More broadly, Bartholomew interpreted homeostasis to mean maintaining dynamic stability in heat production and heat loss, even if that resulted in fluctuating body temperatures. Bartholomew's perspective on homeostasis intentionally blurred the distinction between environmental regulators and conformers. This approach embraced the engineering perspective of Schmidt-Nielsen that emphasized mechanical analogies and sophisticated quantitative analysis of size using allometry. At the same time, it deepened the comparisons by highlighting the ecological idea of niche overlap and an evolutionary appreciation for how natural selection causes convergent evolution. Physical size constraints on small animals, regardless of phylogenetic relatedness, often called forth similar physiological responses, even though the anatomical machinery involved was very different. Thus, heterothermy served as an important bridging concept that Bartholomew used to unify a spectrum of adaptive responses to temperature, in contrast to previous comparative physiologists who had recognized only the extremes.

Several years later Bartholomew returned to the physical constraints that size placed on the ability of small animals to conserve heat. In an essay entitled "A Matter of Size," he wrote, "It is only a slight overstatement to say that the most important attribute of an animal, both physiologically and ecologically, is its size."[8] Although mammals ranged from tiny shrews to enormous whales, there seemed to be a lower size limit beneath which it was impossible to maintain a constant body temperature. The smallest warm-blooded animals—shrews and hummingbirds—weighed only two grams. Balancing metabolic heat production and heat loss in such small animals required that they feed almost constantly. Indeed, Bartholomew noted that most of these creatures were heterothermic, maintaining constant body temperature only when active and well fed.

The size of the smallest birds and mammals overlapped that of large insects. Bartholomew's interest in comparing these distantly related groups was stimulated, in part, by a flurry of studies by various physiologists demonstrating that large flying insects generated heat internally (endothermy) and maintained constant body temperature (homeothermy), at least during flight. This research itself built on a long history of recording elevated temperatures in beehives and attempts to measure the internal body temperature of insects.[9] Early studies were plagued by the technical difficulties of miniaturizing thermometers, and perhaps also by a prejudice that insect thermoregulation was primitive and uninteresting compared to the exquisite homeostatic abilities of birds and mammals. Using tiny thermocouples embedded in the bodies of large moths, Bartholomew and his student Bernd Heinrich accurately recorded internal temperatures during rest, warm-up, and flight. Tethered by the wire connected to the thermocouple, experimental moths flew in place within their laboratory containers. What made the studies particularly compelling were comparisons with small hummingbirds: anatomical, behavioral, and ecological.[10]

Superficially, sphinx moths (family Sphingidae) are so similar to hummingbirds that casual observers often confuse them. The moths and hummingbirds are about the same size, exhibit similar hovering flight, and sometimes feed from the same tubular flowers. Hovering requires rapid movement of the wings, is energetically expensive, and is dependent upon maintaining a high body temperature, 35–38°C. The energy cost and temperature dependence pose an adaptive problem for the sphinx moths, which are nocturnal. Being active at night means that they cannot use solar radiation to raise body temperature by basking in the sun, the way some diurnal insects do. Instead, sphinx moths must rely completely upon metabolic heat production to warm their bodies in a desert environment where nightly temperatures are too cold for flight. The remarkable "biological engineering" that allows this highly efficient internal heat production is quite different from the anatomy and physiology of the hummingbird and constitutes a compelling example of evolutionary convergence.

Bartholomew and Heinrich described the two sets of muscles located in the thorax of the moth as both a motor for flight and a potent heat generator. These flight muscles had a higher rate of metabolism than any tissue previously reported in the physiological literature. A layer of hairlike scales covering the thorax trapped the heat generated by the flight muscles as effectively as the insulating feathers of hummingbirds. In contrast to the thorax, the moth's abdomen was uninsulated and acted as an efficient heat radiator. Even though the thoracic temperature increased with muscular activity, the abdomen remained close to the environmental temperature. The two body segments, working together, maintained a balance of heat production and heat loss. Blood circulation facilitated the transfer of heat between the thorax and abdomen, and controlling this flow provided a homeostatic mechanism for precise regulation of thoracic temperature when the insect was flying. Heinrich and Bartholomew described this anatomy, so different from that of birds and mammals, as a true, self-regulating system. It not only produced heat internally (endothermy) but also maintained a relatively constant thoracic temperature during flight (homeothermy).

At rest the sphinx moth remained an ectothermic poikilotherm whose body temperature matched the surrounding environment. This meant that before flying, the moth required a preflight warm-up period. The insect accomplished this by rapid and simultaneous contraction of both sets of flight muscles. Contracting together, the muscles generated large amounts of heat without moving the wings. According to Heinrich and Bartholomew, this process was the eco-physiological equivalent of shivering in birds and mammals. The length of the warm-up period depended upon the ambient temperature. In a warm laboratory setting, moths were ready to fly within a minute of preflight activity, but at the cooler nocturnal temperatures of the Mojave Desert, the warm-up took up to fifteen minutes. Indeed, nightly temperatures in the desert were always too cold for flight without some preflight warm-up.

Bartholomew and Heinrich admitted that applying the concept of heterothermy to moths and making broad comparisons between moths and hummingbirds was unconventional. Nonetheless, it seemed justified from an energetic perspective. When food was unavailable, small birds and mammals relaxed control of body temperature, temporarily allowing it to fluctuate with the surrounding environment. This adaptive hypothermia was sometimes seasonal, but it was also often part of the daily life of small warm-blooded animals facing cold nights as well as hot days. The "rapid, drastic, and repeated changes in body temperature" that characterized heterothermy were not simply passive responses to a hostile environment but part of a broadly adaptive self-regulation employed by small birds and mammals.[11] In the sphinx moth, "facultative endothermy" provided a supplement to an otherwise ectothermic existence. The moth used internal heat production to forage actively during the cool desert nights when hovering flight would otherwise be impossible. Internal heat generation came at a steep energetic cost that the insect could not maintain at rest. Both hummingbirds and sphinx moths actively maintained a dynamic but stable internal environment by regulating heat production and loss, despite labile body temperatures. Thus, from a broad energetic perspective heterothermy constituted an exquisitely adaptive form of homeostasis that bridged the traditional distinction between warm- and cold-blooded animals.[12]

Being heterothermic was not simply making the best of a bad situation. Indeed, according to Bartholomew, heterotherms had the "best of both worlds."[13] Depending upon environmental conditions, they could exploit the advantages of both homeothermy and poikilothermy. Thus, heterothermy was not an evolutionary novelty but a broad adaptive strategy found in many small animals in diverse taxa. It was, Bartholomew argued, a convincing demonstration of how natural selection caused convergent evolution in distantly related species facing similar environmental challenges.

Bartholomew was sensitive to the potential criticism that his detailed accounts of life in the desert were merely adaptive stories: a misconception that he tried to counter. The concepts of heterothermy and convergent evolution provided a unifying theoretical context for comparisons of small mammals, birds, and flying insects. Allometric analyses provided a rigorously quantitative demonstration that from an energetic perspective, metabolism, heat production, and insulation were comparable in diverse animals, even though the anatomical structures were different. From this physical perspective, sphinx moths and hummingbirds had converged on an energetically expensive method of rapid hovering flight that was adaptive in certain ecological contexts but not necessarily in others. Hovering flight was closely tied to feeding on nectar from tubular flowers, a behavior that was common to hummingbirds during the day and sphinx moths at night. Starting from very different anatomical foundations, natural selection had fashioned convergent forms "approaching the biological limit of aerodynamic performance."[14] However, this ecological niche was only one of many available to flying insects. Other nocturnal moths that Bartholomew studied were not facultative endotherms, and even while flying their body temperatures remained as cool as their surroundings. Of necessity, these ectothermic moths utilized a very different flight dynamics that relied on the combination of a light body build and the slow beating of large wings.

Heterothermy was not a necessary adaptation for life in the desert, but it turned out to be much more common than biologists had previously recognized. Bartholomew's contributions to this area of research illustrated several important axioms that he considered essential for the study of adaptation from the perspective of physiological ecology. The first involved understanding the external environment as experienced by the organism. This required the keen eye of a field naturalist and the quantitative and experimental insights from ecology and laboratory physiology. Secondly, following a tradition that he traced to Claude Bernard, the physiological ecologist needed to approach the organism as an integrated whole and to recognize that it was inseparable from its external environment. Finally, the physiological ecologist relied on the Darwinian understanding of adaptation as a temporary and imperfect solution to environmental challenges. Adaptation was always constrained by the history of a species, which constituted a kind of "phylogenetic trap" that limited the raw materials on which natural selection operated. Immediate reproductive success rather than long-term survival of a species also constrained the ability of natural selection to mold anatomical structure and physiological function. Although adaptations such as heterothermy might result in the seemingly exquisite fit of organism and environment, this fit was an ephemeral solution entailing only adequacy, rather than perfection.

In retrospect, Bartholomew saw a tight and natural fit between the physiological ecology that he helped to create and the modern evolutionary synthesis that established itself at the same time. He noted how reading the prominent books of Huxley, Mayr, Dobzhansky, and Simpson had influenced him during his formative years as a graduate student after World War II. His experiences in various museums, laboratories, and field sites strongly shaped his perspective as a self-described "experimental naturalist," providing a bridge between traditional physiology and evolutionary biology. He suggested that his comparative studies of populations, species, and much more distantly related taxa—often conducted in the controlled setting of the laboratory—were attempts to answer evolutionary questions posed by the modern synthesis, such as divergence, convergence, fitness, and biogeographic variation. Yet, although variation and natural selection formed threads running through his writings, Bartholomew rarely cited the literature of the modern synthesis. That the connections between physiological ecology and evolutionary biology were more apparent to Bartholomew at the end of his career only emphasizes the challenges that biologists at midcentury faced in reconciling physiological and evolutionary concepts of adaptation and self-regulation, the inseparability of the organism and its environment, and the relationship between internal and external environments. Highlighting these difficulties were controversies over the validity of Bergmann's rule and other biogeographical generalizations widely supported by leading figures of the modern synthesis but challenged by prominent physiological ecologists.

Bergmann's Rule, Heterothermy, and Arctic Adaptations

Bergmann's rule states that the size of animals tends to increase as one moves from the equator to the poles. A similar generalization, Allen's rule, named after the nineteenth-century naturalist Joel Allen, states that the extremities of animals such as ears and limbs tend to get shorter in colder regions. Evolutionary biologists considered these clines important for demonstrating natural selection, although both methodological and theoretical ambiguities plagued the generalizations. For example, some prominent museum biologists such as Ernst Mayr insisted that the rules were valid only when making comparisons among populations of the same species (i.e., subspecies). Despite this proviso, other proponents liberally applied the rules to species or genera and, at least by implication, birds and mammals in general. The fact that naturalists often used museum specimens to demonstrate the rules also raised the practical problem of accurately measuring size from preserved skins. This was a particular problem, because most discussions of Bergmann's rule at least implicitly assumed a thermoregulatory explanation based on the ratio of surface area and volume. Because volume increases as the cube of linear dimensions, while surface area increases as the square, larger animals have a smaller surface area/volume than their more diminutive relatives. In contrast to the regular geometric solids, which are often used to demonstrate the relationship between surface area and volume, complex organisms usually cannot be reduced to simple dimensions of length, width, or circumference. Attempts to measure surface area and volume of specimens, particularly preserved museum specimens, posed vexing methodological problems.

Heat generation within the body and loss through the body surface might provide a plausible theoretical explanation for the biogeographical rules, but critics scoffed at the notion that it could account for very small differences along a cline. The explanation was also weakened by the fact that Bergmann's rule applied not only to mammals and birds but sometimes to poikilotherms that do not maintain constant body temperatures.

The Scandinavian American physiologist Per Scholander aired all of these criticisms in a short article published in *Evolution* in 1955.[15] Scholander had spent much of World War II conducting military research in Alaska, and afterward he continued studying the diverse adaptations of arctic plants and animals with his father-in-law, the physiologist Laurence Irving. Their research on birds and mammals emphasized the importance of insulation against arctic cold but also the prevalence of heterothermy in both large and small animals, countercurrent exchange systems to retain heat, and the general adaptability of animals (including humans) to extreme temperatures. Scholander's critique of Bergmann's rule brought a sharp rebuttal from Mayr, one of the major proponents of clines and biogeographical rules. The ensuing exchange among physiologists, evolutionary biologists, and physical anthropologists was controversial, and it raised important issues that continue to challenge physiological ecologists.[16]

Together with the research of the Schmidt-Nielsens and Bartholomew, the "Irving-Scholander legacy" in artic biology was a crucial episode in the development of physiological ecology after World War II.[17] Like the Schmidt-Nielsens, both Irving and Scholander drew inspiration from the Danish Nobel laureate August Krogh. A disinterested medical student who never actually practiced medicine after earning his degree, Scholander turned first to the taxonomy of lichens and then the physiology of diving mammals.[18] Several years before Schmidt-Nielsen arrived, Scholander joined Krogh's zoophysiology laboratory to continue his studies of the diving reflex in harbor seals. At the time, Irving, who was a decade older than Scholander, was chair of the Biology Department at Swarthmore College. He invited Krogh to present a series of public lectures at the college, a fortuitous decision that had several important consequences for the later development of physiological ecology. The extended visit to Swarthmore cemented a warm personal and professional relationship between the two scientists. Krogh reworked his public lectures into an important book on respiratory mechanisms. His ideas on the organization of research in comparative physiology served as a model for Irving's later development of an arctic research institute at the University of Alaska after World War II. Perhaps most importantly, Krogh's visit laid the foundations for Scholander's immigration to the United States at the start of the war to work with Irving using the support of a Rockefeller Fellowship. The partnership initially centered on the scientists' common interest in diving physiology, continued with wartime research in applied physiology, and led after the war to the broader study of arctic physiology of plants and animals. As already discussed, the relationship between Krogh and Irving was also responsible for the eventual immigration of Knut and Bodil Schmidt-Nielsen, leading to their pioneering research with kangaroo rats.

Like the Schmidt-Nielsens and Bartholomew, Scholander and Irving combined elements of field ecology, laboratory physiology, natural history, and behavior within a broad evolutionary context. Unlike Bartholomew, who performed experiments in the modern laboratory of a major research university, Scholander and Irving began their postwar research in a converted Quonset hut on a remote naval base in Point Barrow, Alaska (figure 8). Perched above the Arctic Circle, the makeshift laboratory provided few amenities, but the frigid environment posed interesting questions about thermoregulation. Why, Scholander wondered, did local sled dogs sleep exposed on the snow even when shelter boxes were available? Similarly, Irving noted that the subzero temperatures seemed to have little effect on the behavior of caribou and other native species. Social behavior, including play, continued to be a normal part of animal life even during the harsh winters. This seemed to be true of humans, as well. Irving observed that native Alaskan children were just as likely to shake off their mittens during outdoor play as children living in milder climates. These casual observations highlighted the more serious point that thermoregulation in arctic animals meant more than staying warm and that losing heat was not always a threat to survival.[19] In fact, thermoregulation involved a dynamic balance of heat production and heat loss mediated by the interplay of metabolism, insulation, thermal radiation, and behavior. Given these broad and flexible mechanisms, Bergmann's and Allen's claims about minor differences in the size of the body or its appendages seemed unlikely to be very important for thermoregulation in arctic animals.

Consider the sled dog asleep on the snow in subzero weather. When sleeping, the dog maintains a basal metabolic rate (BMR) just as a tropical or temperate mammal. This basal metabolism maintains a relatively constant body temperature in all of these animals over a range of environmental temperatures referred to as the thermal neutral zone. The ability to survive intense cold is due to the sled dog's thick coat of insulating hair that traps heat and

Figure 8. The physiology laboratory of the Arctic Research Laboratory in Point Barrow, Alaska. Scholander is standing in the doorway (fourth from right) next to Irving (third from right). Laurence Irving Papers, Archives of the University of Alaska Fairbanks. Reprinted with permission from the University of Alaska, Fairbanks.

maintains the steep temperature gradient between internal and external environments. The dog unconsciously manipulates the insulation by fluffing or compressing the hair and thereby greatly extends the range of environmental temperatures over which basal metabolism can maintain body temperature. This thermal neutral zone is much wider in arctic mammals than in their tropical or temperate counterparts. In all cases the animals must expend additional energy to maintain constant body temperature outside the thermal neutral zone.

Scholander, Irving, and their coworkers measured metabolic rates by oxygen consumption in a variety of mammals at different environmental temperatures using a metabolic chamber that they constructed at the laboratory at Barrow.[20] The makeshift device could not accommodate very large animals, but the team tested mammals ranging from lemmings to foxes and polar bear cubs (figure 9). Through Irving's military connections, the group used army aircraft for expeditions to Cuba and Panama to test tropical mammals, including raccoons, coatis, sloths, and monkeys.

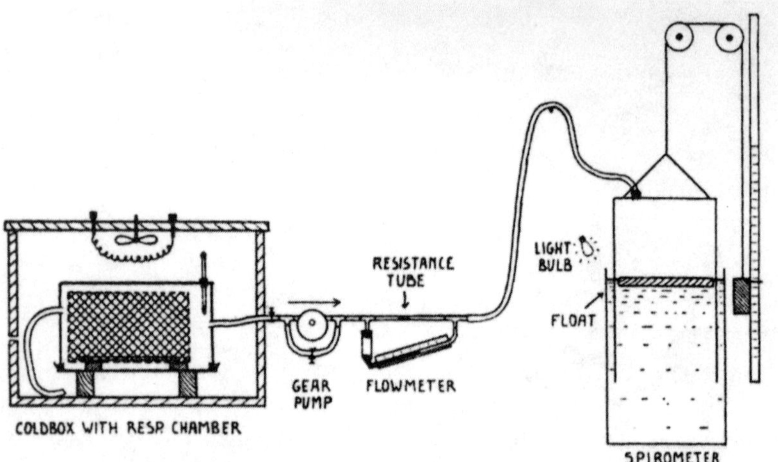

Figure 9. Schematic diagram of the metabolic chamber used at the arctic physiology laboratory. The experimental animal was caged inside a sealed respiration chamber that could be temperature controlled. Respiratory gases were collected in the spirometer for measurement and analysis. P. F. Scholander, Raymond Hock, Vladimir Walters, Fred Johnson, and Laurence Irving, "Heat Regulation in Some Arctic Mammals and Birds," *Biological Bulletin* 99 (1950): 237–58.

The experiments yielded some surprising results. Compared to their tropical relatives, arctic mammals had broader thermal neutral zones that allowed maintaining body temperature over a wider range of environmental temperatures without expending extra energy (figure 10). With their low metabolic rates and poorly insulated bodies, sloths would actually shiver when environmental temperatures dropped a few degrees during the tropical nights. In stark contrast, even in Point Barrow temperatures never got low enough for the researchers to determine the lower critical temperature of the thermal neutral zone for dogs and foxes. When they tested an arctic fox in a more elaborate metabolic chamber at the Naval Research Laboratory in Washington, DC, the animal finally began to shiver at -70°C. If foxes and other mammals and arctic birds could maintain body temperature under such extreme conditions while resting, it seemed likely that overheating might be a serious problem when animals were active. Finding that large animals such as caribou had less insulation than much smaller dogs and foxes also suggested that this might be the case. Indeed, broader comparative studies indicated that there might be no significant correlation between body size and length of fur for animals larger than foxes.

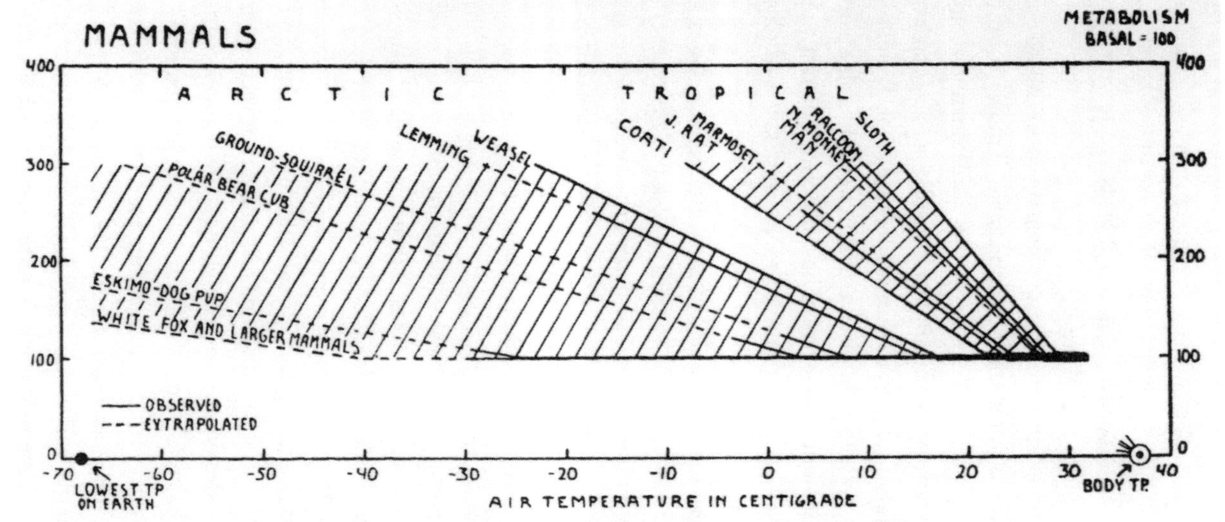

Figure 10. Heat regulation and temperature sensitivity of arctic and tropical mammals. The basal metabolic rate for each animal is represented by 100 on the vertical axis, and relative changes in metabolism are indicated by the sloping lines beyond the lower critical temperature. P. F. Scholander, Raymond Hock, Vladimir Walters, Fred Johnson, and Laurence Irving, "Heat Regulation in Some Arctic Mammals and Birds," *Biological Bulletin* 99 (1950): 237–58.

Mayr would later criticize Scholander for "all or none" typological thinking about insulation, but in fact the physiological ecologist was well aware of evolutionary trade-offs. This was particularly true when comparing how small and large mammals coped with cold temperatures using combinations of physiological and behavioral adaptations. Because of the physics of body size, heat loss was a major challenge for small birds and mammals. Insulation might be adaptive for thermal regulation, but it also compromised mobility. For mice and other small mammals, keeping warm would require a coat of fur so thick that it would make it impossible to move. For these animals there was a compromise between effective insulation and rapid movement necessary to escape predators. By itself insulation was apparently inadequate for the extreme cold of Alaskan winters based on the observation that mice, lemmings, weasels, and other small arctic mammals spent much of the winter burrowing under the snow. Thus, like kangaroo rats and desert ground squirrels, small arctic mammals used a combination of physiology and behavior to maintain homeostasis in a harsh environment.

Larger animals were also constrained by the physics of size, but in a different way. Insulation that maintained warmth at rest, threatened overheating when running. Partly for this reason, and partly to make locomotion more efficient, the limbs of large animals were sparsely covered with hair. Similarly, the function of eyes, ears, and noses required minimizing or eliminating insulation. The sleeping sled dog curled up to protect these exposed parts, but when running the lack of insulation actually acted as a radiator to ensure efficient loss of heat. Altering blood flow to the limbs regulated heat loss and retention, and in some animals, a countercurrent exchange of heat between arteries and veins in the limbs provided another homeostatic mechanism. The warm outgoing blood in the arteries warmed the returning cooler blood in the veins. Thus, through an integrated combination of physiology and behavior, the animal regulated heat retention and loss during both activity and rest.

Despite this remarkable engineering for temperature regulation, Irving described mammalian "heat machines" as thermolabile or heterothermic rather than truly homeothermic.[21] For example, during the winter the limbs of a caribou sometimes operated at a temperature of 9°C, almost 30° below the body core. The fat found in the legs of these large, arctic mammals tended to have a significantly lower melting point than fat in the rest of the body. Peripheral nerve transmission was unaffected by low temperatures, even though it was disrupted in the limbs of temperate animals exposed to the same temperature regime. Thus, the complex adaptations to cold led not to constancy of body temperature, but rather to a broad and flexible adaptability that allowed the body to function with different parts operating at different temperatures. This perspective, Irving claimed, had important implications for both the theory and practice of comparative physiology. Unlike artificial machines, organisms exhibited pervasive adaptation. This adaptation might be genetic, but it also involved phenotypic plasticity and the adaptive modifications to environmental changes associated with acclimation. For Irving, adaptation also meant adaptable behavior in all animals, including material and social cultures in humans. Unlike a thermostatically controlled house, organisms—whether warm-blooded or cold-blooded—were adaptively thermolabile. Variations in body temperature were not defects of a control system, but rather adaptive responses to complex, varying environments. Viewed from this perspective, the idea of homeostasis as maintaining constancy was misleading. According to Irving, such constancy would be both "inoperable" and "unadaptable" under the challenging natural conditions of the arctic, which required homeostatic mechanisms that were both flexible and adaptable.

Physiologists working with rats or mice in the artificial confines of the laboratory too easily ignored the adaptability of animals living in extreme environments. Simple distinctions between warm-blooded and cold-blooded animals also obscured the fact that arctic fish and invertebrates were just as active as their tropical relatives.[22] The idea of warm-blooded birds and mammals also led to a misleading emphasis on homeostatic constancy, rather than emphasizing the adaptive flexibility of self-regulatory systems. According to Scholander and Irving, this rigid perspective on thermoregulation led to the pernicious emphasis on surface area and volume that had preoccupied physiologists during the late nineteenth and early twentieth centuries, and had likely influenced Bergmann's original interest in geographic clines in body size. For Irving and Scholander, studies of surface area had been an "incubus" plaguing physiology and, in the form of Bergmann's rule, had become "dogma" among ecologists and evolutionary biologists.[23] The dogma was all the more dangerous because prominent physical anthropologists applied the rule to human populations making broad claims about the evolution of racial differences.

In his short note published in *Evolution* in 1955, Scholander summarized the arctic research that he and Irving had conducted during the preceding decade and used this as a basis for criticizing Bergmann's and Allen's rules.[24] Animal size was important, but the minute subspecific differences involved in the biogeographic rules were inconsequential compared to the physiological and behavioral mechanisms for thermoregulation studied by the arctic researchers. The rules were plagued by exceptions and methodological difficulties of accurately measuring surface area, particularly in museum skins. Finally, the fact that Bergmann's rule sometimes applied to invertebrates and other cold-blooded organisms cast doubt on the claim that thermoregulation was the basis for the biogeographic clines, even if they did exist.

After publication Scholander sent a reprint to Mayr. The evolutionary biologist replied courteously, but he was emphatic in dismissing Scholander's claims. According to Mayr, Scholander misunderstood the subtlety with which natural selection could shape adaptations. Piqued that Scholander had published this "misunderstanding" in the journal that Mayr helped establish, the evolutionist wrote, "I am very much tempted to point this out in a note to *Evolution*. No doubt you realize that if one challenges a widely accepted conclusion one exposes oneself to questioning."[25] Three weeks later Mayr sent Scholander a draft of his note asking for comments. Scholander's response was uncompromising in its challenge to the evolutionary validity of the biogeographical rules. Remarking sarcastically on the misuse of Bergmann's and Allen's rules, Scholander wrote Mayr: "Evolutionists must beware of physiological interpretations involving factors so subtle that they can neither be measured nor even discussed in physiologically relevant terms. Evolutionists and physiologists meet in the concept of adaptation, and must listen to each other. The trouble with the rules is that they are statistically weak, they cannot be appraised thermally, and their interpretation as the result of temperature adaptation is conspicuously contradicted on the species level."[26] Scholander, who at the time, was working at the Marine Biological Laboratory at Woods Hole, suggested organizing a joint seminar at the MBL to air their differences in public. Although this meeting apparently never occurred, *Evolution* published Mayr's critical note and a detailed rebuttal from Scholander, followed by commentaries written by Irving and the physical anthropologist Marshall Newman. Mayr continued his critique of the physiological ecologists in his big book *Animal Species and Evolution*, published in 1963.[27]

In his earlier writings, Mayr placed great weight on Bergmann's rule, both as a useful taxonomic tool for classifying subspecies and as evidence for natural selection.[28] For Mayr, the biogeographic correlation between body size and climate summarized by Bergmann's rule was a powerful, albeit indirect, proof of natural selection. It was a particularly well-studied example of the more general phenomenon of ecotypic variation that evolutionary biologists had documented in geographically widespread species composed of overlapping populations, each adapted to local environmental conditions. Heat conservation and natural selection for optimal surface area provided a well-accepted theoretical explanation for Bergmann's rule that remained legitimate until proven false. From Mayr's critical vantage, Scholander failed to overturn the rule, even though he demonstrated the complexity of thermoregulation. Exceptions to the rule were to be expected, Mayr claimed, because heat conservation was only one of many selection pressures affecting body size. He pointed out that population geneticists had demonstrated that natural selection could increase the frequency of genes, providing an advantage of only a fraction of a percent. Thus, even if body size had only a minor effect on thermoregulation, it could still be important. According to Mayr, Scholander was guilty of a misguided typological thinking that posited "all or none solutions" to adaptation, rather than recognizing that organisms were evolutionary compromises resulting from multiple, conflicting selection pressures.

In his rebuttal Scholander restated several lines of argument against the validity of Bergmann's and Allen's rules. First, the body size differences involved were too small to have any measurable effect on thermoregulation. Insulating fur or feathers that were grown or shed, raised or lowered, provided a far more effective mechanism for regulating body temperature. So did countercurrent exchange systems in blood circulation and behavioral mechanisms. The arctic researchers had rigorously demonstrated these physiological and behavioral adaptations by experiments and careful measurement. By contrast, Scholander claimed, Bergmann's and Allen's rules rested heavily upon much less precise measurements of museum specimens. Scholander claimed that his reasoning rested theoretically on Newton's law of cooling, which applied to heat transfer between bodies, whether animate or inanimate. This well-accepted physical law provided a far more secure foundation for thermoregulation than the so-called surface law relating surface area and volume used by Bergmann's followers. For Scholander, this biogeographic generalization was nothing more than a vague rule of thumb. The fact that it sometimes applied to invertebrates and other animals whose body temperatures conformed to the environment further undercut the thermoregulatory argument for the biogeographical rules.[29]

Biogeographical Rules and Human Adaptations to Cold Environments

Scholander's critique of biogeographical rules and his response to Mayr were broadly evolutionary, but his line of reasoning had little in common with the population thinking that formed the basis for Mayr's support of Bergmann's rule. The contrast became obvious in the contentious issue of racial differences in human evolution. Carleton Coon and other leading anthropologists turned to the modern synthesis for a new evolutionary perspective. In Coon's case, using the synthesis to explain how human populations evolved dovetailed with an emphasis on genetic differences among human groups. Notoriously, he claimed that races constituted different human subspecies. Although he did not completely endorse Coon's claims about race, Mayr admired the anthropologist's work and saw him as an ally in promoting the modern synthesis. By contrast, Scholander dismissed Coon's claims about "climatic engineering" of human characteristics, particularly the anthropologist's application of biogeographic rules to human populations. Contrary to Coon's claims, Scholander argued that there was no evidence for a reduction in length of extremities or overall body size of Alaskan Natives compared to other ethnic groups. Scholander was deeply interested in human thermoregulation, but he considered the success of indigenous arctic people to be primarily cultural rather than biological: "In the Eskimo the main adaptation lies not in physiology, but in an age-long experience and technical skill in ducking the cold. They conquered the arctic not by submitting to it but by surrounding themselves successfully with a little piece of the same tropical microclimate upon which we also depend."[30] Warm clothing and shelter, rather than natural selection, adapted humans to the arctic. Irving agreed, writing: "In the case of the Eskimos they have suited their physiological adaptability to arctic cold by an ingenious and highly developed material and social culture which has secured their racial existence during ten centuries which have seen the disappearance of most of the societies and many of the populations living in milder lands."[31]

World War II and the Korean War focused attention on the physiological responses of humans to cold conditions. Many of these studies assumed a biological basis for human races.[32] Nonetheless, the extent to which adaptations reflected genetic characteristics or were the product of physiological acclimation, behavioral flexibility, and cultural innovation remained open questions. The opposition of Scholander and Irving to Coon's views reflected their rejection of biogeographical rules but also their admiration for native Alaskan culture.[33]

Using military funding from both the United States and Canada, as well as support from the Rockefeller Foundation and various universities, Scholander helped organize international teams of physiologists to study human responses to cold temperatures in a variety of geographical locations.[34] Coon joined the physiologists on one of these expeditions to Tierra del Fuego. The results of the study of the Alacaluf and other indigenous groups highlighted the difficulty of separating the effects of genetics, acclimation, and culture. Coon's interpretation of the results was strikingly different from those of Scholander and the other physiologists who actually conducted the research.

The ability of the native Fuegians to swim in near-freezing water and to live with minimal clothing was dramatically described by Darwin in *The Voyage of the Beagle*. When the physiologists visited the region in the late 1950s, they found that the Fuegians maintained body temperature through increased metabolism. Coon considered this cold tolerance to be a unique genetic adaptation of this "unmixed race" of Fuegians.[35] In contrast, Scholander claimed that this physiological response was shared widely, even in urban Europeans. He later managed to convince a group of Norwegian college students to sleep unclothed in near-freezing conditions. Although they found it unpleasant at first, the students fairly quickly acclimated to the cold and were able to sleep comfortably by significantly increasing metabolic heat production.[36]

Australian Aborigines of the Pitjandjara tribe exhibited a different physiological response to cold when they slept largely unclothed in freezing temperatures with only small fires for heat.[37] Measuring oxygen consumption, Scholander and his team found that the Aborigines' metabolism did not increase when they slept. However, the temperature of their limbs dropped to about 10°C. On awakening they quickly rewarmed their arms and legs by physical exertion. Coon concluded that this was another unique racial adaptation. The physiologists who actually conducted the research took a more skeptical attitude toward the racial implications of the studies. When interviewed by a reporter, Scholander's protégé Robert Elsner cautioned against drawing hasty conclusions about a genetic basis for racial differences in thermoregulation, stating that they could be due to "psychological or social conditioning."[38] Coon's University of Pennsylvania colleague Ted Hammel cautioned against any broad evolutionary conclusions based on the physiological research.[39]

Although he was not directly involved in the physiological research, Coon used the studies in Tierra del Fuego and Australia to support his claim that human races are biologically real, each with its own suite of adaptations shaped by natural selection. Bergmann's and Allen's rules were a critical part of Coon's argument that differences in stature, facial structure, and length of extremities were adaptations for thermoregulation based on surface area and volume. Responding to Scholander's criticism, Coon aligned himself with Mayr and other evolutionary biologists who defended the biogeographical rules. For Coon, the physiological studies of thermoregulation in indigenous people in varied environments provided important insights into human evolution. In his *Origin of Races*, Coon concluded, "Once man's inventive genius had made it possible for him to live in extreme environments previously barred to him, a new burden was placed on his physiology because he could not, with his incipient skills, overcome all climatic obstacles. We must expect to see the results of genetic responses, through natural selection, to differences in environment, and we must know how to interpret them, for the patterns they take will tell us much about the early history of our genus and species."[40] According to Coon, the different adaptations to cold found in Fuegians and Australian Aborigines were examples of genetic, physiological, and anatomical differences between human subspecies.

Irving shared Scholander's skepticism about such claims, and he was equally committed to the view that humans are capable of quickly acclimating to life in cold climates. In a brief article, Irving reported the results of cold tolerance tests that he had performed on two students at the University of Alaska and on a young airman who had been stationed for two years in the arctic.[41] The students were members of the Fount of Venta, a cult that had recently relocated from California to Alaska. The group's adherants wore light robes and went barefoot even during the Alaskan winter. Irving found that the students could sit comfortably in a temperature-controlled cold room for ninety minutes at freezing temperatures. Irving recorded oscillations in temperature of the students' fingers and toes, dropping to as low as 9°C during the experiment. The students reported little discomfort and retained feeling in their digits throughout these cycles. Even after an hour of inactivity, they experienced only mild shivering. At the end of the experiment, the students reported no pain as their toes and fingers rewarmed. When Irving repeated the experiment with the airman, he found that the subject quickly began shivering despite being dressed in military fatigues that were heavier than the robes worn by the students. His fingers and toes did not get as cold as the students', although the airman reported pain in the extremities. Because of the airman's discomfort, Irving cut the experiment short after forty minutes

According to Irving, the difference between the students and airman was due to the members of the Fount of Venta acclimating to the cold through deliberate exposure with only light clothing. Like the Norwegian students studied by Scholander, the members of this group illustrated a latent adaptability found in all humans that could become highly developed given proper motivation and practice. Through exposure to cold conditions, the students exhibited some of the same physiological responses found in indigenous populations inhabiting cold climates for millennia. Irving suggested that the students probably had not reached the limits of adaptability and that members of the cult who habitually worked outdoors might be even more tolerant of cold than those who spent their days in the classroom.

The criticisms of applying the biogeographical rules to human races put anthropologists on the defensive. This problem was highlighted by Marshall Newman, who published an anthropological response to the exchange of articles in *Evolution* by Scholander and Mayr. Newman's left-leaning politics, particularly on issues of racial equality, were well known. His defense of racial minorities antagonized his superiors at the Smithsonian Institution and led to his surveillance by the FBI and naval intelligence during the Cold War.[42] Like Mayr, Newman rejected extreme claims about a biological basis for human races, although he wrote a favorable review of Coon's book *Races*.[43] Sharing Coon's commitment to the modern synthesis, Newman had conducted research on the application of Bergmann's and Allen's rules to human populations.[44] For Newman, humans provided a "test species" for studying geographic variation, because human populations existed in virtually every terrestrial habitat on Earth. Although he acknowledged Scholander's claim that clothing, shelter, and other cultural innovations were largely responsible for human success in the arctic, Newman hoped that combining anatomy and physiology within an evolutionary context would lead to "more and better controlled studies on native peoples."

Mayr avoided the issue of human races in his response to Scholander in *Evolution*. In *Animal Species and Evolution*, he took a nuanced approach to the adaptive significance of racial characteristics. From the perspective of evolutionary theory, humans were no different from other animal species, and Mayr used research by Coon and other physical anthropologists to argue that variation among human populations was adaptive and sometimes followed the biogeographical rules.[45] At the same time, continual gene flow among populations had likely reduced this ecotypic variation in modern humans.

Historians have analyzed the fraught relationship between physical anthropologists and evolutionary biologists after World War II.[46] Like other leading evolutionary biologists, Mayr was sharply critical of how early anthropologists had used typological concepts of species and subspecies to justify racist conclusions. Mayr claimed that the idea of pure races was "sheer nonsense" and that the differences among human populations were smaller than those of many other polytypic species.[47] Nonetheless, he admired Coon's work, accepted his claims about the adaptive significance of human characteristics particularly in relation to climate, and used these claims to support Bergmann's and Allen's rules.[48] For Mayr, Coon was an important ally for bringing the modern synthesis to physical anthropology but also for challenging Scholander's critique of biogeographical rules as applied to humans.[49]

Mayr completely ignored his disagreements with Scholander when he wrote his monumental history, *The Growth of Biological Thought*. He continued to argue that the biogeographical rules provided important evidence for natural selection, but he omitted any reference to thermoregulation as an explanation, or to the objections raised by physiological ecologists. For his part, Scholander also ignored the episode in his autobiography, although he reveled in several other scientific controversies that marked his career. In retrospect, it might seem that Mayr and Scholander were simply talking past each other, even though both men expressed an interest in finding a meeting place between physiology and evolutionary theory. Indeed, despite his claim that many fruitless biological controversies resulted from failing to distinguish between proximate and ultimate causation, Mayr was quite consciously looking for ways to use homeostasis to explain how thermoregulation contributed to the evolutionary phenomena codified by Bergmann's and Allen's biogeographical rules. Mayr based his support for the rules on an expansive understanding of homeostasis broadly based on both physiology and genetics.[50]

The controversy between Scholander and Mayr highlights the indistinct boundaries between functional biology and evolutionary biology, as well as the difficulty of clearly separating proximate from ultimate causation. The uncertain status of Bergmann's rule and continued disagreements about its causes reinforce the historical conclusion that biologists working in interdisciplinary fields often do not adhere strictly to these philosophical distinctions. The arctic research of Scholander and Irving was based on a broad foundation of evolutionary thinking combined with deeper interests in natural history, ecology, physiology, and behavior. Their legacy inspired a generation of younger scientists in physiological ecology and related areas of organismal biology.[51] For his part, Mayr took a lively interest in physiology, particularly in using a broad understanding of homeostasis that combined proximate and ultimate causation—with little attempt to consistently distinguish the two. Indeed, Mayr and other evolutionary biologists routinely applied homeostasis at the level of both organisms and populations to explain adaptation and speciation.

8

Physiological Teamwork, Homeostasis, and Coadaptation

In his monumental *Growth of Biological Thought*, written toward the end of his long and distinguished career, Ernst Mayr stated: "It is now clear that a new philosophy of biology is needed. This will include and combine the cybernetic-functional-organizational ideas of functional biology with the populational-historical program-uniqueness-adaptedness concepts of evolutionary biology. Although obvious in its essential outline, this new philosophy of biology is, at the present time, more of a manifesto of something to be achieved than a statement of a mature conceptual system."[1] The comment is striking, because although Mayr spent much of his later career on the project, he still considered it to be a "manifesto," rather than a mature philosophy. His controversial exchange with Per Scholander over Bergmann's rule demonstrated the difficulty of achieving this philosophical goal, and also the challenges of combining functional and evolutionary approaches to complex questions involving self-regulation, adaptation, and geographical distribution. Both Mayr and Scholander moved back and forth over the indistinct boundary between proximate and ultimate causation. This was not an isolated incident in Mayr's career. Indeed, he wrestled with countervailing demands, dichotomizing functional and evolutionary approaches in biology while at the same time seeking a synthesis of the two.

Mayr's philosophical distinction between "proximate causation" and "ultimate causation" was a major contribution to philosophy of science. This distinction, as several historians have argued, arose largely as a defense of organismal biology against the perceived threats from molecular biology and a philosophical reductionism that looked to physics as the model for scientific reasoning.[2] Erecting this defense required dichotomizing biology into two "largely separate fields" of functional and evolutionary biology.[3] At the same time, Mayr intended to unify biology on the evolutionary foundation provided by the modern synthesis. To do this required integrating the two largely separate fields that he so carefully distinguished. This integration proved to be a daunting challenge, but even as he argued with Scholander about thermoregulation, Mayr was trying to combine functional and evolutionary biology using the concepts of homeostasis and adaptation to explain both proximate and ultimate causation.

Integrating proximate and ultimate causation had other important philosophical consequences, particularly for eliminating two bugbears of modern biology: vitalism and teleology. Indeed, about half of Mayr's article, "Cause and Effect in Biology," was devoted to these two issues. Turning to cybernetics and information theory to make his point, Mayr described the goal of functional biology as the decoding of genetic information carried in DNA. However, this information had a history and was in constant flux, which required the understanding of evolutionary biology. The fact that mutations were unpredictable and natural selection was creatively opportunistic resulted in a living world characterized by a high degree of indeterminacy. Even those processes that appeared goal-directed were not teleological in the traditional sense, but rather *teleonomic*: the product of highly complex, interacting, and self-regulating genetic programs. The self-regulation inherent in teleonomic systems might result in a degree of predictability, but these systems retained some indeterminacy. Mayr claimed, "Every organic system is so rich in feedbacks, homeostatic devices, and potential multiple pathways that a complete description is quite impossible."

"Cause and Effect in Biology" stirred considerable interest, including an exchange of letters to the editor of *Science* between Mayr and British biologist Conrad Hal Waddington.[4] Waddington was a scientist with broad interests. Originally trained as a paleontologist, he became interested in embryology, genetics, and evolutionary theory.[5] Although Mayr and Waddington differed sharply on some issues, they shared a deep interest in philosophy of biology and particularly in using ideas of self-regulation in evolutionary biology. During the early 1960s, Mayr considered Waddington an ally in creating a new philosophy of biology. Indeed, their correspondence over "Cause and Effect in Biology" led the two biologists to organize one of the first conferences devoted to philosophical issues in biology.[6] Reprints of Mayr's article circulated before the meeting and served as a centerpiece for discussion, although Mayr later complained that participants took his ideas less seriously than he had hoped. Mayr's dour assessment of the meeting was that Waddington's somewhat different philosophical perspective had carried the day.[7]

Waddington generally agreed with the distinctions that Mayr made in his article but argued that the idea of teleonomy needed to be refined to emphasize what he referred to as "quasi-finalistic" phenomena in development and evolution.[8] Basing his argument on the same cybernetic analogies that Mayr had used, Waddington pointed out that development was the outcome of both a genetic program and a complex system of negative feedback controls that steered the embryo toward the adult stage. By analogy he also argued that the genetic program itself was shaped not only by natural selection (a nonpurposive process) but also by at least two types of negative feedback that made evolution itself a quasi-finalistic process. First, behavior influenced the outcome of natural selection because an animal always had some choice of the environment in which it lived. Second, the evolution of developmental processes had resulted in redundant systems that were highly stable. Waddington's developmental and behavioral feedback systems combined proximate and ultimate causation to explain self-regulation. Describing both development and evolution as similar quasi-finalistic processes involving cybernetic feedback seemed to undermine the distinction. In his letter to the editor of *Science,* Waddington ignored the proximate–ultimate and functional–evolutionary dichotomies that Mayr considered to be so important.

In his response to Waddington's critique, Mayr was remarkably conciliatory, claiming that he had no disagreements with Waddington over the facts or even the British biologist's emphasis on the evolutionary role of the development but merely what counted as deterministic.[9] For Mayr, natural selection was opportunistic, and combined with random mutation it resulted in an evolutionary process far too capricious to be considered "quasi-finalistic," as Waddington claimed. Indeed, one might have expected Mayr to sharply criticize such thinking as an example of discredited orthogenesis and for failing to clearly distinguish between functional and evolutionary explanations. Waddington, for his part, wanted to integrate development and evolution by emphasizing the similarities between the processes. He was less willing than Mayr to see natural selection as a creative process and hence looked to the adaptability of organisms and developmental feedback systems to account for evolutionary novelties. Substantial as these differences might appear, the meeting ground lay in the abiding interest that Mayr and Waddington shared for ideas of homeostasis and cybernetic self-regulation at the levels of both individuals and populations. Waddington had explored these issues in *The Strategy of the Genes* (1957). Mayr's attempt to bring homeostasis into evolutionary biology was most evident in his highly influential *Animal Species and Evolution* (1963), published two years after his article on cause and effect in biology. Although not so evident in the book itself, in his private correspondence Mayr went out of his way to emphasize points of agreement with Waddington on gene interaction and the integration of the genome.[10]

Populations, Species, and Physiological Teamwork

Mayr spent more than a decade writing *Animal Species and Evolution.*[11] His earlier book *Systematics and the Origin of Species* (1942) had been written explicitly from the perspectives of a field naturalist and museum taxonomist. By contrast, *Animal Species and Evolution* took a more expansive and inclusive perspective on the evolution of populations and species. Most notably, Mayr incorporated insights from population genetics that emphasized gene interaction and the integration of the genome, and departed significantly from the earlier "beanbag genetics" that he harshly criticized.[12] Less obvious, Mayr explored a physiological perspective on species that looked to Walter Cannon's *Wisdom of the Body* for inspiration. Although also deeply indebted to the genetic homeostasis that had been promoted and popularized by I. Michael Lerner and other geneticists, Mayr crafted his own understanding of populations and species as self-regulating entities that intentionally bridged functional and evolutionary thinking in a novel way. For Mayr, Cannon's homeostasis explained adaptation of individual organisms and, by extension, the cohesion holding populations and species together. The temporary breakdown of this homeostatic balance was a necessary condition for the fragmentation leading to speciation.

Mayr rarely discussed specific examples of organismal homeostasis in detail, but he considered these "rather complex" self-regulatory systems to be integral characteristics of a species, both physiologically and genetically.[13] Homeostatic mechanisms were species-specific, and as such, they set the physiological limits of tolerance for each species. Each species had an optimal environment, usually near the center of its range, with diminishing environmental conditions extending out to the periphery where the species finally met its limits of tolerance. However, there was also a tendency for populations to become closely adapted to local conditions. Therefore, most species were composed of more or less distinct subpopulations or ecotypes. Part of the complexity of homeostatic systems was the balance that every ecotype faced between being adapted to local conditions, while at the same time sharing the "heritage" of species-specific physiological mechanisms that provided unity to the entire species.

Self-regulation was an organismal response to the environment that required coordinated responses from multiple organs. For this reason, the genetic basis for homeostatic mechanisms was always polygenic and might involve the entire genotype of the organism. Mayr's broadly conceptual description of this genetic program relied heavily upon metaphors of cohesion, harmony, integration and teamwork. Effective homeostatic responses were not so much the result of the individual performance of superior genes as the effective teamwork of otherwise average genes. The entire genotype was a "physiological team" based on cooperation rather than the outstanding performance of particularly talented "soloists." Comparing genes to athletes or musicians, Mayr wrote: "The total genotype can be considered a 'physiological team,' an analogy that has considerable illustrative value. Some of the best-known athletes are poor team players, or might star as members of one team but not of another. Some musical virtuosos, unexcelled as soloists, are only mediocre in an ensemble. Genes are never soloists, they always play in an ensemble, and their usefulness, their 'selective value,' depends on the contribution to the goodness of the product of this ensemble, the phenotype."[14]

Mayr's physiological model employed ideas of feedback and regulation at multiple levels from molecules to populations.[15] Although the same genes existed in all of the cells of the body, they were active only in some. Furthermore, gene activity could be turned on and off during the process of development. The products of one gene generally affected the activity of numerous other genes to form balanced biochemical pathways. Nearly all genes were pleiotropic (affecting more than one characteristic), and most important characteristics were polygenic (influenced by multiple genes). These intricate feedback systems often corrected deficiencies and imbalances during the development of an organism (developmental homeostasis) or replacement of alleles through natural selection in populations (genetic homeostasis). Eventually well-adapted populations had genotypes that were "closely knit" and functionally integrated.[16]

This emphasis on feedback regulation and stability might seem to preclude change, and Mayr admitted that if a well-integrated genetic system came into perfect balance with its environment, evolutionary change would be impossible.[17] Evolution required that populations escape from a too-rigid system of genetic homeostasis, although the process of rejuvenation was itself risky and often failed. Speciation, in particular, involved revolutionary changes in homeostatic systems. This typically occurred in small populations that encountered new niches, often at the periphery of a species distribution. In retrospect, Mayr considered this idea of "genetic revolutions" to be his major contribution to evolutionary biology.[18] An important corollary for the success of this revolution was the rapid reorganization of the genotype to restore homeostatic balance.

Because homeostatic systems were part of the shared "heritage" of the species as a whole, they were an "immensely conservative force" that held each species together and put a "brake" on evolutionary change. However, most species were polytypic, being composed of several more or less distinct geographical races or ecotypes. Ecotypic variation was a kind of overlay on the homeostatic system. It allowed adaptation to local environments but also acted as a "centrifugal evolutionary force" that potentially pulled a species apart.[19] The great stability of the homeostatic system, together with gene flow, counteracted this tendency for species to fragment, but it also meant that local populations were in a dynamic state of readjustment between adaptation to the demands of local environments and maintaining overall homeostasis that required an integrated, species-specific response of populations and their members. This view of populations and species as homeostatic systems had two important consequences. First, speciation was a rare event, but when it happened it usually occurred in small, peripheral populations that entered a new ecological niche. Second, speciation was a revolutionary event that upset the finally tuned genetic system. In many cases the breakdown in homeostasis was catastrophic and led to extinction of the local population. Those rare populations that successfully survived did so through rapid readjustment by evolving harmonious genotypes leading to new homeostatic systems—both physiological and genetic.

I. Michael Lerner, Theodosius Dobzhansky, and Genetic Homeostasis

The idea that populations are highly integrated evolutionary units regulated by homeostatic mechanisms was widely shared by geneticists during the 1950s and early 1960s. Perhaps the most influential presentation of this population-level self-regulation was I. Michael Lerner's short book entitled *Genetic Homeostasis,* published in 1954.[20] Lerner was an émigré, raised in a Russian enclave in China. His father was a prosperous merchant whose business fortunes worsened after the Russian Revolution. Rather than wait for a visa to the United States, Lerner took advantage of a Canadian program that accepted immigrants who agreed to work toward college degrees in agriculture. Thus, although initially he had no particular interest in either science or agriculture, Lerner studied poultry breeding at the University of British Columbia. There he met Theodosius Dobzhansky when the prominent population geneticist visited the university for a month. The two men became good friends, and Dobzhansky served as an informal mentor who used his influence to get Lerner into the doctoral program in genetics at the University of California at Berkeley. Lerner spent the rest of his career there and played an important role in bringing modern genetics into the practice of animal breeding. It was within this agricultural context that Lerner developed his ideas about genetic homeostasis, which he introduced in his *Population Genetics and Animal Improvement* (1950) and developed more fully in *Genetic Homeostasis*, written three years later during a year in Italy on a Guggenheim Fellowship.

Despite the agricultural context of his books, Lerner claimed that the origin of feedback mechanisms was central to evolutionary biology because it focused attention on the evolution of adaptation.[21] The evolution of populations, whether by artificial or natural selection, involved a fundamental tension between maintaining stability and promoting change. From the practical perspective of animal breeding, success depended upon overcoming the stabilizing action of genetic homeostasis to produce new and useful breeds. From the perspective of evolutionary theory, the success of populations involved a balance between maximizing average fitness and maintaining sufficient genetic variability for adaptive change in a constantly fluctuating environment.

In *Genetic Homeostasis*, Lerner emphasized his own originality, but at the same time placed his ideas in the mainstream of biology. He freely admitted his intellectual debt to Walter Cannon and *The Wisdom of the Body*. He readily acknowledged that several other geneticists and evolutionary biologists shared similar ideas about self-regulation, citing Dobzhansky, Cyril Darlington, Kenneth Mather, I. I. Schmalhausen, H. J. Muller, Sewall Wright, and C. H. Waddington. Aside from his explicit transfer of Cannon's physiological ideas of homeostasis into genetics and evolutionary theory, Lerner claimed that his own innovations included drawing close ties between stabilizing mechanisms in individual development and the evolution of populations, in attributing this stabilization specifically to overall heterozygosity, and in presenting novel experimental evidence from the agricultural breeding of chickens to support his broader evolutionary claims.

In artificial selection experiments that he conducted with his Berkeley colleague Everett Dempster, Lerner tried to increase the length of the lower leg or shank in chickens. Initially selection was successful, but after a few generations, loss of fitness hindered further progress in the selected line. When selection was relaxed, the average shank length decreased in the population, but not to the original value. A new equilibrium was established with a longer average shank length in the population. Resuming selection after several generations resulted in further increase in shank length without loss of fitness. Other scientists reported similar results in selection for the number of bristles on the abdomen of *Drosophila* in laboratory populations, and British geneticists Cyril Darlington and Kenneth Mather coined the term "genetic inertia" for a population's ability to resist change. Lerner preferred the more dynamic term "genetic homeostasis," because he believed not only that populations resisted change but, more importantly, that they "auto-regulated" their genotypes. Lerner's genetic homeostasis linked the population-level phenomenon to Walter Cannon's physiological ideas on self-regulation in several important ways.

Lerner noted with approval that Cannon had used the term "homeostasis" to denote the general property of self-regulation but also to describe specific mechanisms that organisms used to maintain constancy in the face of a fluctuating environment. Cannon's mechanistic approach was important because, although ideas of self-regulation dated back to Hippocrates, Lerner identified earlier ideas with mysticism and a lingering vitalism that Cannon had eliminated with his careful studies of nervous and hormonal control. More importantly, by linking physiological homeostasis to social homeostasis, Cannon had suggested that self-regulation was a general property of all organized systems. Following this suggestive lead, Lerner used homeostasis to bridge the organism-population boundary in two ways. First, he claimed that populations were self-regulating entities and that this genetic homeostasis could be explained mechanistically using the well-established principles of Mendelian genetics. Second, he drew connections between this population-level homeostasis and the developmental homeostasis that helped ensure normal ontogeny. According to Lerner, both the resistance of populations to sudden genetic changes and the buffering of development in embryos against environmental perturbations were a result of heterozygosity. Genetic variation at multiple loci ensured that heterozygous individuals were better protected against developmental perturbations than were homozygous individuals. Populations with high levels of heterozygosity were more stable than populations with higher homozygosity.

Other theories supported by prominent population geneticists such as R. A. Fisher and H. J. Muller emphasized the uniformity of populations, which exhibited only low levels of variation in the form of rare recessive genes. Evolutionary change was largely due to natural selection acting at individual loci to replace less-well-adapted alleles with better-adapted ones. In contrast, Sewall Wright and Theodosius Dobzhansky had emphasized locally adapted populations that were genetically diverse and tightly integrated entities maintaining this genetic diversity in a dynamic balance with the environment. Natural selection was not simply a sieve that eliminated less advantageous alleles. It was a creative process for molding harmonious, coadapted groups of genes. These gene combinations worked well together in individual organisms but also gave the population a degree of structure and stability that transcended individual variation.

Consider the artificial selection for shank length in chickens or bristle number in *Drosophila* that Lerner used to introduce the concept of genetic homeostasis. According to Lerner, the genes responsible for these characteristics were parts of "equilibrated systems" composed of many genes. The strong directional selection on single characters that the experimenters employed disrupted or destroyed these delicately balanced systems. This breakdown of homeostasis led to decrease in average fitness.[22] Relaxing selection restored balance, but at a slightly different level. This new balance was the result of natural selection that operated on the physiology of the entire individual, rather than just the characteristic that had been the target of the artificial selection. This "balancing selection" restored and maintained the integration of the genotype. Lerner described problems of artificial selection familiar to all animal breeders but explained them in novel terms of genetic homeostasis, integrated genotypes, and balancing selection.

As Lerner admitted, other genetic mechanisms might account for the equilibrated system of the integrated genotype. Although not completely rejecting epistasis, or interaction among a few genes, Lerner emphasized the importance of overall heterozygosity. For Lerner, populations were "exceedingly heterogeneous," and genetic variation, in and of itself, was primarily responsible for genetic homeostasis.[23] A population had "reserves of genetic variation" that were maintained by balancing selection. In his experiments these genetic reserves provided the raw material for artificial selection to work—often shifting the mean value of a characteristic by several standard deviations. However, as the genetic tolerance limits were exceeded by continued selection, extreme deviates appeared in the population that were incapable of reproducing. At this point, natural selection and artificial selection were in an antagonistic relationship that increasingly prevented further evolutionary changes in the characteristic. Restoring genetic equilibrium when artificial selection was relaxed was not simply due to mutation or crossing over at the loci directly responsible for the characteristic under selection. Rather, it was a system-wide phenomenon resulting from a new array of well-integrated genotypes in the population created by balancing selection.[24]

This highly integrated genetic structure of the population was a property that emerged from the selection of individuals.[25] According to Lerner, balancing selection favored heterozygous individuals that were phenotypically intermediate, and it eliminated more extreme phenotypes that arose sporadically in every population through recombination. Heterozygosity also provided a mechanism for buffering or channeling development against environmental perturbations. Through the action of complex genetic feedback, the embryo tended to develop normally over a range of environments and in the face of environmental perturbations. Lerner wrote, "It should be clear that the basis of buffering must reside in some sort of cybernetic model, where alternative pathways are available to the organisms, depending on the variation in its genetic and environmental milieu."[26]

His emphasis on the importance of heterozygosity placed Lerner at the center of a major controversy in population genetics. Although Lerner presented his model of genetic homeostasis as part of the broad mainstream of evolutionary thinking, some prominent population geneticists, such as the Nobel laureate Hermann J. Muller, dismissed claims about the benefits of heterozygosity.[27] According to Muller and his supporters, most loci in a genotype were homozygous for well-adapted alleles, and natural selection acted primarily to remove rare, deleterious mutants. In those unusual cases in which a novel allele proved adaptive, directional selection quickly increased its frequency in the gene pool generally leading to elimination of the previously established allele. Muller scathingly rejected the idea that genetic diversity, in and of itself, was a good thing.

Lerner shied away from controversy, but Dobzhansky and his student Bruce Wallace engaged in acrimonious exchanges with Muller over the importance of heterozygosity and genetic homeostasis. Richard Lewontin, another Dobzhansky student who was equally enamored with homeostasis during the 1950s, later wrote extensively on the problem of genetic variation and the issues dividing the Muller and Dobzhanksy groups. Never fully resolved, the disagreements between what Lewontin referred to as the "classical" and "balanced" schools of thought were shaped not only by the technical difficulties of testing alternative hypotheses but also by important implications of the competing views for a range of social issues—including eugenics and testing nuclear weapons—during the 1950s and 1960s. Homeostasis, borrowed directly from Cannon's *Wisdom of the Body*, as well as reformulated in terms of "genetic homeostasis" by Lerner, played a crucial role in the thinking of Dobzhansky and his students.

Dobzhansky had conceptualized populations as highly integrated entities governed by processes that constituted a form of "population physiology" two decades earlier when he published *Genetics and the Origin of Species* in 1937. These processes, including mutation, natural selection, genetic drift, and migration, maintained and regulated the variation in populations. Describing the consequences of the Hardy-Weinberg equilibrium, Dobzhansky wrote: "The maintenance of the genetic equilibrium is evidently a conservative and not a progressive factor. Evolution is essentially a modification of this equilibrium. We shall proceed now to show that agents that tend to modify the equilibrium actually exist in nature. A significant fact is that each of such agents is counteracted by another of opposite sign, which tends to restore the equilibrium. A living population is constantly under the stress of the opposing forces; evolution results when one group of them is temporarily gaining the upper hand over the other group."[28] Although he did not initially use homeostasis to describe this balance of forces, the description neatly encapsulated Dobzhansky's idea of a physiological approach to studying evolution. Particularly noteworthy were the interplay of internal and external regulatory processes and the tension between stability and evolutionary change. The tension between homeostasis as the maintenance of stability and a more evolutionary view of homeostasis as a balance between stability and lability was an issue that Dobzhansky and his students wrestled with during the 1950s and early '60s.

Dobzhansky and Wallace wrote glowing reviews of Lerner's *Genetic Homeostasis* in high-profile journals, and even before the book appeared, they embraced genetic homeostasis.[29] While acknowledging the important insights of Walter Cannon, Dobzhansky and Wallace emphasized that homeostasis had a genetic basis that demanded an evolutionary interpretation, writing "The 'wisdom of the body' is an outcome of the molding of the genetic structure of the species by natural selection in the process of evolution, and it cannot be understood outside this evolutionary context."[30] Successful populations were adapted because individuals were physiologically capable of self-regulating within the range of environments usually encountered by the population but also because most populations were polymorphic, with different individuals optimally adapted to slightly different environments. Thus, homeostasis was a matter of both individual adaptability and genetic variation among individuals within the population as a whole.

Although admitting that one could not categorically conclude that heterozygotes were superior to homozygotes, Dobzhansky and Wallace claimed that their experiments with *Drosophila* demonstrated that both individual and population homeostasis were largely due to overall heterozygosity. Homeostasis was a matter of degree, but heterozygous individuals tended to be more successful than homozygotes at facing the challenges of a fluctuating environment. This homeostatic property of individuals provided the basis for understanding high levels of heterozygosity and its consequences at the level of the population. According to Dobzhansky and Wallace, natural selection formed gene pools with alleles that were likely to form "harmonious combinations" with other alleles both at the same locus and epistatically with alleles at other loci. Extreme variants not "mutually compatible" with other alleles were eliminated by natural selection. Consequently, successful populations not only exhibited high levels of genetic variation, but this variability formed a "store of mutually compatible alleles" that made the population a coadapted genetic system.[31]

For Lewontin, physiological homeostasis was not so much maintaining stability in a static sense as it was balancing stability and lability.[32] Discussing Cannon's *Wisdom of the Body* and Lerner's *Genetic Homeostasis*, he noted that to maintain stability of some processes, other processes inevitably fluctuated. Similarly, the dynamic equilibria among genotypes within a population shifted to meet the demands of an environment fluctuating in space and time: "A genotype is homeostatic if individuals of this genotype can so adjust their physiology and morphogenesis as to survive and leave offspring in a wide variety of environments. A population or species is homeostatic if its genotypic or phenotypic composition can be so adjusted as to assure its survival in a variety of environments."[33] Thus, while Lerner had emphasized the ability of populations to resist sudden changes in genotypic composition, Lewontin warned that too much resistance would be maladaptive. It might impede adaptive change, perhaps even leading to extinction. Instead, Lewontin pointed toward a more dynamic evolutionary process that used genetic variation in populations to produce increased physiological homeostasis at the individual level. For Lewontin, evolution was a process for converting the collective homeostasis of populations into the physiological homeostasis and adaptation of individuals. The gene pool of a population included alleles that worked well together in a variety of combinations. Balancing or stabilizing selection maintained the homeostatic structure of the population and, as a result, favored individuals with harmonious combinations of genes that promoted self-regulation.

For Dobzhansky, Wallace, and Lewontin, heterozygosity provided flexibility for both populations and individuals. Although experiments with *Drosophila* demonstrated that in any particular environment some homozygotes were more viable than heterozygotes, when compared in the full range of environments encountered by a population, heterozygotes always outperformed homozygotes. Homozygotes were "narrow specialists," while heterozygotes were flexible generalists.[34] The maintenance of balanced polymorphisms in a population was therefore the result of variable and fluctuating environments that placed a selective premium on adaptability. Without environmental variation, homeostasis—whether at the individual or population level—would provide no benefit.

C. H. Waddington, Canalization, and Genetic Assimilation

One of Lerner's major goals in *Genetic Homeostasis* was to demonstrate that the same basic mechanism that buffered populations against rapid changes also buffered embryos to ensure normal development over a wide range of environmental conditions. For Lerner, developmental homeostasis depended upon the ability of heterozygotes to remain within normal developmental pathways against environmental perturbations that tended to disrupt development in poorly buffered homozygotes.

C. H. Waddington had discussed similar ideas, although he preferred the term "canalization" to "homeostasis," because he wanted to emphasize the process of developmental change rather than stability of the outcome. He was also skeptical of Lerner's insistence on the importance of heterozygosity for buffering development. Nonetheless, he applauded *Genetic Homeostasis* for focusing attention on the important connections bridging developmental biology, population genetics, and evolutionary theory.[35] He found it "refreshing" that a population geneticist would take embryology seriously, and he found the book to be a "stimulating discussion" of topics on the forefront of biology.

Despite his ambivalence toward homeostasis, Waddington embraced cybernetic ideas of networks and feedback regulation. He conceptualized development as a ball (the embryo) rolling down a kind of epigenetic landscape of valleys and ridges.[36] This landscape canalized development because, other things being equal, the ball followed a particular series of valleys leading to the normal adult phenotype. However, environmental factors sometimes nudged the embryo over a ridge into a neighboring developmental pathway leading to an altered phenotype. The topography of this epigenetic landscape—the heights of the ridges and depths of the valleys—resulted from a subterranean system of guywires corresponding to the network of interacting genes making up the coadapted genotype. During the 1950s, biologists knew enough about the relationships among genes, enzymes, and biochemical pathways to make the conceptual scheme plausible, and Waddington's epigenetic landscape has had an enduring attraction for some developmental biologists and geneticists.[37]

Closely associated with Waddington's model of the epigenetic landscape was his idea of genetic assimilation, which formally linked genetics and developmental biology with evolutionary theory, albeit in a somewhat unorthodox fashion. First published in a short article during the middle of World War II, genetic assimilation was Waddington's attempt to explain how phenotypic traits resulting from environmental influences could later come under genetic control.[38] According to Waddington, a trait originally triggered by an environmental influence during development might later become established in the population if its effect was mimicked by a mutation. In such a case, an environmentally induced phenotype became "assimilated" by the new form of a gene.

One could conceptualize how this might occur using the epigenetic landscape. The downward movement of the ball might be shifted from one valley to another by a lateral force representing some environmental disturbance. A mutation might duplicate this effect if it lowered the height of the ridge separating the two valleys, thus reducing the barrier between the two developmental pathways. What had been a "phenocopy," to use a term coined by I. I. Schmalhausen, was now a genetic trait inherited from one generation to the next. This was not just speculation, because Waddington conducted experiments that seemed to demonstrate the effect. When *Drosophila* pupae were administered a heat shock, some of the adults later lacked a crossvein in their wings. After several generations of combined treatment and selection, the crossveinless trait began to show up even in the absence of the heat shock.

Genetic assimilation was an enigmatic contribution to evolutionary biology. Nobody doubted the experimental results that Waddington offered in support of the theory, but for more orthodox evolutionary biologists, genetic assimilation was little different from earlier ideas of organic selection and the Baldwin effect.[39] Genetic assimilation was not false, but it probably was not very common and was likely due to natural selection for genes, each with a small effect that eventually reached a collective threshold. In the title of his earliest article on genetic assimilation, Waddington rhetorically linked the concept to the discredited Lamarckian idea of inheritance of acquired traits. This unnecessarily provocative implication served as a poor strategy for gaining acceptance of the idea during the heyday of the modern synthesis. In this regard Waddington's theoretical work was in marked contrast to the efforts of Lerner, Dobzhansky, Wallace, and Lewontin to place developmental and genetic homeostasis squarely within an orthodox neo-Darwinian context of Mendelian genetics and natural selection. It seems likely that Waddington's unconventional suggestion was partly responsible for the less-than-enthusiastic responses of critics such as Mayr and George Gaylord Simpson.[40] Waddington's later supporters have sometimes viewed this historical episode as an unnecessary narrowing of the focus of evolutionary biology and a missed opportunity to bridge embryology and population genetics.[41] But Waddington's emphasis on the interaction of the developing organism and its environment, including the important interactions between self-regulating or cybernetic genetic systems and environmental forces, was also part of a broader integrating effort by biologists, including Mayr, who explicitly endorsed genetic homeostasis even though rejecting genetic assimilation. As such, Waddington's cybernetic ideas became an important bridging element in attempts to unify evolutionary and functional approaches by combining proximate and ultimate causation. It also provided a point of departure for discussing human social issues using perspectives that combined functional and evolutionary ideas of balance, self-regulation, and development.

Genetics, Evolution, and Social Homeostasis

The idea of genetic homeostasis may have developed within the immediate context of experiments on fruit flies and chickens, but the implications for human society were never far in the background. During the 1950s and 1960s, Mayr, Dobzhansky, Lerner, Waddington, and other evolutionary biologists found homeostasis useful for discussing the social implications of evolutionary theory, including such hot-button issues as race, intelligence, eugenics, and ethics.

In 1954 Dobzhansky became the first biologist to deliver the Page–Barbour Lectures at the University of Virginia. Established in 1907, the public lectures aimed to present fresh ideas from any field in the arts and sciences in a form understandable to a general audience. Hailed as a leading figure in evolutionary biology, the visiting Dobzhansky received extended coverage by both the student newspaper and the local press. He later expanded and modified his public lectures as *The Biological Basis of Human Freedom*.[42] Although he did not mention local issues involving race, the broad implications of Dobzhansky's discussion of the importance of human genetic variation would have been hard to miss within the context of Virginia's politics and the segregated university at which he spoke. Although the university accepted a few women and minorities into graduate programs during the 1950s, moves to integrate the undergraduate population at the university did not begin for another decade. Efforts to integrate public high schools near the university were a key part of litigation in the broader case of *Brown v. Board of Education of Topeka*. That Supreme Court case was reaching its conclusion when Dobzhanksy visited the university. The ruling announced three months later quickly led to the Massive Resistance to integration instigated by Virginia senator and Democratic Party boss Harry F. Byrd.[43]

The Biological Basis of Human Freedom provided Dobzhansky with an opportunity to present evolutionary theory, his ideas on the importance of genetic variation, and the human implications of this research in an expansive way not possible in his more technical writing. Although "the racist hypothesis" was his main target, Dobzhansky also criticized "diaper anthropologists" who denied genetic variation and attributed all human differences to nurture rather than nature.[44] The more nuanced perspective that Dobzhansky presented to his audience balanced heredity, environment, and culture in arguing against scientific racism and genetic determinism, more broadly. While acknowledging the importance of variation in humans, as well as other species, Dobzhansky repeatedly used the idea of "phenotypic plasticity" to emphasize the flexibility of human development and to undermine claims about race, particularly applied to complex characteristics such as intelligence. Phenotypic plasticity also resonated with an evolutionary interpretation of homeostasis that was a centerpiece of Dobzhansky's lectures.

In a chapter entitled "Who Is the Fittest?" Dobzhansky elaborated his ideas on the evolutionary importance of homeostasis. According to Dobzhansky, "The 'wisdom of the body' is, then, not a mysterious gift or an inherent property of life. This 'wisdom' has been built slowly and painfully in the long process of evolution and is controlled by natural selection."[45] The marvelous physiological mechanisms for maintaining stability that Walter Cannon described were examples of unique adaptations, rather than manifestations of some intrinsic property of living matter. Historically, Dobzhansky claimed, ignoring this important distinction about the "wisdom of nature" had misled some earlier biologists to embrace vitalism and the inheritance of acquired traits. By redirecting attention to the interaction of genes and environment in producing the phenotype, evolutionary theory highlighted the creativity of natural selection in the evolution of adaptive homeostatic capacities of organisms. Offspring did not inherit fixed characteristics such as temperature regulation or antibody production, but rather what Dobzhansky referred to as "reaction patterns" that led to the development of those homeostatic characteristics that were adaptive in environmental conditions that might be highly variable. The observation that all humans can regulate body temperature suggested that this characteristic was so broadly adaptive as to be essential. On the other hand, dark skin pigmentation might be an adaptive homeostatic mechanism in environments with intense solar radiation, but not necessarily in less sunny habitats. More importantly, skin color was an example of broad human variation incapable of compartmentalization into biologically meaningful categories of race. Its adaptive significance was context dependent and in modern society probably irrelevant.

Although emphasizing the evolution of homeostasis as a gradual process, Dobzhansky also elaborated three distinct stages in this process. Lack of sexual reproduction deprived bacteria and other asexual organisms of a powerful means of generating genetic variation through recombination. As a result, bacterial populations were essentially clones with limited genetic variation arising completely through mutation. In a fluctuating environment, most individuals might perish, although a few well-adapted mutants could quickly repopulate due to the extremely high reproductive rates exhibited by microbes. Dobzhansky cited recent research on antibiotic resistance to illustrate this point. In contrast to this rather inefficient process of adaptation, Dobzhansky wrote, "In higher organisms, every individual is too valuable to the species to be wasted in the relatively inefficient process of adapting to environmental changes by natural selection."[46] Genetic recombination that accompanied sexual reproduction produced large storehouses of genetic variation. Given this variation, natural selection became more than an agent of elimination. Natural selection now became a creative process producing flexibility in adapting to changing environments through physiological and developmental homeostasis but also through increasingly sophisticated behavior. In humans, the unique process of cultural evolution greatly enhanced this adaptive flexibility and allowed human societies to adapt to a rapidly changing world more efficiently than through natural selection alone. Through cultural evolution, the "wisdom of nature" and the "wisdom of the body" were largely, though not completely, supplanted by human wisdom.

The interaction between natural selection and cultural evolution was a topic of considerable debate during the decades following World War II, particularly as it related to intelligence, ethics, and other seemingly unique human characteristics.[47] If big brains had been a product of natural selection, what would be the consequences if selection were relaxed? Pessimists feared that intelligence was, in fact, declining because of civilization. Without the continual action of natural selection operating through competition between individuals (or races), such a decline seemed inevitable.[48] The behavioral biologist Curt Richter went so far as to compare humans to the overly domesticated rats bred for life in the laboratory but incapable of survival in the wild.[49] The rise of the welfare state only contributed to indolence and degeneration. Mayr fretted that current tax policies discouraged "superior individuals" from having large families, and he scorned social leveling that confused social uniformity with equality of opportunity.[50] Mayr's social commentary seemed at odds with his homeostatic claim that genetic teamwork and cooperation were more important to the genotype than the special contributions of particularly talented "soloists." The fact that he had so enthusiastically analogized the action of genes to the teamwork of athletes or cooperation of musicians performing in an ensemble might have led his readers to conclude that such teamwork was just as true of human societies. Ordinary or average team players might be the best-adapted genes in an integrated genotype, but Mayr seemed unwilling to embrace the broader social analogies of that claim.

Dobzhansky was much less conflicted about problems of social equality and intelligence than Mayr.[51] He pointed to the paucity of scientific data supporting declines in IQ, and he was openly skeptical about attempts to measure intelligence. Emphasizing the importance of variation, Dobzhansky claimed that intelligence was not a single characteristic but included a spectrum of talents equally important for society. Educational reform was likely to have a greater impact on this broadly defined human intelligence than misguided attempts to select for an enhanced genetic endowment. Dobzhansky was scathing in his condemnation of Muller's continued support for utopian programs of genetic engineering and artificial insemination using sperm from particularly talented men.[52]

Dobzhansky's interest in homeostasis, particularly as applied to human evolution, wavered in his later general writings. Even at the height of his interest in the 1950s, he noted that attempts to find a biological basis for ethics suffered from "mechanical oversimplification."[53] By the 1960s, he was particularly critical of attempts by Julian Huxley and Waddington to create an evolutionary ethics. Waddington, in particular, was using ideas of homeostasis and cybernetics to discuss how "biological wisdom" evolved through a feedback system involving both evolution and epigenetic development.[54] Avoiding his earlier effusive praise of Cannon, Dobzhansky responded to Waddington's suggestions for an evolutionary ethics by noting, "'Wisdom of the body' and 'wisdom of evolution' are good metaphors, but they are not synonymous with wisdom which is the source and validation of ethics."[55] This was not to deny that genes and natural selection were irrelevant to understanding the *origin* of ethics, but human values and wisdom were both products of cultural evolution irreducible to processes involved in biological evolution.

If homeostatic ideas led to mechanical oversimplification in ethics, they also seemed unnecessary and unhelpful in considering other social issues. Dobzhansky was opposed to various forms of biological determinism used to justify distinctions based on race or socioeconomic class. In *Mankind Evolving*, Dobzhansky used "feedback" in the subtitle of a section on the relationship between genes and culture, but the metaphor remained undeveloped and appeared to mean little more than that both were involved in human development.[56] One seemingly did not need homeostasis or cybernetics to emphasize the importance of genetic variation in human populations, the idea that equality was not synonymous with uniformity, or the possibility that educational opportunities and social programs could ameliorate genetic differences in intelligence.

By the late 1960s, Lerner also found homeostasis too conservative and inadequate as a metaphor for discussing social issues. He had initially been attracted to Cannon's foray into social homeostasis because it suggested that self-regulation occurred at all levels of organization, rather than for any particular insight into human social regulation. For Lerner, genetic homeostasis was a mechanism by which populations resist rapid change, but social change required guidance rather than resistance. Technological innovation, the accumulation of scientific information, future prospects for genetic engineering, not to mention problems of human population growth and environmental degradation were accelerating at what seemed to be an exponential pace. In his textbook for nonscience majors, *Heredity, Evolution, and Society*, Lerner presented genetics and evolution as a necessary foundation for making informed decisions about social issues arising from the looming technologies of genetic engineering.[57] Expressing a common post–World War II attitude among liberal biologists, Lerner dismissed older eugenics movements in the United States as racist pseudoscience rooted in reactionary Bible Belt politics and bigotry. Nonetheless, he cautiously embraced other, more recent, suggestions for scientific planning for the future of humanity. In the chapter "Management of the Human Genome," he argued for new eugenic techniques, such as genetic counseling combined with "euphenic" approaches to improving phenotypes through both preventative and therapeutic medical techniques. Rapid advances in neurophysiology, immunology, and molecular biology would undoubtedly encourage a "feedback" between eugenics and euphenics, although this interaction would presumably involve positive rather than negative feedback.[58] Lerner consciously avoided using the language of social homeostasis to describe this scientific trend. In the introduction to the book, he cautioned that "in traditional society, culture, including ethics and religion, acted as a homeostatic stabilizing force. Now culture has become an instrument of rapid change. Reasoned decisions have become much more difficult in the absence of historical guidelines."[59] Although Lerner could call for the development of an "effective and just machinery for collective ethical decision-making," mechanistic ideas of self-regulation or social homeostasis seemed inadequate, even as metaphors.[60]

More generally, ideas of genetic homeostasis lost much of their force by the end of 1960s. One of the early reviewers of Lerner's *Genetic Homeostasis* pointed out the disanalogy between the internal control by hormones and nerves that characterized physiological homeostasis and the external role of stabilizing selection in maintaining genetic homeostasis.[61] Despite his enthusiasm for applying various homeostatic concepts to evolutionary theory, Lewontin also realized the danger in analogizing too broadly from Cannon's physiological idea. In the context of the 1950s, Lewontin thought homeostasis was an apt way of thinking about adaptation, but he claimed that he would abandon the idea if it led to confusion.[62] By the time that he wrote *The Genetic Basis of Evolutionary Change*, in 1974, Lewontin ignored his earlier homeostatic perspective, even though he retained basic commitments to the importance of genetic variation and adaptation as an interactive process between organism and environment. The controversy between the balanced and classical schools of population genetics seemed intractable, even though Lewontin and Jack Hubby had demonstrated widespread genetic variation at the molecular level using electrophoresis. The adaptive significance of this variation remained an open question because much of this variation appeared to be selectively neutral.

In retrospect, Lewontin was also skeptical that the traditional theoretical understanding of fitness and other evolutionary parameters was adequate for dealing with the multidimensional interactions among numerous genes, each with many alleles. The idea of the coadapted genotype made up of mutually compatible, interacting genes was still attractive, but Lewontin ignored the idea of genetic homeostasis in his 1974 book. Other evolutionary biologists argued that it was possible to visualize genes, organisms, and populations from a much more atomistic perspective. The selfish genes popularized by Richard Dawkins might be the components of "gene machines," but they were autonomous and in competition with one another for reproductive success.[63] An integrated combination of genes might be formed by genetic recombination, random mutation, and natural selection, much as a successful crew might be assembled by randomly shuffling alternative rowers in and out of a racing shell while testing each combination for speed.[64] Dawkins was certainly intrigued by this blind process of evolutionary creativity, but he was even more focused on reducing causation to the lowest levels of biological organization. For Mayr, the interaction of genes was comparable to the cooperation of members of a musical ensemble who forgo the individual virtuosity of soloists to produce a harmonious group performance. For Dawkins, genes were "Chicago gangsters" who ruthlessly acted for personal gain. Of course, gangsters may cooperate in criminal mobs, and ensemble musicians are not always high-minded in their cooperation, but the contrasting metaphors suggested strikingly different relationships among genes, organisms, and the populations to which they belong. Mayr's deep commitment to organismal biology was rooted in a holism and organicism that invited physiological analogies, while dismissing a reductionist "beanbag genetics" focused on individual genes.[65] Although also deeply interested in organisms, Dawkins viewed them as less fundamental than Mayr did. Introducing *The Selfish Gene,* Dawkins wrote, "I shall argue that the fundamental unit of selection, and therefore of self-interest, is not the species, nor the group, nor even, strictly, the individual. It is the gene, the unit of heredity."[66] From this perspective, organisms and populations served as weapons in the struggles among gene lineages. Like his precursor and ally George C. Williams, Dawkins was intent on purging biology of unnecessary or misleading concepts. As critics, they didn't need to single out the genetic homeostasis of Mayr or

Dobzhansky to cast doubt on the idea that populations are physiological entities comparable to organisms. Easier targets than the prominent leaders of the modern synthesis were readily available for criticism, as we shall see in chapter 10. Regulation did not necessarily imply self-regulation, nor did it necessitate thinking explicitly in terms of homeostatic checks and balances acting at superorganismal levels of organization.

9

Limits of Tolerance, Adaptation, and Speciation

Thinking physiologically in terms of integration, balance, and self-regulation became an important, though certainly not dominant, feature of the modern synthesis. Ernst Mayr, Theodosius Dobzhansky, and other evolutionary biologists turned to homeostasis and the metaphor of physiological teamwork to explain both the cohesion that held species together and the centripetal tendency of species to fragment when ecotypes adapted to local environmental conditions. This was part and parcel of the biological species concepts of Mayr and Dobzhansky that defined species as populations of interbreeding individuals separated from other species by various physiological and behavioral mechanisms maintaining reproductive isolation.[1] Each species was in a constant tug-of-war between local adaptation and the homogenizing effect of gene flow. Thus, species formed evolving lineages that persisted over time but occasionally split, producing daughter species increasingly separated by the evolution of reproductive isolating mechanisms. From this perspective species were not simply categories created by taxonomists but real entities whose functional and evolutionary attributes could be studied in the field and laboratory by ecologists and geneticists—and perhaps also by physiologists.

In the introduction to his highly influential *Comparative Animal Physiology*, University of Illinois professor C. Ladd Prosser wrote, "Foremost among general principles which emerge from a study of comparative physiology is the functional adaptation of organisms to their environment. The distribution of a species is determined through natural selection by its limits of tolerance."[2] Combining the ideas of "functional adaptation," "limits of tolerance," and natural selection provided a conceptual framework for thinking physiologically about how populations confronted the challenges of a heterogeneous landscape of different ecological niches, even though the genetic bases for most complex physiological processes were unknown. It also provided an entry point for discussing how physiology could help explain how populations evolve and how speciation occurs. During the 1950s and early 1960s, Prosser actively promoted a "physiological species concept" and outlined a theory of how acclimation played a crucial role in adaptive evolution and speciation. Prosser argued that each species had a unique constellation of homeostatic responses that defined its "limits of tolerance" and determined its geographic range and environmental preferences but also allowed local populations to inhabit marginal habitats, and preadapt individuals to extend a species range.[3] If a mutation mimicked this phenotypic change, the preadaptation might become an adaptation in the Darwinian sense.

Prosser claimed that his physiological species concept complemented the new biological species concepts advanced by Dobzhansky and Mayr based on ideas of genetic continuity within species and genetic isolation between species. According to Prosser, "Species may be considered as reproductively isolated populations which constitute natural units in taxonomy. To the comparative physiologist species are not rigidly fixed populations; rather the individuals within a species show a range of potentialities with respect to environmental stresses. By ascertaining which adaptations are physiological and which genetic it may be possible to picture the evolution of small groups—the origin of varieties and subspecies."[4] Enticing as this prospect seemed, it faced the problem not only of reconciling functional and evolutionary concepts of adaptation but also of finding good examples of how physiological adaptations contributed to genetic divergence among populations, the evolution of reproductive isolation, and ultimately speciation.

The centenary celebrations of Darwin's *Origin of Species* provided Prosser with a number of opportunities to share his nascent ideas with other leading evolutionary biologists during the late 1950s.[5] In preparation for these presentations, he corresponded with Mayr both to share ideas and to get advice. Mayr was skeptical about the evolutionary importance of preadaptation and genetic assimilation that so interested Prosser at the time.[6] Nonetheless, he was enthusiastic about finding common ground between comparative physiology and the modern synthesis. In particular, Mayr directed Prosser's attention to the work of John Alexander Moore, a developmental biologist who was interested in the geographic distribution, temperature adaptations, and reproductive isolation among various populations of the leopard frog, *Rana pipiens*, and its close relatives. Moore, who was eight years younger than Prosser, had completed his PhD at Columbia in 1940. During his graduate studies, Moore interacted with Dobzhansky and Mayr, both of whom influenced his interests in speciation and reproductive isolation. Both of the eminent evolutionary biologists actively promoted Moore's research on frogs, presenting it as a classic study of adaptation, evolutionary divergence, and reproductive isolation. For Prosser, Moore seemed to provide a bridge between the species concepts articulated by Dobzhansky and Mayr and the new physiological species concept that he was developing. Moore's studies of the geographically far-flung leopard frogs provided a natural case study of how physiological responses to temperature expanded the limits of tolerance of a species, leading to adaptive evolution and incipient speciation.

John Alexander Moore: Laboratory Biologist and Museum Taxonomist

During his long career, Moore distinguished himself as a researcher, educational reformer, and historian.[7] Despite an impoverished childhood and education in a one-room school in rural Virginia, Moore published a note on the short-eared owl in *Auk* when he was only fifteen years old. His precocious interest in collecting and identifying organisms expanded when he volunteered as a high school student compiling specimen lists of birds at the American Museum of Natural History after moving to New York with his mother. Despite what he later described as serious academic deficiencies, Moore matriculated at Columbia University. The zoology department at Columbia had built a distinguished reputation through the early leadership of Edmund Beecher Wilson and Thomas Hunt Morgan, and continued to offer outstanding programs in genetics and embryology. Moore began working as a dishwasher in the laboratory of the embryologist Lester Barth and later completed a PhD under Barth's direction. At the time, Barth was conducting experimental research on embryonic induction using a variety of inorganic and organic chemical agents. Barth's emphasis on the nonspecificity of substances capable of causing induction sharply contrasted with the search for a unique inducer led by the British biologists Joseph Needham and C. H. Waddington.[8] Barth's Rockefeller Foundation grant in chemical embryology supported Moore's early research on frogs, but his interests moved in a distinctly different direction from his mentor's program. While freely acknowledging the encouragement and direction that he received from Barth, Moore never coauthored a publication with Barth, and he rarely cited the older embryologist's research in the articles that he wrote on the embryology and systematics of North American frogs.

Despite forays into physiology, genetics, ecology, museum taxonomy, and biogeography, Moore always considered himself an embryologist. His early laboratory experiments must be viewed within the historical context of embryological experiments on the effects of temperature and other environmental factors on the early development and metamorphosis in frogs, experiments begun in the late nineteenth century by Frank Lillie, Oscar Hertwig, August Krogh, and other researchers in Europe and the United States. These quantitative studies demonstrated the temperature dependence of development and the close correlation between rates of development and metabolism. They also showed that the temperature tolerances of eggs and embryos often followed environmental gradients, with early-breeding species tolerating colder temperatures and developing more rapidly than late-breeding ones. What set Moore's research program apart from these earlier studies was the way that he combined the functional understanding of environmental effects on development with an evolutionary perspective focusing upon adaptation, population distribution, and reproductive isolation. Dobzhansky and Mayr shaped these interdisciplinary interests, both through their writings and through direct personal interactions with Moore at Columbia and the American Museum of Natural History. Moore had just started his graduate studies when Dobzhansky published his influential *Genetics and the Origin of Species* (1937). Mayr's *Systematics and the Origin of Species*, published five years later (1942), proved even more influential for Moore's developing views on evolutionary systematics.

In his first major article, published in *Ecology*, Moore compared rates of development of the five species of frogs in the genus *Rana* living in the New York area.[9] Although adult frogs of all five species were broadly temperature tolerant, embryos were much more sensitive. Each species had a distinctive range of temperatures for normal development and metamorphosis (figure 11). Temperatures falling outside these zones were usually fatal to the developing embryo. The developmental range for each species closely correlated with the season when breeding occurred and the geographic range of the species. Embryos of *R. sylvatica*, the most northerly species and an early spring breeder, required water temperatures 2–24°C for normal development, while the embryos of the more southerly, summer-breeding *R. clamitans* could survive only between 12 and 35°C . Rates of development for all species were temperature dependent, increasing with temperature. However, at any given experimental temperature within the tolerance range, *R. sylvatica* developed significantly faster than *R. clamitans*. Other characteristics also followed this

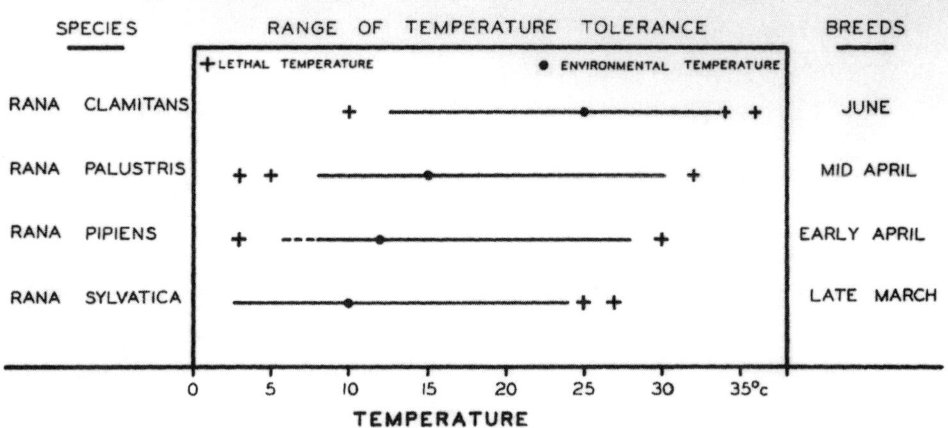

Figure 11. The relationship between temperature and breeding habits in four species of frogs from the northeastern United States. Horizontal lines indicate temperature ranges for normal development, crosses represent lethal temperatures, and dots represent average water temperature when eggs were laid. John A. Moore, "Temperature Tolerances and Rates of Development in the Eggs of Amphibia," *Ecology* 20 (1939): 459–78.

pattern, including the morphology of egg masses laid by females. For example, cold-adapted species deposited eggs in compact clusters, while the warm-adapted species laid eggs in thin films that were less prone to oxygen starvation at higher temperatures.

Both Mayr and Dobzhansky applauded these early studies, which had important implications for intraspecific variation as well as variation between related species. In the second edition of *Genetics and the Origin of Species*, Dobzhansky concluded his discussion of Moore's early experiment with this statement: "Results such as these ought to encourage further comparative studies on physiology of geographical races and ecotypes, a field which was almost completely neglected until recently."[10] Noting that zoologists were "far outdistanced" by botanists in studying the physiological basis of geographic variation in ecotypes of the same species and similar ecotypes in closely related species, Dobzhansky stated that "every class of ecotype is like a specially equipped detachment evolved by a species to colonize a certain type of habitat." As Dobzhansky was well aware, Moore had already started investigating ecotypic variation and its taxonomic consequences in leopard frogs.

Shortly after his initial publications, Moore directed his attention toward the far-flung species of leopard frogs *R. pipiens* and the systematic challenges that he referred to as the "*pipiens* problem." Unlike the geographically circumscribed and anatomically homogeneous *R. sylvatica* and *R. clamitans*, *R. pipiens* was a jigsaw puzzle of distinctive populations that ranged over much of North America, from Canada to Panama. Understanding this unusually broad range and diversity posed ecological, evolutionary, and taxonomic questions that seemed to be amenable to Moore's interdisciplinary approach to research, which included museum studies, genetic crosses, and embryological experimentation.

This research program, which consumed most of Moore's career, invites interpretation from three related perspectives. In discussing Moore's work, Dobzhansky stressed the similarities in approach with the highly successful interdisciplinary approaches that botanists were using to study adaptation and geographical distribution of ecological races or ecotypes of plants. This "experimental taxonomy" employed a combination of techniques from genetics, ecology, physiology, and taxonomy to study populations of plants both in controlled gardens and in native habitats.[11] Often conducted by teams of specialists, this interdisciplinary approach shed considerable light on how geographically widespread species evolved partially isolated populations adapted to unique local conditions of elevation and temperature. In contrast, Moore worked largely by himself, and although he used a diverse set of methods, his research was constrained by his independent style. In particular, he generally relied on frogs collected by commercial suppliers and conducted little research in the field. Although he acknowledged the importance of Dobzhanksy, Moore always allied his research more closely to the perspective of Ernst Mayr and museum systematics. From this zoological perspective, the study of leopard frogs found parallels in the clines found in birds and mammals codified by Bergmann's rule and similar biogeographical generalizations. Assuming that Bergmann's rule reflected adaptations for thermoregulation, Moore noted that the parallel between the frogs that he studied and warm-blooded vertebrates was not exact. Nonetheless, the temperature responses of different species or populations of frogs were indisputably physiological adaptations that not only provided a fit with the environment but also isolated some populations from others.[12] The extent of this reproductive isolation became the principal focus of Moore's research. From C. Ladd Prosser's perspective, Moore's research held the promise of investigating the role of acclimation in preadapting populations and setting the stage for speciation. It provided the bridge between a physiological species concept based on local adaptation and a biological species concept based on reproductive isolation.

For Moore, *R. pipiens* posed both a taxonomic problem and an evolutionary paradox.[13] Because of the broad geographic distribution and phenotypic variation, taxonomists had long disagreed whether leopard frogs consisted of a single species or a complex of closely related, geographically contiguous species. Considering leopard frogs to be a single species, as he did, Moore faced the evolutionary paradox that the northern and southern races of *R. pipiens* exhibited reproductive isolation and differences in temperature response that were comparable to the more well-established, homogeneous species *R. sylvatica* and *R. clamitans* that he had studied earlier. Unlike the morphological characteristics that taxonomists traditionally used for classification, the frogs' developmental responses to temperature were physiological characteristics that had obvious implications for adaptation and reproductive isolation.

Moore's entry to these problems took two forms: experimental testing of species boundaries using genetic crosses and a monographic study of variation within species of leopard frogs inhabiting eastern North America. In a shipment of frogs from a dealer in the Midwest, Moore found several unspotted individuals. Most taxonomists had considered these unusual frogs to be a color variation of *R. pipiens*, but two decades earlier the naturalist Alfred Weed had given the name *R. burnsi* to these frogs, which inhabited northern Iowa and southern Minnesota. Admitting that his designation was unorthodox, Weed wrote: "There may be some question as to the propriety of describing species based on color characteristics alone, especially in a group where the color variation of any particular individual may be so great as in the frogs. However, the author feels that the differences shown are so constant and of such a degree as to warrant their receiving a name and that their exact status whether specific, subspecific or varietal must be determined by future examination, for which he has neither the time nor the equipment. It seems probable that the final decision as to the validity of these species must rest on the result of carefully controlled breeding experiments."[14]

Weed's new species was not widely accepted by other taxonomists, but Moore took up the challenge of determining the genetic basis of the unspotted frogs. Using frogs sent to him by a commercial dealer, Moore conducted a series of breeding experiments among spotted and unspotted frogs. The results suggested that the color variation associated with *R. burnsi* might be due to a single, dominant gene. Using rough estimates provided by collectors of the relative numbers of *R. burnsi* compared to typical *R. pipiens*, Moore calculated Hardy-Weinberg frequencies for the three hypothetical genotypes. Because *R. burnsi* typically made up less than 1% of leopard frog populations in

Minnesota, Moore estimated the frequency of homozygous dominants (BB) to be about 0.025. In other words, nearly all of the unspotted frogs were heterozygous. When typical *R. pipiens* were crossed, they always showed spots as adults, but when *R. pipiens* was crossed with Weed's *R. burnsi*, about half the offspring were spotted and half unspotted. When *R. burnsi* were crossed, the offspring showed a range of coloration, but about 25% were unspotted. Moore concluded, "*R. burnsi* should not have the status of a species or subspecies but should be reduced to synonomy with *R. pipiens* and be referred to as the 'burnsi mutant.'"[15]

Moore's foray into the population genetics of leopard frogs was a significant initial step toward unraveling taxonomic and evolutionary relationships, although ultimately the genetic basis for the coloration pattern turned out to be more complex than he had hypothesized.[16] He admitted the difficulty of classifying coloration, both because of the individual variation in spotting patterns and the fact that the patterns sometimes changed as frogs matured. Nonetheless, the very fact that *R. burnsi* and *R. pipiens* freely crossed in the lab without any indication of defects in the offspring provided strong evidence that they were members of the same species. In a broader way, Moore's experimental discrediting of Weed's taxonomic claims for species status of *R. burnsi* demonstrated the potential of laboratory hybridization for determining evolutionary relationships and degrees of isolation among various populations of leopard frogs. Variations on this experimental method, especially combined with embryology, became a major feature of Moore's research program during the 1940s and 1950s.

His critique of the new species identified by Weed based on color patterns provided a microcosm of the broader "*pipiens* problem" that began to interest Moore during the early 1940s. Noting the historical interest in the taxonomy of meadow frogs, Moore pointed out the lack of agreement about the group. Naturalists had proposed several alternative classification systems during the early twentieth century. Some taxonomists preferred splitting leopard frogs into northern (*R. pipiens*) and southern (*R. brachycephala*) groups, although there was disagreement about whether these constituted true species or only subspecies. Other taxonomists had further divided the frogs, recognizing another group (*R. sphenocephala*) restricted to the southeastern United States. There was considerable disagreement over the geographical boundaries of these groups as well as their taxonomic status. Conservative taxonomists lumped leopard frogs into a single, highly variable species (*R. pipiens*). From Moore's perspective, these disagreements were more than a parochial taxonomic issue, but had a broader significance for evolutionary biologists who studied polytypic species with broad geographic ranges.[17]

Moore's careful study of geographic variation based on the extensive collections of frogs at the American Museum of Natural History reflected taxonomic ideas often associated with the new systematics and modern synthesis that were gaining influence during the 1940s. In particular, Moore had read Mayr's recently published *Systematics and the Origin of Species.* The book provided a general intellectual framework for explaining variation, although not a fool-proof "recipe" for classifying frogs.[18] In contrast to what he considered the prevailing taxonomic method based on studying a limited number of specimens, Moore treated species as populations that required careful sampling and statistical analysis of morphological characteristics. According to Moore, solving the "*pipiens* problem" depended upon this quantitative approach.[19]

Comparing means and standard deviations for measurements of morphological characteristics and plotting the results as frequency distributions provided a basis for deciding whether differences in populations from different localities were significant. Although Moore's approach might appear quite elementary, it represented a new perspective on quantitative zoology advocated by Mayr and other leading evolutionary biologists.[20] Not widely used in earlier systematics, these statistical methods were quite innovative during the 1940s when Moore began his work. Comparing populations drawn from the geographical ranges of *R. pipiens, R. brachycephala,* and *R. sphenocephala* showed broadly overlapping distributions for all of the morphological characteristics and ratios that Moore measured. Based on his morphological study of over 500 museum specimens and freshly killed frogs, Moore concluded: "It does not appear possible to recognize three species or subspecies of meadow frogs on the basis of differences in body proportions or pigmentation. Therefore, the meadow frogs of eastern North America should be known as *R. pipiens* Schreber. *R. sphenocephala* (Cope) and *R. brachycephala* (Cope) should be reduced to synonyms of *R. pipiens* Schreber."[21]

Moore's conservative taxonomic conclusions found a theoretical justification in the evolutionary interpretations of species and speciation championed by Mayr and Dobzhansky.[22] Species were evolutionary units held together by the exchange of genes and separated from other species by various reproductive isolating mechanisms that prevented gene flow. Mayr's polytypic species concept was particularly useful for making sense out of the variability found in leopard frogs. According to Mayr, many species were composed of local populations or ecotypes adapted to particular environmental conditions. These populations might evolve unique sets of traits, but gene flow with adjacent populations limited their divergence. Thus, there was an uneasy balance between local adaptation fragmenting a species into highly differentiated populations and conservative gene flow holding the entire species together. The idea of reproductive isolating mechanisms that evolved to maintain the cohesion and integrity of species fit particularly well with Moore's early embryological studies. The different rates of development found among species of frogs might well prevent gene flow through hybridization. Within *R. pipiens,* populations from different geographic localities might also exhibit different rates of development that could provide the basis for incipient speciation. Both of these hypotheses seemed amenable to experimental testing in the laboratory. For Moore, the "*pipiens* problem" provided a model system for exploring reproductive barriers and testing competing hypotheses about the evolution of isolating mechanisms in a well-controlled setting. Of course, leopard frogs had long served as a model organism for laboratory research, even though experimenters were rarely aware of differences between populations of their experimental subjects. Therefore, understanding the evolutionary relationships of this diverse model organism might have far-reaching implications for research that went well beyond Moore's own interests.[23]

During the late 1940s, Moore expanded his hybridization experiments into an ambitious program that integrated genetics and embryology in novel ways. The comparative studies proceeded in two related directions aimed at both interspecific and intraspecific variation. First, Moore studied the fate of embryos from hybrids between well-established but closely related species of frogs. In most cases, this did not involve actual mating. Moore manually stripped eggs from gravid females and fertilized them with sperm obtained by chopping testes in Ringer's solution. In the artificial crosses between *R. pipiens* and the wood frog *R. sylvatica,* fertilization was uniformly successful and early development proceeded fairly normally until gastrulation.[24] In most cases the hybrid embryos failed to complete this complex developmental process, although some survived for several days as "arrested gastrulae." On one hand, these experiments investigated embryological questions of competency and induction that continued to interest developmental biologists during the 1940s. On the other hand, they suggested a general mechanism for reproductive isolation that was both physiological and ecological. Moore's earlier studies had shown that the more northerly *R. sylvatica* bred earlier in the spring and in colder water than *R. pipiens.* In the laboratory, the rate of embryonic development in *R. sylvatica* was faster than in *R. pipiens* regardless of the water temperature. Assuming that the two species were adapted to slightly different climatic conditions, one could explain the failure of gastrulation in hybrids as a consequence of genetically controlled rates of development that differed in *R. pipiens* and *R. sylvatica*. It seemed likely that hybridization under natural conditions would rarely, if ever, result in gene flow between the two species.

Moore reached the same general conclusion about intraspecific variation among different populations of *R. pipiens*, which he considered a single, polytypic species.[25] Broadly distributed from the northern United States to Central America, different populations bred at different times of year and at different water temperatures. Populations from similar temperature zones tended to be completely interfertile, even when they were from such geographically distant locations as Vermont and Colorado. However, hybrid embryos from crosses between frogs from Vermont and Florida (or Mexico) failed to develop and usually died in early stages of gastrulation or neural tube formation. The developmental fates of these defective embryos depended upon the maternal population. Hybrids derived from northern eggs developed enlarged heads and defective circulatory systems, while hybrids from southern eggs had reduced heads, fused olfactory pouches, and no mouths.

The concept of the ecotype, which Moore sometimes used, provided an evolutionary explanation for how various local populations of leopard frogs combined to form the polytypic species *R. pipiens*. Originally used by botanists to describe and classify local populations, the ecotype emphasized genetically based physiological adaptations to local environments.[26] It infused the idea of geographic variation with an ecological emphasis on such causal factors as temperature, moisture, or altitude. In the case of the leopard frogs, populations from Canada to Panama had evolved distinctive physiological adaptations exemplified by the different rates of embryonic development. These differences in developmental rates adapted each population to a narrow range of water temperatures characteristic of different localities and, as a by-product, served as isolating mechanisms that might form the basis for incipient speciation. Indeed, populations of leopard frogs living in Florida and Vermont were reproductively isolated both by distance and by the fact that hybrid embryos failed to develop. If not for the intergrading series of populations along the eastern United States, Moore concluded, there would be no question that the extreme north and south populations were separate species. Because of limited gene flow throughout the entire complex, Moore continued to argue, taxonomically *R. pipiens* was a single species—although quite unique in its extensive range and complex population structure.[27]

Moore's Critics and the Demise of Polytypic *R. pipiens*

By the end of the 1960s, Moore was no longer publishing original research on leopard frogs, although he remained professionally active as an educational reformer and a historian of biology. During these later years, evidence began to accumulate suggesting that speciation within the *R. pipiens* complex had proceeded further than Moore had earlier claimed. Rather than a single, highly variable, polytypic species, it increasingly appeared, *R. pipiens* might be composed of many morphologically similar but reproductively isolated sibling species. The most compelling evidence came from field studies of small, localized populations by behavioral ecologists who claimed that mating calls might serve as effective prezygotic isolating mechanisms. Moore had never seriously considered the mating behavior of frogs. He conducted most of his experiments with frogs shipped to him by animal suppliers, and he used artificial fertilization rather than allowing frogs to freely mate in the laboratory. The casual observations that he made when he collected his own frogs in areas around New York led him to believe that mating was indiscriminate and that choice of mates was relatively unimportant. By contrast, field biologists claimed that male frogs from different populations had distinct mating calls. The use of sonograms, a technology previously exploited by ornithologists, provided compelling visual representations of these differences in vocalization, despite the lack of equally compelling evidence that female frogs actually responded to these differences.[28]

The idea that adjacent or overlapping populations of leopard frogs might be reproductively isolated found further support from the molecular evidence of gel electrophoresis, introduced to evolutionary biology during the mid-1960s.[29] Ironically, one of Moore's students, Stanley Salthe, published electrophoretic evidence that helped undermine Moore's evolutionary interpretation of the *R. pipiens* complex. According to Salthe, patterns of variation in enzymes such as lactase dehydrogenase and serum proteins such as albumin closely correlated with the geographic distribution of different mating calls.[30] By the early 1970s, critics confidently claimed the demise of Moore's idea that *R. pipiens* was a single, polytypic species. A steady increase in named species, some of which were morphologically indistinguishable, reflected the wide acceptance of the competing claim that *R. pipiens* was a complex of sibling species.[31]

In important ways Moore was a victim of his own success but also of his somewhat idiosyncratic approach to research. He had conceived his research program within the intellectual framework of the burgeoning modern synthesis and had enjoyed the early and continuing support of Mayr and Dobzhansky. Indeed, even as criticism of Moore's work mounted, Dobzhansky hailed his research as a "classic" example of reproductive isolation through hybrid inviability.[32] Authors of popular general biology and evolution textbooks highlighted Moore's work on leopard frogs as the best example of incipient speciation and the polytypic species concept. Moore's high profile had the effect of overshadowing early criticism of the polytypic species concept but also of intensifying the criticism after he retired from research on leopard frogs, during the late 1960s. Critics pointed to the popularization of Moore's research in textbooks as an important reason for correcting a widely accepted but misleading claim about the evolutionary status of the *R. pipiens* complex. For some critics, this was not simply a matter of taxonomic revision but a revolutionary rejection of Moore's "pseudo-subspecies."[33]

Moore's extreme individualism and independent style of research was also partly responsible for the fate of his life's work on leopard frogs. Virtually all of his publications were single-authored. Although he trained several graduate students at Columbia University, he also encouraged them to publish independently. Thus, in contrast to scientists who establish cohesive research teams or schools that build legacies that outlive the founder, Moore's influence was more diffuse.[34] When critics challenged his ideas on the evolutionary dynamics and taxonomy of leopard frogs late in his career, Moore lacked institutional resources for responding effectively. Indeed, the work of his own students sometimes helped to undermine Moore's position. Perhaps not surprisingly, as he neared retirement, Moore himself characterized his life's work as a paradigm under siege.

In his 1974 presidential address to the American Society of Zoologists, Moore described the history of systematic research on *R. pipiens* as a series of paradigms.[35] His nearly three decades of research beginning in the late 1930s had contributed importantly to replacing a purely morphological concept of species with an evolutionary concept that emphasized populations, genetic variation, and physiological adaptation to a complex external environment. He had placed his results within the explanatory framework of the modern synthesis, and his studies served as an exemplar of the polytypic species concept championed by Ernst Mayr. He had skillfully combined a variety of laboratory methods to explore both functional and evolutionary aspects of reproductive isolation within a single unified research program, although he missed the critical role that behavior played in separating populations. When he gave the address, Moore had not published any original research on leopard frogs for several years, and he acknowledged that a new paradigm based on the recognition that *R. pipiens* was a collection of often highly localized, sibling species was gaining ascendancy among evolutionary biologists interested in the group.

Although both Moore and his critics emphasized discontinuity by describing this episode using a Kuhnian framework, one might question the historical interpretation of species complexes and polytypic species forming competing paradigms. Indeed, Ernst Mayr, who claimed to have coined the term "sibling species," viewed it as part of a more encompassing biological species concept based on reproductive isolation rather than morphology.[36] A species might be geographically widespread and morphologically variable, forming relatively distinct subspecific ecotypes, as Moore originally claimed. Groups of species could also be morphologically indistinguishable but reproductively isolated, as his critics later pointed out. According to Mayr, evolutionary systematists should expect to find a full range of possibilities between these two extremes.

Viewed in this broader historical context, the later rejection of Moore's claim that *R. pipiens* was a single polytypic species was not so much a shift in paradigm as a further elaboration of an evolutionary perspective on the systematics of a challenging group that continued to pose evolutionary "puzzles" for systematists.[37] Moore seemed to acknowledge this deeper continuity when he concluded his historical narrative of "changing paradigms." After posing the rhetorical question of whether local populations of leopard frogs constituted good species, he wrote: "There is no obvious way that taxonomy can be simple enough to be useful but complex enough to be applicable to all natural variation. Much more interesting questions can be asked about the interrelations of the numerous local populations of the *pipiens* complex. It is probable that one finds here a full spectrum of the stages of speciation from adjacent populations showing but the slightest amount of differentiation to others that are behaving as full biological species. And we must remember that the variety of studies that can be made on these frogs, both in the field and the laboratory, is almost unique among animals. Thus, it is not unreasonable to conclude that, as we learn more and more about the interrelations of the local populations of the *pipiens* complex, the more we will know about the dynamics of speciation."[38] Moore's bold foray into the borderlands between physiology and evolutionary biology provided an impressive body of work on geographic variation and physiological adaptation that later critics could not ignore. Despite the rejection of his taxonomic conclusions, Moore's experimental studies provided a firm foundation for revision of a group important both as a model organism for laboratory research and a perennial source of evolutionary questions.

10

Adaptation, Natural Selection, and Homeostatic Populations

The idea that populations are self-regulating physiological entities was widely popular after World War II. C. Ladd Prosser's physiological species concept emphasized how natural selection, a broadly conceived homeostatic regulation, and the close adaptive fit between organism and environment set the limits of tolerance for populations. He was confident that this dynamic theoretical understanding of species as populations of adaptable organisms harmonized with the biological species concepts championed by Ernst Mayr and Theodosius Dobzhansky. For their parts, Mayr and Dobzhansky employed homeostasis to discuss both stability and change in evolving populations. Although this interdisciplinary sharing of physiological, ecological, and evolutionary ideas seemed promising during the 1950s, it later encountered strong criticism, particularly when applied to population growth. The idea that populations were capable of self-regulating numbers to match environmental resources was widespread during the post–World War II period, but it encountered stiff resistance from critics set on purging biology of unnecessary theoretical impediments during the 1960s.

The Problem of Self-Regulating Populations

After World War II there was an important, although incomplete, shift from older forms of organicist thinking in biology and the social sciences toward broader systems theories, notably cybernetics. Vague organismal analogies gradually gave way to seemingly more rigorously objective explanations in terms of feedback and other cybernetic concepts. This shift was evident in the work of Norbert Wiener, who acknowledged an intellectual debt to Walter Cannon's homeostasis.[1] For Wiener, self-regulation in machines, organisms, and society found a common explanation in the mathematical language of cybernetics. The incompleteness of the intellectual shift was obvious, at least in retrospect, in the Macy Conferences, a series of interdisciplinary meetings held during the late 1940s and early 1950s to explore new ideas of cybernetics and other forms of systems thinking. Organized by the neurophysiologist Warren McCulloch, the Macy Conferences had a strong orientation toward living systems (both organismal and social), but there was an unresolved tension between the general, and sometimes loose, analogies encouraged by interdisciplinary sharing of the participants and the formal theories developed by Wiener and a few other mathematically adept participants.[2] Although the titles of later conferences explicitly referred to cybernetics, the earliest meetings identified self-regulation with "teleological mechanisms" and "circular causal systems." For the organizers, these general terms were equivalent to the more formal concepts of feedback, self-regulation, and servomechanisms that were rapidly developing as core concepts in the new science of cybernetics.[3] However, the earlier and more ambiguous terms never became widely established in biology or other fields influenced by systems thinking.

Participating in the second Macy Conference in January 1946, the British émigré and Yale ecologist G. Evelyn Hutchinson presented his ideas in an influential paper titled "Circular Causal Systems in Ecology."[4] In keeping with the general theme of the meeting, Hutchinson discussed cases of self-regulating systems in ecology, although he was ambivalent toward the idea that these constituted "teleological mechanisms." He did not describe ecological self-regulation in terms of feedback, nor did he refer to the nascent literature of cybernetics or to Wiener's mathematics. Although his mathematical applications were somewhat novel, Hutchinson used equations developed earlier in theoretical ecology and already well accepted by ecologists. Typical of the conference presentations, Hutchinson's remarks about ecology were broadly synthetic, abstract, and speculative. His mathematics, synthetic perspectives, and emphasis on stability in biological systems paid homage to Alfred Lotka, whom Hutchinson cited in several different contexts in his article. As Sharon Kingsland and Peter Taylor have carefully documented, Lotka's underappreciated goal of creating a synthetic physical biology had a profound effect on Hutchinson and his students.[5]

Hutchison's article was particularly noteworthy for its attempt to forge connections between seemingly disparate ecological phenomena using what he referred to as "formal analogy." Although the processes that he discussed seemed unrelated, Hutchinson discussed them in terms of shared mathematical formulations. Hutchinson divided ecology into two convenient categories that he termed *biogeochemical* and *biodemographic*. The biogeochemical approach focused on the cycling of nutrients and other chemical substances between the living and nonliving worlds. The biodemographic approach focused on interactions of individuals within and between species, particularly phenomena of population growth, competition, and predation. This classification neatly mapped Hutchinson's eclectic mix of research interests, as well as the divergent directions taken by his prominent students, notably the population and community ecologists Robert MacArthur and Lawrence Slobodkin and the ecosystem ecologist Howard T. Odum.[6]

The "formal analogy" that Hutchinson used to bridge the seemingly unrelated phenomena of biogeochemical cycling and population regulation was the logistic equation $\frac{dN}{dt} = Nb\,\frac{K\text{-}N}{K}$ and its graphical representation, the sigmoid growth curve. In the equation, *b* represented the reproductive rate, in other words, the difference between birth and death rates in a population. With no regulating factor, a positive reproductive rate would cause the population size (*N*) to grow exponentially. *K* represented the saturation level of the environment beyond which a larger population was unsustainable. When the population density was low, $\frac{K\text{-}N}{K}$ approximated 1, and growth was close to exponential. However, as *N* increased, $\frac{K\text{-}N}{K}$ decreased, eventually approaching zero. Theoretically, population growth would approach the saturation level asymptotically. Real populations rarely fit this model exactly, but in at least some cases the size oscillated or fluctuated around the saturation level. Hutchinson emphasized that, similar to other self-limiting systems, the precision of the self-regulatory mechanism depended upon the speed with which the population responded to changes in size. Long time lags resulted in potentially destabilizing oscillations or population cycles.

Hutchinson readily acknowledged the earlier theoreticians who had explored the logistic equation to study population growth and regulation, notably the nineteenth-century mathematician Pierre Verhulst and his twentieth-century successors Raymond Pearl, Alfred Lotka, Vito Volterra, and Georgii Gause. Under carefully controlled laboratory conditions, the growth of populations of aquatic microorganisms and insects sometimes approximated the sigmoid curve. According to Hutchinson, this type of growth also occurred in the productivity of aquatic communities during succession from a pond to a marsh, even though the populations and species changed during the historical process. Similarly, Hutchinson claimed that photosynthetic production and organic decomposition in the global carbon cycle had followed this pattern with carbon assimilation and liberation currently balanced around a saturation level. The sigmoid curve also modeled the growth of many multicellular organisms. More broadly, Hutchinson speculated that the logistic equation could represent economic growth in capitalist societies (an insight that he attributed to Lotka) and the historical development of schools or styles of art. He concluded, "It is legitimated to regard the term $\frac{K-N}{K}$ as formally describing a self-regulatory mechanism. There can be little doubt, biologically, the mechanism can take a great variety of forms."[7]

Indeed, even in the restricted case of population regulation, Hutchinson listed a number of possible factors that might limit growth. These included extrinsic factors such as lunar cycles or meteorological events, as well as factors intrinsic to biological communities, notably food supply, competition, predation, and disease. In providing this comprehensive list of causes, Hutchinson sidestepped the controversies over the relative importance of "density-dependent" and "density-independent" factors for regulating animal numbers.[8] Complicating matters was the realization that populations were never isolated units but interacted with one another in complex ways. In this sense no population was truly "self-regulating." Although Lotka and Volterra had studied competitive and predatory interactions theoretically, and Gause, Thomas Park, and other ecologists had performed some compelling laboratory experiments, Hutchinson noted that the regulation in natural populations remained largely unknown. Population cycles in rodents and fur-bearing mammals and their predators studied by Charles Elton and his students had garnered interest among ecologists, although Hutchinson considered such cycles highly unusual cases and probably derangements of the self-regulating mechanisms found in typical populations. Presumably, most populations evolved precise mechanisms for minimizing oscillations in size, or they went extinct because of the destabilizing fluctuations.

Hutchinson did not delve very deeply into exactly how population regulation had evolved, although he assumed that it was a common feature of most natural populations. The idea that at least some species could self-regulate was widely shared, and Hutchinson took it for granted that this self-regulation was adaptive and that it had evolved through natural selection.[9] Because population size was a group attribute, Hutchinson reasoned that selection must have occurred by the replacement of less successful populations by those better able to regulate numbers. In a rather off-hand conclusion, Hutchinson commented, "Selection of this type is of course commonly recognized in modern evolutionary theory."[10] This comment may be jarring to modern ecologists, but it reflected mainstream biological thinking about natural selection after World War II. Later critics would agree that the idea of group selection was widespread, but they dismissed it as detrimental to evolutionary thinking.

The fundamental ideas in "Circular Causal Systems in Ecology" influenced prominent group selectionists, and these ideas faced intense scrutiny. The fact that they were not directly attributed to Hutchinson helped him to avoid becoming a primary target of the contentious group selection controversy that ensued during the 1960s. Richard Lewontin, who himself had been enthusiastic about homeostasis at the population level, dismissed self-regulation of population size as an unscientific notion. In an influential paper on the units of natural selection, he wrote "self-regulation of numbers, on which most of the argument for group selection is built, simply does not occur. It is an aesthetic prejudice of ecologists that animals regulate their own numbers so as somehow to achieve 'optimal' numbers."[11] Lewontin was echoing the critique of George C. Williams, who denied that negative feedback was a valid concept for explaining changes in population size. He repeatedly argued that ideas of self-regulation reflected aesthetic and anthropomorphic assumptions of optimality. Other critics more broadly attacked idea of equilibrium or "balance of nature" in ecology. Thus, the group selection controversy of the 1960s served as a watershed in twentieth-century ecology. It focused scrutiny on adaptation and natural selection, calling into question how prominent ecologists had employed these concepts particularly when discussing populations, communities, or ecosystems. More broadly, it opened debate about what constituted a legitimate evolutionary explanation. Atomistic conceptions of individuals in groups eclipsed ideas of group homeostasis but never completely eliminated this physiological perspective.

Group Selection, Biotic Adaptations, and Superorganismic Communities

When George C. Williams wrote his trenchant critique of group selection, he took particular aim at Warder Clyde Allee and his colleagues at the University of Chicago. Allee had been dead for a decade by the time that Williams's *Adaptation and Natural Selection* appeared in 1966, but the attenuated influence of the deceased ecologist continued to shape the field—or so Williams suspected. Williams accused Allee of promoting "biotic adaptation," the idea that certain behaviors or physiological characteristics evolved for the benefit of the population or community (i.e., the biota) rather than the individual. Carrying this line of thought to its extreme, which Williams claimed the

Chicago ecologists did, the entire biosphere constituted an adapted unit. As authors of a prominent ecology textbook that espoused such views, Allee and his colleagues provided a prominent target for Williams's extended argument against the existence of group adaptations and group selection. According to Williams, group selection and group adaptations might be logical possibilities, but in the real world they did not exist.

Williams introduced his argument around the Darwinian topic of the soil-forming activity of earthworms. Darwin had marveled at the prodigious ability of earthworms to decompose organic detritus and aerate the soil. As Williams pointed out, although the impact of a single worm was negligible, taken together earthworms modified the soil in ways that benefited other members of the community, particularly plants. Based on this beneficial result, Williams claimed, Allee believed that the feeding behavior of earthworms should be designated a "soil improvement mechanism" that constituted a biotic adaptation of the soil community as a whole. A simpler explanation, according to Williams, was that soil formation was a fortuitous side effect or statistical summation of the individual feeding behavior of earthworms. Close examination of this behavior and the physiology of the earthworm's digestive tract adequately explained causes of soil alteration in terms of individual nutrition. A causal explanation invoking biotic adaptations at the level of the community or natural selection at the level of populations was gratuitous and violated the principle of parsimony.

The attack on Allee's account of soil formation was polemical, and Williams made no attempt to contextualize the Chicago ecologist's claims beyond the bare skeleton of erroneous evolutionary assumptions. Indeed, his summary of Allee's ideas was a caricature.[12] Both the introduction and conclusion of *Adaptation and Natural Selection* provided just enough details of Allee's research to make the idea of group selection and biotic adaptation seem implausible, if not preposterous. While earthworms had never been the central focus of Allee's work, his account of soil formation reflected his deep interests in group behavior, which had interested him from the beginning of his career. Earthworms were part of an "aggregation" of soil organisms that worked together in a largely harmonious and mutually beneficial way to create and sustain a living community. The "grouping tendency" that brought diverse animals together, regardless of genetic relatedness, had evolved from an automatic and pervasive predisposition to provide mutual aid.[13] There was no need to consider the interactions cooperative in the sense of conscious behavior, but the "protocooperation" or "automatic mutualism" among microorganisms and small soil animals was a rudimentary but tightly knit form of sociality. For example, large groups (regardless of relatedness) provided mass defense against predators or protection against harmful or toxic substances. According to Allee, combined activities of living organisms "conditioned" the environment in complex ways that might regulate population growth or enhance stability of the community as a whole. Changes wrought to the surrounding environment by organic activity also facilitated the orderly development of the community through the process of succession. Therefore, the soil itself was a kind of organism or "superorganism" whose parts operated harmoniously to promote homeostasis.

For the Chicago ecologists, homeostasis was as much an ecological principle governing communities and ecosystems as a physiological concept applied to individual organisms.[14] Self-regulation and self-correction leading to stability, balance, and orderly change were characteristic of all biological and social systems. In his detailed history of the Chicago school of ecology, Gregg Mitman emphasized how physiology and development shaped Allee's early thinking much more than Mendelian genetics and Darwinian evolutionary theory.[15] The developing embryo provided a metaphor for understanding both evolution and social development as orderly and progressive. Similarly, Allee's early conception of populations evinced a holistic, organicist perspective on ecological communities that emphasized the processes of integration and division of labor rather than the sifting of alleles by natural selection. From this perspective, even simple aggregations of different populations were analogous to ethnic neighborhoods in a city, which though distinct, still combined to form a larger organic whole.[16] As a discipline bridging the biological and social sciences and providing a naturalistic basis for ethics and politics, Allee's ecology emphasized organization, cooperation, and stability.

Allee's later collaboration with Alfred Emerson, beginning in the late 1930s, brought a greater emphasis on genes and natural selection to the Chicago ecologists.[17] But, although Emerson was a respected evolutionary biologist, his primary interest in termites and other social insects infused his thinking with organicist metaphors and "significant analogies" that went well beyond those employed by Ernst Mayr, Theodosius Dobzhansky, and other major figures in the modern synthesis. Populations of social insects were superorganisms linked by close genetic relatedness of individuals. This relationship provided a mechanism for the evolution of specialization, division of labor among castes, and replacement or regeneration of the population's parts. The tight integration of these groups also allowed the colony to maintain homeostasis within the nest by tightly controlling temperature and humidity. The superorganism formed by populations of social insects was the primary target of natural selection, but to a lesser extent this was true of all populations and communities. Homeostasis was an adaptive characteristic of all of these biotic entities. Natural selection, acting primarily at these higher levels, was a creative process primarily responsible for the progressive evolution of the self-regulating and self-correcting ability of populations, communities, ecosystems, and human societies.

Although they claimed originality for their ecological and evolutionary ideas, Allee and his colleagues also argued that they were part of mainstream biological thinking. In their *Principles of Animal Ecology*, the Chicago ecologists cited George Gaylord Simpson, Sewall Wright, and G. Evelyn Hutchinson as supporters of an evolutionary role for group selection acting on competing populations. They referred repeatedly to Hutchinson's "Circular Causal Systems in Ecology" to support their claims about group adaptations, homeostasis, and population regulation. Circular causal systems were the "rule" in evolution. According to Allee and his colleagues, "evolutionary trends are in the direction of increased homeostasis within the organism, the species population, and the ecosystem."[18]

The idea of group selection and the problem of levels of selection sharpened during the mid-1960s, but when *Principles of Animal Ecology* appeared in 1949, the relationship between adaptation and natural selection was comparatively broad and amorphous. That natural selection acted on populations and that this was an important evolutionary mechanism was part of the mainstream of biology. That certain characteristics were group adaptations that evolved for the good of the population, species, or community was also widely accepted. In framing their evolutionary perspective, the Chicago ecologists relied not only upon Darwinian natural selection but also on biological traditions that emphasized progressive evolution, development, cooperation, integration, and stability. Allee, in particular, wove this congeries of biological ideas together with a liberal political philosophy that emphasized social harmony and peaceful coexistence.

From George Williams's critical perspective, the reasoning used by the Chicago ecologists amounted to "cryptic opposition" to Darwinian theory. While claiming to be evolutionary biologists, Allee and his colleagues supported unnecessary and erroneous ideas about adaptation and natural selection that were foreign to Darwin's original thought. Williams took it as a "doctrine" that adaptation was an "onerous concept" that should be applied only when warranted by compelling evidence that natural selection had produced a characteristic. Evolutionary biologists should attribute adaptation only to the lowest levels of biological organization.[19] For Williams, this meant alternative alleles, individual organisms, or perhaps closely related family groups, but never populations, communities, or ecosystems. His book aimed to "purge" biology of unnecessary theoretical distractions. If Allee and the Chicago ecologists were guilty of a cryptic opposition to Darwinism uncritically shared by many other biologists, the British naturalist V. C. Wynne-Edwards had more recently proposed a detailed mechanism for group selection that constituted an overt challenge to Williams's reductionist and parsimonious stance. Homeostasis was a key element of Wynne-Edwards's evolutionary mechanism.

Group Selection in Homeostatic Populations

Wynne-Edwards's name has become indelibly linked with the concept of group selection. Where a clear distinction had not previously existed, Wynne-Edwards sharply differentiated between natural selection acting on individuals and populations. He argued strongly that individual selection could not account for density-dependent regulation of population size, and in his massive *Animal Dispersion in Relation to Social Behaviour*, Wynne-Edwards compiled 650 pages of examples from diverse animal taxa to support group selection. The book ignited a firestorm of criticism that toppled Wynne-Edwards's claims, and group selection became anathema to most evolutionary biologists beginning in the mid-1960s. Mark Borello has carefully documented the history of this episode in evolutionary biology, particularly in relation to the controversy over levels of selection.[20] He also examined the close and underappreciated intellectual connections between Wynne-Edwards and the modern synthesis. Wynne-Edwards quickly grasped the evolutionary importance of populations, and his idea of population physiology shared important similarities with claims made by Mayr and Dobzhansky.

Wynne-Edwards's emphasis on cooperation and group selection also shared considerable common ground with the Chicago school of ecology. My account focuses more closely on the relationship between Wynne-Edwards's views on population regulation and those expressed by other supporters of density-dependent regulation. In particular, I show the important roles that the general idea of population physiology and more specifically the concept of homeostasis played in Wynne-Edwards's argument for group selection and group adaptation. Homeostasis formed a subtext that was often ignored by Wynne-Edwards's critics. Despite its importance early on, the idea was also abandoned by Wynne-Edwards in his later writings about group selection. But, as we shall see, other ecologists used ideas of homeostasis to discuss self-regulation of population size without invoking group selection in the way that Wynne-Edwards and the Chicago school did.

It is significant that the index of *Animal Dispersion in Relation to Social Behaviour* has more entries for "homeostasis" and "feedback" than for "group selection." Introducing the topic of population regulation, Wynne-Edwards wrote, "To build up and preserve a favourable balance between population-density and available resources, it would be necessary for the animals to evolve a control system in many respects analogous to the physiological systems that regulate the internal environment of the body and adjust it to meet changing needs. Such systems are said to be homeostatic, and it will be convenient for us to use the same word."[21] Such a regulatory system required two parts. First, the population needed an input of information about population size and available resources, and according to Wynne-Edwards, complex social behaviors had evolved for exactly this purpose. Members of a population were constantly using a variety of behavioral cues, or "epideictic displays," to assess these two parameters. Second, the population needed a mechanism for using this information to equilibrate population growth at a sustainable level. Rather than maximizing their reproductive success, Wynne-Edwards argued, individuals adjusted reproductive effort to optimize population size. In effect, individuals sometimes sacrificed self-interest for the common good. The only way that this could occur, according to Wynne-Edwards, was if poorly regulated populations were outcompeted by populations that regulated numbers more precisely. In effect, natural selection acted at the level of groups rather than individuals.

The idea that populations could self-regulate size through a type of homeostasis was diametrically opposed to an earlier explanation provided by David Lack. Rather than the internal self-regulatory mechanism envisioned by Wynne-Edwards, Lack argued that external factors in the environment such as food supply, predation, and parasitism controlled population size. Introducing the disagreement at the beginning of his book, Wynne-Edwards juxtaposed extrinsic controls with intrinsic behavioral regulation: "It will suffice for the present to say that these external checks, while they may sometimes be extremely effective in preventing population-density from rising are on the whole hopelessly undependable and fickle in their incidence, and not nearly as perfectly density-dependent as has often been imagined. They would in most cases be incapable of serving to impose the ceilings found in nature;

what is more, experiment generally shows that they are unnecessary, and that many if not all the higher animals can limit their population-densities by intrinsic means. Most important of all, we shall find the self-limiting homeostatic methods of density-regulation are in practically universal operation not only in experiments, but under 'wild' conditions also."[22] Later, contrasting his idea of limiting reproductive output with Lack's idea of individuals maximizing reproductive success, Wynne-Edwards wrote: "The apparent alternative to Lack's hypothesis is that the recruitment rate is the dependent variable, and can be continually modified as part of the homeostatic process by which an optimum population-density is maintained."[23]

The controversy between Wynne-Edwards and Lack centered upon a number of important issues that were related but not necessarily interdependent. At least in retrospect, the major point of contention was the level at which natural selection operated. By taking an extreme position on the necessity of group selection, Wynne-Edwards raised important questions not previously examined by most evolutionary biologists. How common was group selection? What happens if group selection and individual selection are in opposition? The responses to these questions by Lack, and later by Williams, Lewontin, Richard Dawkins, and other evolutionary theorists had a profound impact on the direction of evolutionary biology during the late twentieth century. Before Wynne-Edwards's book it was relatively unproblematic for evolutionary biologists to discuss adaptations that existed "for the good of the species," but after his critics had their say, such language became very problematic, indeed.

Other issues raised by Wynne-Edwards's book were equally important, at least to ecologists who had long argued about the causes of population regulation. By contrasting Lack's emphasis on predation and mortality with his own emphasis on limiting birth rates, Wynne-Edwards highlighted issues of population regulation in an evolutionary context that had not always been an important part of population ecology. The controversy surrounding his book helped usher in a new "evolutionary ecology" explicitly based on natural selection acting on genes and individuals.[24] Because homeostasis played such a central role in Wynne-Edwards's discussion of population regulation, it also focused critical attention on earlier ways of thinking about populations widely accepted by ecologists and evolutionary biologists. If social behavior acted to regulate population growth, was it a form of negative feedback control? Were populations the sort of entities that truly exhibited homeostasis, or was this a superficial and misleading analogy? If populations were, in fact, homeostatic, what roles did cooperation and competition play in regulating population numbers? For some ecologists, Wynne-Edwards was largely correct about the importance of social behavior, but not necessarily about group selection. At least in retrospect, these ecologists complained that the group selection controversy was a "rather fruitless controversy" over an "untenable mechanism" that resulted in diverting attention from the more important question of self-regulation of population size that Wynne-Edwards raised in his book.[25] One could accept population regulation, even self-regulation, without necessarily adopting group selection arguments.

Stress and Self-Regulation in Individuals and Populations

When John J. Christian won the George Mercer Award from the Ecological Society of America in 1957, he told a reporter, "I have been interested for some time in bringing the highly developed techniques of physiological research to bear on field and ecological (population) problems."[26] Later in life, he would express some exasperation that other ecologists considered him a laboratory endocrinologist, although he thought himself, first and foremost, a field naturalist. His eclectic career encompassed both areas, and although his claims about the role of stress in population regulation met skepticism from some quarters, the link between hormonal feedback at the organismal level acting upon birth and death rates at the population level has had an enduring attraction for other population ecologists. His award-winning article on the effects of population growth on adrenal activity in laboratory mice reported doctoral research that he had done under the direction of the ecologist David E. Davis at Johns Hopkins University. Davis was a leading authority on the population ecology of rats and other rodents, and had spent a brief stint on a rat eradication program in Baltimore after World War II. The unusual collection of characters involved in that project—medical researchers and ecologists—has generated considerable interest from historians.[27] Christian's indirect connection to this group via Davis was an important part of his intellectual development, but more than any of the other figures associated with the rat eradication project, Christian combined the skills and perspectives of both laboratory physiology and field population ecology.

In a retrospective, Christian recalled a lifelong interest in natural history, beginning with birdwatching during his childhood in Scranton, Pennsylvania. His nascent interest in natural history found focus during a six-week high school trip to Yosemite Park for a naturalist program, during which he trapped small mammals, attended lectures, went on nature tours, and typed his teacher's notes. Christian's interest in endocrinology began during his undergraduate training at Princeton University and developed further in medical school at Columbia. However, partway through the medical program, Christian left to study aeronautical engineering at the University of Michigan. Still interested in natural history, he took a part-time job as an illustrator for the noted mammologist William H. Burt, who introduced him to population ecology by recommending Charles Elton's book *Voles, Mice, and Lemmings*. Both engineering and natural history were partly suspended when Christian was drafted into the navy to serve as the skipper of a PT boat in the Pacific, although he continued to collect and draw birds in his spare time during World War II. After the war he held a variety of research positions in vertebrate physiology with the Pennsylvania Game Commission, the Wyeth Medical Research Institute, the Penrose Research Laboratory, and the Naval Medical Research Institute. Christian continued as the head of the vertebrate physiology laboratory of the Naval Medical Research Institute while working on his doctoral research under Davis in the vertebrate ecology division of the Johns Hopkins School of Hygiene and Public Health, where he earned an ScD in 1954.

Even before working with Davis, Christian had a well-developed theory of population regulation that integrated individual reproductive physiology and behavior with the control of birth and death rates at the population level.[28] The theory had gone through a short but significant gestation while Christian briefly pursued graduate studies at the University of Pennsylvania before moving to Johns Hopkins. During his time in Philadelphia, he worked with Herbert L. Ratcliffe, the director of the Philadelphia Zoo, but Christian also held research positions at Wyeth Labs and with the Pennsylvania Game Commission, where he was involved with studying sudden die-offs in muskrats that presumably were the result of epidemic disease. This unusual combination of work environments allowed him to combine interests in endocrinology, natural history, and ecology. Christian was particularly intrigued by reports of "shock disease" in snowshoe hares described by a group of wildlife biologists in Minnesota led by the microbiologist Robert Gladding Green.[29] Both in the wild and in captive populations, apparently healthy animals would sometimes die quickly and in large numbers. When it occurred the disease seemed to affect entire populations. According to Green, an expert on animal diseases, intensive pathological investigation of dead hares provided no evidence of bacterial infection, but the animals exhibited atrophied livers and severe hypoglycemia. Green reported that he could sometimes temporarily alleviate the convulsions that were a characteristic symptom of shock disease by glucose injections. He concluded that the "finely balanced" carbohydrate metabolism of snowshoe hares was fatally disrupted by stress, including overcrowding, and he posited shock disease as the cause of the regular population cycles characteristic of the species.

Susan Jones has detailed Green's important role in a network of researchers associated with Charles Elton. Despite his medical training, Green's first love was ecology, and Jones describes how he pursued "a double professional life as a top-flight biomedical scientist and active ecologist."[30] The extensive correspondence between Green and Elton during the 1930s suggests a warm relationship based on mutual respect. After Green visited Oxford to present his research on shock disease, Elton wrote, "You certainly have made us sit up, and everyone here is tremendously impressed with your work (and you)."[31]

Christian read Green's reports within the context of Elton's broader perspective on population cycles presented in *Voles, Mice, and Lemmings*. He also brought insights from recent developments in endocrinology, particularly Hans Selye's ideas about the role of the adrenal gland in response to stress. Selye defined stress as a nonspecific response of the body to an external stimulus or threat. Regardless of the nature of the stimulus, the body's response followed a distinctive series of stages. The initial alarm or shock phase, lasting between six and forty-eight hours, was marked by a decrease in blood pressure and muscle tone, and depressed neural activity. For Selye, this constituted a "generalized call to arms of the body's defensive forces."[32] If the animal survived the initial threat, it entered a stage of resistance to the stress that might last for several weeks. This stage involved both the short-term, specific homeostatic mechanisms described by Cannon (e.g., secretion of adrenaline by the adrenal medulla), but also a nonspecific, longer response mediated by hormones secreted by the adrenal cortex. Regardless of the type of stressful stimulus, these corticosteroids acted in a variety of ways on nearly all systems of body: suppressing immunity, mobilizing energy reserves, increasing blood pressure, and inhibiting sexual behavior and lactation. As a result, the organism "adapted" to the stressor, not by returning to its original state through homeostasis but by accommodating to the stressor through a process that Selye referred to as heterostasis. In essence, the body formed a kind of symbiotic relationship with the stressor by resetting to a new, abnormal equilibrium. This new state of adaptation might resolve itself if the stressor was removed, but the new equilibrium might also be lost during prolonged stress. In such cases of unrelieved stress, the body entered a third stage of exhaustion. Depleted of its "adaptive energy," the body became subject to chronic, degenerative conditions that Selye referred to as "diseases of adaptation." Ironically, these pathological effects of stress were often the result of the body's own defense mechanisms overcompensating in a nonhomeostatic manner. Selye's ideas met with some skepticism from other physiologists, including Cannon, but as Mark Jackson recently demonstrated, Selye broadly and profoundly influenced both stress physiology and popular perceptions of the social effects of stressful encounters.[33] Within this broader context, Christian's enthusiasm for Selye's general scheme, as well as its obvious connection to Green's idea of shock disease, was quite unremarkable.

Working with caged mammals exposed to even minor stress convinced Christian that shock disease was a general phenomenon and that it corresponded to the exhaustion phase of Selye's general adaptation syndrome.[34] Christian reasoned that as a population grew, the effects of crowding led to the type of pathological conditions characteristic of Selye's heterostasis. This contributed both to increased mortality and decreased reproduction. Stress diverted energy not only from reproduction but also from parental care, leading to reduced survival of young. In addition, Christian argued that overcrowding led to a host of other dangers, including increased fighting, inadequate food, nutritional deficiencies, susceptibility to disease and parasites, increased exposure to predators, and inadequate shelter from the elements. This complex combination of increased mortality and decreased reproduction would affect all populations that approached the carrying capacity of the environment, but particularly in rapidly growing populations of small rodents that often exhibited population cycles. Christian was able to marshal a substantial body of circumstantial evidence from both the laboratory and field to support his claim, but he acknowledged that it remained hypothetical: "We now have a working hypothesis for the die-off terminating a cycle. Exhaustion of the adreno-pituitary system resulting from increased stresses inherent in a high population, especially in winter, plus the late winter demands of the reproductive system, due to increased light or other factors, precipitates population-wide death with the symptoms of adrenal insufficiency and hypoglycemic convulsions."[35]

Christian's hypothesis became prominent for its emphasis on intrinsic regulation of population growth mediated by the stress response, but his early statement of the hypothesis is revealing for several reasons. First, the decline that he described was population-wide rather than selective. Shock death apparently affected otherwise healthy animals regardless of their age, sex, or social status. Second, although adrenal insufficiency and other intrinsic factors were critical for regulating population growth, extrinsic factors such as weather, disease, and predation also played important roles in Christian's scheme. Finally, despite his interest in population growth and decline, Christian did not explicitly discuss this phenomenon within the context of mathematical population ecology. A decade later, the logistic equation would become a central part of his explanatory framework of stress regulation, but in his 1950 paper, published when he was a graduate student, Christian made only general references to exponential growth and carrying capacity.[36] Perhaps not surprisingly, in his earliest writings on population regulation, Christian was not thinking in terms of negative feedback controls. Later, as he came to place greater emphasis on intrinsic control of population size and as he distanced himself from the nonselective effects of shock disease, he adopted the language of cybernetics by presenting the stress response as a negative feedback mechanism operating differently on various individuals exposed to crowding.

Pivotal to the development of Christian's later ideas was his loose affiliation with an ambitious program in urban ecology growing from attempts to control rat populations in Baltimore during World War II. The Rodent Ecology Project combined practical concerns of public health and animal control with theoretical issues of population growth. The interests of the scientists involved combined medically oriented stress physiology, animal behavior, wildlife biology, experimental psychology, and population ecology. Working on the periphery of this eclectic mix, Christian continued to investigate the relationship between crowding, pituitary-adrenal activity, and stress in laboratory populations of rodents. However, the ecological studies of natural populations of Norway rats in an urban environment sharpened his emphasis on intrinsic regulation of population size based on behavioral differences among individuals.

The Rodent Ecology Project originated through the chance discovery of a highly effective rodenticide. Searching for chemical compounds that he could use in his studies of taste aversion in laboratory rats, Curt Richter, a brilliant but eccentric behavioral biologist at Johns Hopkins University, found that alpha napthyl thiourea (ANTU) was highly lethal to the rodents, but less toxic to humans and other mammals.[37] He convinced city officials to use the compound to eradicate rats that had infested neighborhoods, particularly around the harbor in Baltimore. He also secured financial support from the Rockefeller Foundation for ecological studies on rat population dynamics by John T. Emlen, and later by David E. Davis and John B. Calhoun.

Emlen had conducted waterfowl surveys with Aldo Leopold for the US

Biological Survey during the 1930s before taking a teaching position at the College of Agriculture at Davis, California. Raised as a Quaker, he registered as a conscientious objector when the US entered World War II and was assigned to the rat control project for his alternative service.[38] He later considered this experience to be a brief interlude before the "second half" of his life as a successful behavioral ecologist at the University of Wisconsin, but he had a rare talent for careful field studies that led to a series of publications on the population dynamics of rats in Baltimore. Davis was also a behavioral ecologist with a particular interest in questions of applied ecology and epidemic diseases. He completed his PhD at Harvard in 1940 with research on avian social and reproductive behavior of the smooth-billed ani in Cuba.[39] During the war he conducted research on yellow fever in Brazil for the Rockefeller Foundation. He joined the Rodent Ecology Program when Emlen left at the end of the war. Unlike Emlen, who had the rather undignified designation of rat control officer, Davis held a faculty position in the School of Hygiene and Public Health at Johns Hopkins from 1946 to 1959. Also unlike Emlen, who took no particular interest in rats, Davis became a leading authority on the population ecology of *Rattus*.[40] Davis's understanding of the natural history and population dynamics of these highly adaptable rodents had an important influence on Christian's thinking.

Although the Norway rat does not exhibit the population cycles characteristic of rodents such as lemmings and voles, the rodent control program in Baltimore allowed ecologists to study populations decimated by poisoning, to follow their recovery, and to compare these populations with untreated populations both in the city and in adjoining rural areas. Early in the program, Emlen and Richter designed highly effective box traps for catching rats, and Emlen modified established field trapping methods to estimate the size of urban rat populations.[41] It was a daunting challenge, made somewhat more tractable by the fact that rats usually avoided crossing streets and other exposed areas. Thus, the urban environment was subdivided into semi-isolated populations, each inhabiting a city block. During a nine-month period during 1943–1944, Emlen conducted exhaustive trapping campaigns in eighty-nine city blocks. Each campaign typically lasted two to three weeks and utilized 100–300 live traps. As the trapping continued, the daily catch decreased exponentially, eventually approaching zero, while the accumulated catch reached an asymptote, which provided a rough estimate of population size. Emlen independently evaluated the effectiveness of trapping using relative measures of rat abundance such as counting droppings, rat burrows, and live sightings of animals.

Impoverished urban neighborhoods constituted a favorable environment for rats, providing abundant shelter and adequate food. Estimated populations of over one hundred rats per block were typical in Baltimore neighborhoods sampled. These numbers were significantly higher than in rural areas adjoining the city. Furthermore, urban populations rapidly recovered after trapping or poisoning and often returned to the original size in a matter of months. Summarizing nearly a decade of studies within the context of theoretical population ecology, Davis concluded that the recovery of decimated rat populations followed the sigmoid growth curve postulated by Pearl and Lotka.[42] Although growth approached an equilibrium, different populations varied greatly in size, and there was considerable fluctuation in the size of any given population. According to Davis, "this equilibrium is a moving thing, constantly being displaced and returning to the original state, and there is a constant flow of individuals through the system. Furthermore, even if the total environment (asymptotic conditions) were absolutely constant, there are slight fluctuations due to the variations among individual rats and the fact that rats are not 'infinitesimally small' units."[43]

Although theoretical population ecology assumed that all individuals were equal, animals were not like the atoms or molecules that combined and dissociated in a chemical equilibrium. The death of a large, dominant rat might affect the population in far-reaching ways when replaced by a smaller individual. More importantly, the environment was never constant. For example, improved sanitation and rehabilitation of neighborhoods in Baltimore improved living conditions for humans but permanently degraded it from the perspective of rats. Unlike the transient effects of poison, urban renewal permanently reduced the number of rats in some neighborhoods.

The striking decline in the rat population in Baltimore during World War II, which Davis attributed to rehabilitation of neighborhoods rather than the use of poisons, involved an interaction of ecological factors that controlled the "flow of individuals through the system."[44] Davis assumed that reproduction remained relatively constant during this period, so the overall decrease in numbers was primarily due to increased mortality. Eliminating suitable habitat and food caused an increase in competition, and to a lesser extent predation. Superimposed on this simple input-output description of population dynamics was a complex set of social interactions that would play importantly into later ideas that Davis and Christian developed about the population effects of stress and adrenal hormones.

Early attempts to study the population effects of competition involved adding rats to stable populations near carrying capacity. John Calhoun investigated such "supersaturated" populations by transferring marked "alien" rats from one block to another in Baltimore.[45] The addition of new individuals destabilized the population, leading to death of both introduced and resident rats, although mortality tended to be higher among the intruders. Emigration also increased, particularly if the introduced rats came from neighboring populations. After a period of weeks, the populations appeared to stabilize once again after surviving aliens integrated into the larger group. Later, Davis and Christian modified this experimental design by first removing a number of resident rats before replacing them with an equal number of introduced rats.[46] In these cases no significant increase in mortality occurred. However, when they used introduced rats to increase the size of rapidly growing populations, they could terminate growth, apparently through a combination of increased mortality, decreased reproduction, and unsuccessful lactation. Reporting these results in the *Journal of Wildlife Management*, Davis and Christian noted the practical implications for the common practice of stocking populations of wildlife. At least under certain circumstances, artificially increasing the number of individuals might do little to change the population size. Indeed, it might lead to a decline.

Richter's earlier laboratory research had established the highly sophisticated behavioral adaptations of both wild and domesticated rats that contributed to physiological homeostasis. By selectively choosing food and drink, rats could compensate for metabolic deficiencies caused by experimentally removing various endocrine glands.[47] The ecologists working with the Rodent Ecology Program described an equally sophisticated complex of social behaviors that integrated members of a population. Populations of rats were not mere collections of atomistic individuals, but cohesive groups organized into well-defined hierarchies. Disrupting this organization by adding strangers or removing residents generally led to decreased reproduction, and increased strife and mortality. Calhoun would later become famous—or infamous—for his studies of social pathology in grossly overcrowded laboratory populations of domesticated rats, but in his early studies of wild rats he described how individuals formed highly integrated social groups and how they adaptively modified their environments. This adaptive "biological conditioning" prevented disharmony, regulated group size, and perhaps even increased the carrying capacity of the environment.[48] Calhoun was so impressed with the rats' ability to manipulate environmental variables for the benefit of the group that he described it as a rudimentary form of culture. He could experimentally disrupt this system of behavior, but rats naturally used social behavior so effectively that the group became a highly integrated unit of interdependent individuals. Calhoun's later speculative account of the implications of rat studies for humans was not necessarily shared by other members of the Rodent Ecology Program, but the emphasis on social behavior as an integrating force in animal populations became a core idea of the group.

Although the studies of natural, urban populations of rats provided few clues on behavioral effects of crowding, John Calhoun's early laboratory studies suggested that crowding had a detrimental effect on female reproduction. Moreover, this effect was not population-wide but disproportionately affected subordinate females. In his earliest experiments, Calhoun constructed large pens one hundred feet on a side. Food and water were abundant, but available only in a small central area enclosed by a fence with four openings.[49] To eat or drink, the individuals living in the outer areas were forced to trespass on territories maintained by individuals in the inner alleys surrounding the food supply. As the population grew, competitive encounters increased, and the stressful effects were strikingly different in the two subpopulations. Compared with females living in the alleys surrounding the food supply, those living in the outer areas weighed less, had more bite wounds and fewer successful pregnancies, and were generally unsuccessful at weaning litters. Later experiments involved much smaller enclosures with much denser populations of laboratory rats, and Edmund Ramsden and other historians have analyzed how the bizarre and aberrant behavior of Calhoun's rats influenced both academic and popular discussions of human population growth, urban crowding, and social pathology.[50] However, Calhoun's earlier and less artificial enclosure experiments also suggested that competition and stress might reduce reproductive success of individuals in natural populations that were much less crowded.

During the late 1950s, Davis and Christian developed a physiological model to explain how competition and stress could regulate populations through a complex feedback system. Like Christian's earlier discussion of population regulation, the later accounts owed much to Selye's emphasis on adrenal activity and the general adaptation syndrome. In other ways Christian's thinking shifted significantly. He largely abandoned his earlier emphasis on the indiscriminate, population-wide effects of shock disease. Instead, stress influenced some members of a population disproportionately, causing decreased reproduction, increased mortality, and emigration. Instead of focusing on population cycles (which Davis dismissed as a "mystical attraction" for population ecologists), Christian turned attention to regulation, more generally.[51] As a result, the idea that there was a feedback loop involving population size, stress, adrenal function, and reproductive success became more closely tied to the logistic curve. He had not discussed this theoretical model in his earlier work, but the emphasis that Davis, Emlen, and Calhoun placed on logistic growth clearly influenced Christian's later thinking.

Christian measured and compared the weights of adrenal glands from both male and female rats that he and Davis collected during population surveys in Baltimore during the early 1950s. At the same time, he was working on his doctoral dissertation at Johns Hopkins studying the effects of crowding on adrenal weights and reproductive success in laboratory colonies of mice. The results were equivocal, but at least in some cases Christian reported a correlation between the number of mice in a cage and changes in the adrenal glands. As expected from Selye's general adaptation syndrome and Calhoun's experiments on social behavior in crowded rat colonies, mice from denser populations tended to have enlarged adrenal glands. Conversely, the weights of sexual accessory glands decreased with crowding. These anatomical changes correlated with the social status of the animals, affecting low-status individuals more than dominant mice. At higher densities these females continued to reproduce but were less successful weaning the litters.

Christian and Davis increasingly presented the results of their studies within the explicit context of negative feedback.[52] Hormones of the adrenal cortex were the key feature of a system that linked individual physiology with a density-dependent regulation of the population (figure 12). Hormones secreted by the adrenal cortex functioned to protect animals from injury or disease, but at the cost of shifting energy away from reproduction. Thus, a secondary adaptive function of these adrenal hormones was to maintain populations at or below carrying capacity by suppressing the reproduction of subordinate individuals, who suffered disproportionately from competition and attendant stress. Increased mortality and emigration of low-ranking members of a population also acted to regulate numbers but were generally not as important as reproductive inhibition. Speaking at a symposium on crowding, stress, and natural selection sponsored by the National Academy of Sciences, Christian concluded, "The evolution of the social regulation of population growth could be considered a marked developmental advance over direct environmental regulation, and coincides with the greater development and

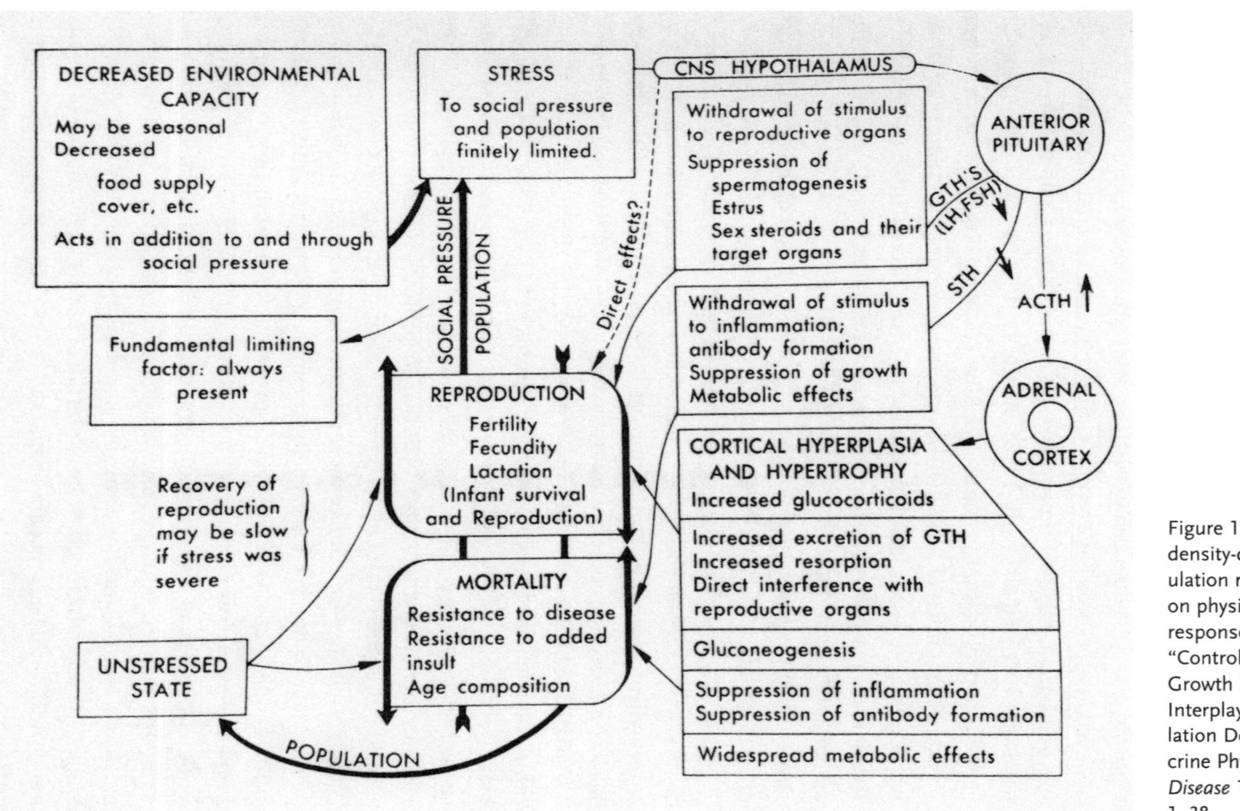

Figure 12. Model of density-dependent population regulation based on physiological stress responses. J. J. Christian, "Control of Population Growth in Rodents by Interplay between Population Density and Endocrine Physiology," *Wildlife Disease* 1 (January 1959): 1–38.

importance of the higher central nervous system and warmbloodedness in mammals in contrast to the lower animals."[53]

The significance of Christian's work was not lost on the organizers of the symposium. In his opening remarks, the anthropologist Carleton Coon raised practical implications for human population growth and the effects of stress in modern, urban life. According to Coon, Christian had opened a new approach to the study of evolution by proposing a novel, non-Malthusian form of natural selection.[54] Survival and reproductive success often went hand in hand, but when there was insufficient energy available to support both, reproduction was sacrificed. This happened when population size outstripped resources, but in Christian's model a behavioral-endocrine feedback mechanism began to shut down reproduction in subordinate individuals well before starvation occurred. This did not occur because of any voluntary sacrifice, but rather through the hormonal side effects of fighting and other stressful encounters with more-aggressive individuals in a crowded environment. Significantly, Christian's explanation did not rely on group selection, which Davis later characterized as a fruitless controversy and distraction for population ecologists.[55]

When pressed on the implications of his work for human populations, Christian was reticent to generalize too broadly from studies of rodents. In a televised interview with the urban architect Ian McHarg, Christian acknowledged that humans shared the neuroendocrine system of other mammals and were equally susceptible to deleterious effects of stress in daily life.[56] Nonetheless, physiological studies of stress in human populations remained rudimentary. Significantly, the likely problem for the future of human population was continued growth, rather than decline. Internal regulation might curtail this population growth at some future point, but Christian found already-existing megalopolises connected by freeways both appalling and traumatic. Nonetheless, despite McHarg's constant prodding, Christian cautioned that his personal feelings about the urban life of humans reflected the perspective of an "unscientist," rather than an expert.

Davis popularized the tight linkages among organismal behavior, physiology, and population dynamics a few years later in a short book appropriately titled *Integral Animal Behavior*.[57] The didactic framework of the textbook highlighted ideas that were more implicit in the technical papers of Christian and Davis. The two scientists made liberal use of negative feedback control systems in population regulation, but in his textbook Davis explicitly explained this self-regulatory mechanism as a form of homeostasis working at both individual and population levels. The first half of the book, devoted to individual organisms, emphasized the integration of behavior and physiology in adapting organisms to the environments in which they lived. This section culminated in a chapter entitled "Emancipation from the Environment," which emphasized how both behavioral and physiological aspects of homeostasis provided freedom from environmental fluctuations but also control over the environment. Although this control had reached its peak in humans, Davis argued that even simpler organisms did not respond passively to the fluctuating environment. Some coral animals had domesticated algae, and this internal symbiosis provided a stable food supply. Indeed, the animals could not survive without their photosynthetic partners. Similarly, fungus-farming ants had domesticated a food supply in a complex way that rivaled human agriculture.[58] These themes of integration, emancipation, and control of the environment formed the foundation for the second half of the book, dealing with population biology.

In a section entitled "Population Survival," Davis imagined a completely unemancipated population totally at the mercy of the environment. Food supply and a variety of other environmental variables determined the size of such a population. Because these variables often acted independently, population size fluctuated randomly, sometimes resulting in local extinction. Well-documented cases, even in some mammals such as voles, provided evidence that such density-independent fluctuations actually occurred in nature. However, successful species reduced the likelihood of extinction both by the evolution of homeostatic mechanisms within individuals and density-dependent homeostasis that dampened fluctuations in population size. Significantly, Davis argued that both individual and population homeostasis were the result of the same physiological and behavioral mechanisms.

According to Davis, the simplest understanding of density-dependent regulation involved an idealized, particulate model analogous to atoms colliding in space. As a population grew, competitive encounters increased at a rate approximately equal to the square of the number of individuals. If each encounter led to the death of one or the other competitor, the effect at the population level would precisely approximate the logistic growth curve. Of course, competition did not usually lead to immediate death. The evolution of territoriality, hierarchies, submissive gestures, and other social behaviors reduced lethal encounters. These behaviors might "soften the impact of competition," but the ultimate effect on populations still approximated logistic growth. The very hormones that helped organisms respond adaptively to toxins, infections, and injuries also curtailed reproduction in chronically stressed individuals. According to Davis, "behavior thus becomes a means of population homeostasis, regulating the population toward an equilibrium with its environment. In this situation the population maintains close adjustment to the resources, and the boom and bust sequence of numbers rarely occurs."[59]

In 1964 Christian and Davis coauthored a review article in *Science*, "Endocrines, Behavior, and Population," in which they laid out their argument for the integration of hormones and social behavior in regulating mammalian populations to a broader scientific audience, while distinguishing it from other theories of density-dependent regulation, particularly ideas advanced by Charles Elton's student Dennis Chitty.[60] Chitty took the stress hypothesis seriously, and Christian's 1950 paper effectively "scooped" his own early thoughts on applying Selye's ideas to population regulation.[61] But Chitty's explanation was less overtly physiological and less aligned with mathematical theory than the stress hypothesis of Christian and Davis. From Chitty's perspective population regulation involved natural selection favoring one set of characteristics at low population densities and another at high population densities. As a population increased and environmental resources decreased, selection would shift from favoring high reproductive ability to aggressive behavior. Although coming to the same general conclusions that Chitty did, Christian and Davis emphasized the "integration" of social behavior and physiology in individuals and the common elements of negative feedback at both the individual and population levels. The same neuroendocrine responses to environmental stresses mediated both physiological self-regulation in the individual and density-dependent regulation of population size. Self-regulation had evolved from simple beginnings in less complex organisms to the intricate homeostatic controls found in mammals that largely emancipated them from fluctuations in temperature and other environmental variables. Similarly, self-regulation of population size, which was dependent on well-developed endocrine and central nervous systems, had reached a peak of efficiency in mammalian species. Both organisms and populations had evolved greater independence from the vicissitudes of the environment.

This idea of population homeostasis maintained by social behavior and stress hormones was one of the "beautiful hypotheses" undermined by "ugly facts" that Dennis Chitty described so vividly in his memoir of population ecology. Other prominent population biologists such as Williams and Lewontin simply dismissed negative feedback models of population regulation as unscientific "aesthetics" that did not rise to the status of hypotheses. A major difficulty for Davis and Christian was convincingly demonstrating elevated stress hormones in members of natural populations, which were always much less crowded than the caged mice that Christian studied in the laboratory. Data collected from natural populations of rats studied by Davis and Christian in Baltimore were suggestive, rather than conclusive. For critics, the stress hypothesis was an artifact generated by the artificially crowded conditions of the laboratory, unlikely to apply to most natural populations.

Chitty and Davis continued to skirmish after they retired, but Chitty's student Charles Krebs took a more encompassing perspective by acknowledging the important positive contributions that both earlier ecologists had made to a broader understanding of population biology and social behavior.[62] Christian and Davis encouraged links to endocrinology and stress physiology, while Chitty's emphasis on natural selection encouraged greater contact between population geneticists and ecologists. From Krebs's perspective, continuing developments in these neighboring fields not only kept both approaches alive but also encouraged attempts to bring all of them closer together. In his efforts to accomplish this, Krebs completely ignored earlier controversies over group selection, focusing instead on how stress affected individual fitness, sometimes over several generations through maternal effects. Nor was he particularly interested in framing his results within the formal theoretical frameworks of cybernetics, although he used general ideas of self-regulation and feedback to discuss individual stress physiology and fluctuations in population size. One might consider the interplay of stress and adrenal hormones to form a feedback loop within an individual, but from Krebs's perspective that did not necessitate—or even encourage—thinking of populations as homeostatic or cybernetic. Krebs emphasized that physiological stress resulted from a range of environmental factors other than crowding, including the threat of predation, disease, limited food, and weather. This complexity implied that regulation of size was not simply a population phenomenon but occurred within the broader framework of the community and ecosystem. Half a century earlier, Hutchinson suggested that population regulation and ecosystem function might be unified by formal analogy involving mathematical equations describing circular causality. Such theoretical unification continued to be enticing, if highly problematic. As we shall see, discussions of homeostatic ecosystems faced some of the same ambiguities of self-regulating populations.

11

Symbiosis and Coadaptation in Homeostatic Ecosystems

Writing in the *American Journal of Medical Sciences* in 1964, the microbiologist René Dubos described the digestive tract as an ecosystem.[1] He based this somewhat unusual claim on research that he was conducting with his postdoctoral fellow Russell W. Schaedler at the Rockefeller Institute. The scientists reported that bacteria and other microorganisms not only inhabited the large intestine but thrived in all parts of the digestive tract, including the mouth, esophagus, stomach, and small intestine. Indeed, the digestive tract was not a single ecosystem; each organ had its own distinctive "microbiota." These microorganisms were not simply contaminants or a random assortment of species. They were integral constituents of the host, interacting with one another and with the animal body of which they were a part. Indeed, some resident bacteria adhered so tightly to the gut wall as to be almost part of the anatomy, and quite possibly the physiology, of the organ. This indigenous microbial "flora" included autochthonous bacteria that were truly symbiotic partners with the body, other bacteria that were ubiquitous in the external environment but only transient inhabitants of the gut and not integral parts of the microbiota, and opportunistic bacteria that flourished only under unusual conditions or situations in which the internal environment was disturbed. This last group was sometimes responsible for disease, but usually its members constituted what Dubos termed "persisters" unable to successfully compete with the better-adapted autochthonous microbes.[2]

The way that Dubos combined ecological, physiological, and medical metaphors for describing the microbial inhabitants of the digestive tract reflected his eclectic training and biological interests but also the ambiguity of the complex living system that he began exploring in the early 1960s. Today researchers who study the so-called microbiome continue to describe it as an ecosystem, an organ within an organ, or even an organ in its own right. As such this complex combination of microbial and host tissue is a site of homeostasis, adaptation, coevolution, and a variety of shifting symbioses.

Dubos claimed that medical researchers were largely unaware of the wealth of bacterial diversity because most gut microbes were difficult to culture in the laboratory. The well-known coliform bacteria such as *Escherichia coli*, generally considered typical intestinal inhabitants, were, in fact, only minor constituents of the natural microbiota of most mammals. Dubos and Schaedler's careful enumeration rested on arduous methods of culturing bacteria under anaerobic conditions that they had perfected during the preceding decade.[3] Dubos and Schaedler also supported the claim about the gut ecosystem by pointing to their research conducted on germfree mice. Raised from birth in sterile enclosures devoid of autochthonous symbionts, the laboratory animals developed abnormally and exhibited a range of anatomical and physiological deformities. The researchers were able to reverse some of these pathologies by introducing autochthonous gut microorganisms. Treating animals with antibiotics for extended periods also had adverse effects on the autochthonous bacterial community and sometimes opened the way for opportunistic pathogens to gain a foothold in an ecosystem to which they were otherwise poorly adapted.

The richly interactive microbial worlds that Dubos described highlighted both the importance and complexity of symbiosis, homeostasis, and adaptation within a medical and ecological setting. The human digestive tract constituted the environment within which various populations of microorganisms interacted, but the microorganisms also constituted an important part of the environment within which humans had evolved. Writing in *Man Adapting*, his most extensive treatment of the subject, Dubos stated that "the microbiota is part of the environment to which man has had to become adapted."[4] The symbiosis between microorganisms and their host resulted in a complex homeostatic balance that was the primary determinant of health. Pathogenic bacteria might cause disease, but only within a broader ecological context in which evolutionary equilibrium between host and microbiota was disturbed in a maladaptive way.

The general themes of symbiosis, homeostatic balance, and adaptation that Dubos highlighted in his discussion of the mammalian digestive tract paralleled similar developments in ecosystem ecology, more broadly. The idea that species interacting within the context of a physical environment constituted a self-regulating system was widely influential. Introduced by the British ecologist Sir Arthur Tansley during the 1930s, the ecosystem concept gained substantial support after World War II, and the study of ecosystems was an established specialty by the early 1960s when Dubos wrote his article.[5] Yet Dubos kept this specialty at arm's length even as he generalized his microbial research to broader philosophical questions of health in humans and nature.[6] Despite early publications in ecological journals and his later prominence as a public intellectual and environmental spokesperson, Dubos remained professionally tied to medical research, and his work existed on the periphery of professional ecology. Despite his prominence and the ecological implications of his research, the ecosystems of the gut that Dubos studied rarely appeared in ecology textbooks.

The Background of Dubos's Microbial Ecology

Dubos was born in 1901 in a small village north of Paris, where his parents ran the local butcher shop.[7] He contracted rheumatic fever as a child, which kept him out of school for a year and caused permanent damage to his heart valves. He suffered another attack of the disease when he was eighteen, preventing him from taking the entrance examination for the École de Physique et Chimie. When he recovered, Dubos successfully completed the entrance examination in economics to gain admission to the Institute National Agronomique. After completing his studies, he worked as an editor at the International Institute of Agriculture in Rome. His duties including abstracting journal articles, which introduced him to research conducted by the microbiologist Serge Winogradsky. Although the two men never met, Dubos claimed that Winogradsky's insistence on studying soil bacteria in their natural environments, rather than in artificial laboratory cultures, had a profound effect on Dubos's developing interests in microbiology.[8] During this period he also met Selman Waksman at an agricultural conference. This meeting led to an offer of a graduate fellowship to work in the prominent microbiologist's laboratory at Rutgers University, where Dubos eventually earned a PhD for work on the decomposition of cellulose by soil microbes. Extending procedures previously developed by Waksman and his students, Dubos compared the digestion of cellulose by soil fungi, actinomycetes, and both aerobic and anaerobic bacteria, under a variety of environmental conditions. He published the results of this early research in the journal *Ecology*.[9]

During a visit to the Rockefeller Institute, Dubos met Oswald Avery, and the two scientists were immediately attracted to each other's research. They spent an afternoon discussing Dubos's work in soil microbiology and Avery's interests in the role of the bacterial capsule in promoting virulence in pneumococci. The meeting led to an offer from Avery to join his laboratory team, although at a much smaller salary than for competing positions that Dubos was pursuing. Despite the low pay, Dubos accepted the position at the Rockefeller Institute, where he spent most of his career. His early work with Avery led to the discovery of a bacterial enzyme that could degrade the protective capsule of pneumococci. Interestingly, soil bacteria synthesized the enzyme only when the capsular material was the only source of carbohydrate provided in the culture medium. Dubos described this phenomenon as "adaptive enzyme formation" and later considered it his most intellectually satisfying discovery.[10] Continued research with soil microorganisms, particularly the bacterium *Bacillus brevis*, led to the discovery of gramicidin, which found limited clinical application as a topical antibiotic, as well as other antibiotic substances. The research brought both professional and popular recognition to Dubos as one of the pioneers of antibiotic research. It also secured him a permanent position as a member of the Rockefeller Institute, as well as election to the National Academy of Sciences.

Despite his early success in discovering antimicrobial substances, Dubos shifted his attention away from this line of research. Partly this was due to his intellectual curiosity and his conviction that pharmaceutical laboratories were better suited for drug development than the Rockefeller Institute. Presciently, he also recognized that bacteria would likely adapt to antibiotics and that resistant strains of bacteria would evolve in response to the spread of antibiotic therapy. His skepticism about antibiotics reflected his evolving ideas both about the nature of disease and pervasive role of adaptation and symbiosis in the natural world. Antibiotics were not magic bullets in a war against infectious disease, and Dubos studiously avoided military metaphors in discussing the relationship between bacteria and humans. Rather, disease was an inevitable result of an unstable, constantly shifting balance among a host, its indigenous microbiota, and potentially pathogenic microorganisms. Where this balance rested at a given time depended as much on environmental influences as the innate characteristics of pathogens. Under most conditions, host and parasite maintained an equilibrium, as evidenced by the fact that apparently healthy humans often harbored pathogenic bacteria. Any one of a number of endogenous or exogenous disturbances could upset this precarious balance, although according to Dubos, "Infection is the rule, and disease is the exception."[11]

This expansive ecological perspective, which led to Dubos identifying the digestive tract as an ecosystem, owed a debt to multiple intellectual influences. His early embrace of Winogradsky's microbial ecology and his PhD research under Selman Waksman provided obvious foundations for Dubos's later thinking about organisms and environments. Equally important, as Mark Honigsbaum cogently argues, was Dubos's post–World War II research on tuberculosis and the ability of the mycobacterium to persist inside host cells even after recovery from the disease.[12] This recognition of the intricate interactions of hosts and parasites, as well as the complexities of internal and external environmental factors in both health and disease deeply affected Dubos's thinking. Honigsbaum emphasizes Dubos's ambivalence toward professional ecology; nonetheless, Dubos interacted closely with other medical researchers who turned to ecology for insights into disease. Dubos's skepticism about the long-term efficacy of antibiotic therapy also reflected an ecological and evolutionary emphasis on the pervasive roles of variation and adaptation in the bacterial world. His views in this regard were shaped both by the groundbreaking work on bacterial transformation that Oswald Avery, Colin MacLeod, and Maclyn McCarty were pursuing at the Rockefeller Institute and by Dubos's own early work on antibacterial enzymes and antibiotics in Avery's lab.

Dubos synthesized his early laboratory experiences within the broader context of microbiological research, both historical and contemporary, in his monograph *The Bacterial Cell* (1945).[13] As he noted in the preface, the book owed much to Avery's influence, and Dubos celebrated ideas circulating in the Rockefeller Institute during the late 1930s. The book also presented his own nascent perspective on the bacterial world that he would develop during the following decades. In a chapter on variation, Dubos contrasted the discredited nineteenth-century theory of pleomorphism, which claimed that diverse bacteria were interchangeable manifestations of one or a few types, with monomorphism, the widely accepted view that bacteria formed distinct species comparable to other forms of life.[14] Although he held out hope for a phylogenetic classification of microbes, Dubos complained that monomorphism had become a dogma, blindly accepted particularly among medical microbiologists. The constancy of characteristics that these microbiologists stressed were a misleading result of pure culture methods and other standardized laboratory techniques. Laboratory workers had essentially domesticated bacteria and "trained" them to act in standardized ways that constrained the broader variability found in natural environments. According to Dubos, bacteria were capable of quickly adapting to new environmental conditions, sometimes changing seemingly fundamental biochemical characteristics.

Dubos acknowledged that in the past this plasticity had misled microbiologists to invoke Lamarckian explanations for bacterial adaptation.[15] He attempted a more orthodox classification of causes based on genetic and Darwinian concepts, although the lack of sexual reproduction in bacteria precluded the Mendelian recombination of genes found in plants and animals. Some adaptive changes in bacteria involved strictly phenotypic acclimation to environmental changes such as salt concentration, temperature, or oxygen levels. In contrast to these transient, reversible changes, other cases involved modifications transmitted from generation to generation. Environmental changes sometimes activated adaptive enzymes to catalyze metabolic reactions not typically part of the repertoire of a particular bacterium. For example, *Escherichia coli* normally did not metabolize lactose but occasionally became capable of digesting this food source when cultured in media enriched with the sugar. This did not happen immediately, and it occurred only in a few colonies, suggesting either the occurrence of a mutation or selection for a rare variant already present in a mixed culture. Dubos's own research on capsule-destroying enzymes in soil bacteria also pointed to the importance of such adaptive enzymes in nature. The work done by Avery, MacLeod, and McCarty demonstrating the likely role of deoxyribonucleic acid in the transformation of nonvirulent to virulent *Streptococci* suggested one possible mechanism for how bacterial adaptation might occur. For Dubos, however, pathogenicity was less a genetic characteristic of particular bacteria than a symbiotic interaction between host and parasite.[16] The difference between a harmless infection and disease lay in the balance between the host and microbe, crucially mediated by environmental conditions affecting both partners.

Ecosystem Research in the Mammalian Digestive Tract

Studying the digestive tract as an ecosystem depended heavily on three important technical innovations: large-scale anaerobic culturing methods, the development of a unique strain of laboratory mice, and the use of germfree animals as a comparative tool for investigating bacterial symbiosis. This broad research agenda rapidly expanded beginning in the late 1950s when Russell W. Schaedler joined Dubos's lab. Schaedler had recently earned an MD from the Thomas Jefferson School of Medicine in Philadelphia. His technical skill with new microbiological methods complemented Dubos's broad intellectual vision of microbes interacting in the ecosystem of the gut.[17]

According to Dubos and Schaedler, medical microbiologists overlooked the diversity of the microbiota by focusing on *E. coli* and other intestinal bacteria

Figure 13. Anaerobic incubator constructed from milk cans and used in Dubos's laboratory at the Rockefeller Institute. Each can could hold over one hundred petri dishes. Russell W. Schaedler, René Dubos, and Richard Costello, "The Development of the Bacterial Flora in the Gastrointestinal Tract of Mice." *Journal of Experimental Medicine* 122 (1965): 59–66.

that could be easily cultured aerobically. Inoculated on petri plates with nutrient agar, these bacteria grew so quickly that they overwhelmed other microbes. However, because the environment of the digestive tract was primarily anaerobic, the so-called coliform bacteria might not be representative of the indigenous microflora of the gut. To determine the relative numbers of various bacteria from the gastrointestinal tract, the researchers used selective media that inhibited coliform growth and grew the mixed cultures of fecal bacteria under oxygen-free conditions mimicking the gut. An anaerobic incubator constructed from a set of stainless steel milking cans connected to a vacuum pump and gas cylinders was capable of culturing several hundred petri plates under an inert atmosphere of nitrogen and carbon dioxide (figure 13).[18] This apparatus was state of the art in the early 1960s, and it allowed Dubos and Schaedler to quantify populations of different bacteria from various parts of the digestive tract and throughout development of mice from newborn to adult.[19] Although they expressed differences as orders of magnitude, the results convincingly demonstrated that lactobacilli and members of the *Bacteroides* vastly outnumbered *E.coli* and other coliforms. These populations were neither uniform nor static. The microbiota developed in a predictable fashion reminiscent of ecological succession described by plant ecologists, and, at any given stage, the digestive tract constituted a variety of microenvironments, each with its own distinctive assemblage of bacteria.

The development of a new strain of laboratory mice at the Rockefeller Institute proved crucial for the ecological studies of the microbiota pursued by Dubos, Schaedler, and other members of the lab. Derived from a long-established strain of Swiss mice originally imported from Europe, the New Colony Swiss (NCS) mice had a number of highly desirable characteristics for microbiological studies.[20] They were free from several important pathogens common in mice. They also apparently harbored no *E. coli* in their intestines; at least the bacteria were undetectable in the feces. The animals proved highly resistant to bacterial endotoxins that were often fatal when administered to other mice. The NCS mice were also capable of growing on nutrient-deficient diets missing key amino acids required in the mouse diet. This suggested to Dubos that the unusual microbiota of NCS mice included unique symbionts capable of synthesizing the essential amino acids for their hosts. This claim found some support from experiments in NCS mice treated with antibiotics, which resulted in an alteration in the species of bacteria recovered from feces and a loss of weight in the mice. Presumably, the antibiotic treatment adversely affected amino-acid-synthesizing symbionts.[21]

The NCS mice were particularly useful when grown in germfree environments. The researchers antiseptically removed fetuses by Caesarian section and raised them in sterile plastic enclosures. Comparisons with mice raised under normal laboratory conditions documented developmental deficiencies and anatomical changes in the digestive tracts of germfree mice. The scientists could reverse some of these pathological conditions by exposing the germfree mice to autochthonous bacteria. This was sometimes accomplished through contact with feces, but in other cases, the mice were administered "Schaedler cocktails" composed of various combinations of bacteria derived from pure cultures.[22] These bacteria quickly and selectively colonized different parts of the digestive tract, often following successional changes similar to those during normal development.

In his historical study of gnotobiotics, Robert Kirk carefully documents the widespread interest in germfree animals in medicine, agriculture, and industry after World War II.[23] Championed by a small group of somewhat eccentric pioneers working on the fringes of academic biology, the goal of creating completely sterile environments led to the construction of a variety of grandiose steel and glass environments but also to smaller plastic enclosures suitable for more-traditional laboratory research. The assumption that bacteria were harmful and that germfree life could lead to better health motivated this research. As Kirk relates, even some critics of the project tacitly accepted the notion of a "war" against microbes but feared that germfree life would lead to weakened immune systems vulnerable to accidental exposure to a pathogen. Dubos's biographer Carol Moberg emphasizes that the Rockefeller microbiologist avoided such military metaphors and that his perspective on the microbial world was distinctly different from those who advocated eradicating bacteria, whether by antibiotics or germfree enclosures.[24] Indeed, experiments conducted in Dubos's lab provided compelling evidence that germfree mice were less healthy than their counterparts with intact microbiota.[25] Indigenous microbes likely provided key nutrients and protected their hosts from disease, whereas germfree animals required exotic, synthetic diets and protection from potential pathogens to survive.

The Environmental Context of Health and Disease

Dubos's research on the intestinal microbiota closely coincided with the publications of his philosophical explorations of health, disease, and the environment: *The Mirage of Health* (1959) and *Man Adapting* (1965). His broadly ecological and evolutionary perspective permeated both of these widely influential books, and he referred liberally to his research on the symbiotic relationships between humans and microbes. Dubos emphasized that health was more than an absence of disease, and he contrasted two opposing schools of thought in medicine, each of which had deep historical roots. The dominant position in modern medicine reflected what Dubos referred to as the myth of Asclepius, which emphasized the role of the physician as healer. The rise of scientific medicine, powerfully symbolized by the germ theory of disease and the discovery of antibiotics, made this the primary foundation for modern medicine. The opposing myth of Hygiea viewed health as a natural state dependent upon a balance of conditions and the physician's role as restoring this balance rather than treating symptoms. Restoring health meant more than simply curing disease, and it required more than the application of scientific methods: "Solving specific problems of disease by discovering causes and remedial procedures is within the grasp of the scientific method, but curing disease is not the same thing as ensuring health and happiness. This requires wisdom and vision that transcends knowledge of treatments and remedies. This problem cannot be solved by medical science or by a naïve return to nature."[26]

Dubos presented Asclepius and Hygiea as polar opposites, and he was critical of both. The search for magic bullets in modern medicine faced the daunting challenge of bacterial resistance to antibiotics, not to mention the difficulties of successfully applying the military metaphor to cancer and mental illness.[27] Although more closely allied with Hygiea, Dubos dismissed naive beliefs in a static balance of nature to which humans might return. Such notions of harmonious equilibrium were a form of Platonic idealism that lacked the "flesh and blood" of real life.[28] Human health and happiness meant adaptation to a constantly changing environment, implying both resiliency and adaptability. By extension, Dubos's perspective on the dynamic interaction between humans and the ecosystems in which they lived provided the basis for environmental health that did not rest on a return to a pristine natural state but rather on a more interactive and creative relationship.

In *Mirage of Health*, Dubos included a chapter subsection titled "Health as Adaptation" that reflected in broad strokes ideas derived from both Darwinian evolutionary thinking and the physiological tradition of Bernard and Cannon.[29] For Dubos, health and happiness were not absolutes but manifestations of a relative biological fitness, entailing constant adaptation to a fluctuating environment. To a certain extent, this adaptation involved populations tracking changes in the external environment through natural selection for alternative genes. Dubos was acutely conscious of the heavy hand of evolutionary history that both shaped and constrained a population's ability to respond to current environmental conditions. According to Dubos, the human form had been "molded and chiseled" by interactions with the physical environment not only on the species level but also in numerous populations adapted to nearly every environment on the Earth.[30] Just as impressive, however, was the resiliency and adaptability of individuals, which allowed them to meet environmental challenges by acclimation and other transient, reversible changes to the phenotype. Indeed, the ability to maintain the internal environment within narrow limits was the determining factor in maintaining individuality and consciousness. For Dubos, Bernard's "visionary guess" about the interplay between the internal and external environment, at all levels of complexity, was a necessary supplement to Darwinian evolutionary adaptation.[31]

In *Man Adapting* (1965) and his later writings, Dubos continued to equate homeostasis with adaptation, although he cast a more critical appraisal on the ambiguities surrounding the two concepts.[32] Adaptation, according to Dubos, was a "treacherous term" that had a wide variety of meanings in biology, medicine, and the social sciences. Darwinian fitness provided an important, but incomplete, definition of adaptation. Homeostasis, acclimation, accommodation to disease, and other physiological responses were also necessary aspects of a more inclusive description of how organisms reacted to environmental change. Particularly when considering humans, social and cultural responses were often more important than purely biological adaptation.

Homeostasis was an important form of adaption, but maintaining internal constancy was an ideal, rarely attained by organisms in the real world. Echoing criticisms raised earlier by his friend the Nobel laureate Dickinson Richards, Dubos described homeostasis as a conservative, static, physiological perspective, reflecting a late Victorian worldview that had influenced Cannon and other followers of Bernard in England and the United States. The concept was overly mechanical and emphasized deterministic reactions to stimuli, rather than more creative, adaptive responses to the environment. Norbert Wiener and other systems thinkers reinforced this earlier mechanical, reactive element of Cannon's thinking in the later cybernetic concept of negative feedback. Like Richards, Dubos criticized Cannon for largely ignoring disease, particularly degenerative diseases, in *The Wisdom of the Body*. In some cases, such as immune responses or the development of scar tissue, processes that might be homeostatic in the short term later led to chronic disorders and disability. Although explicable in Darwinian terms of maximizing fitness, such short-term homeostatic reactions did not adequately capture the full range of creative responses to environmental perturbations. In many cases humans could convert passive homeostatic responses into creative processes, including shaping the environment through deliberate action. Although Dubos continued to discuss homeostasis throughout his career, he sometimes suggested that the concept should be replaced by a more flexible and dynamic "homeokinesis."[33]

The creativity that Dubos emphasized in human responses to the environment also applied more broadly to other organisms, populations, and ecosystems. Symbiosis, which played such an important role in Dubos's laboratory studies of the gut microbiota, applied more generally throughout the natural world.[34] Symbiotic partners often formed mutually beneficial relationships, sometimes to the extreme of totally depending upon one another for survival. Yet these partnerships were the result of long histories of coevolution, and in almost all cases, an element of conflict remained. It was therefore misleading to categorize symbiotic relationships into unambiguous categories of parasitism, competition, mutualism, or commensalism. Symbiotic relationships such as those found in the gut existed in a tenuous balance, with the difference between health and disease often depending upon environmental factors influencing both human host and various bacterial populations.

In his nuanced, Darwinian interpretation of symbiosis, Dubos continued to emphasize creativity. In some cases, symbiotic partners influenced each other to such a degree that the combined association took on the appearance and function of a new organism. Lichens, the symbiosis of fungi with algae or cyanobacteria, was a case in point. Often the partners were distinctly different from their free-living relatives and were difficult to culture separately. The lichen combination could thrive in environments inhospitable to either partner. The combined physiologies of the partners often profoundly altered the environments in which they lived, opening the way for ecological succession. Yet, despite the fact that lichens seemed to be organisms of a higher order, Dubos was skeptical that the relationships between the partners were necessarily mutualistic. One or the other partner might well be more or less dependent upon the relationship. Like the situation in which a gut microbe could be pathogenic, harmless, or beneficial depending upon environmental factors such as nutrition or antibiotic treatment, the partners making up a lichen maintained an uneasy balance forged by both evolutionary and ecological factors. Given that indeterminism, homeostasis required a more flexible definition than maintaining internal constancy.

Symbiosis became the central concept that informed Dubos's ecology and environmentalism. From this perspective, humans formed a partnership with nature. Although human activity often led to destruction of natural areas, it could also lead to creative improvement. The resilience of natural ecosystems impressed Dubos, but so did how these systems could be modified in mutually beneficial ways.[35] Ecosystems were highly resilient entities having enormous powers of recovery and self-healing analogous to the homeostatic processes in organisms. Abandoned farmlands and degraded waterways often returned to their original states when allowed to do so. Human creativity, appropriately applied, could modify ecosystems in ways that were economically productive and aesthetically pleasing, while still maintaining the ecological integrity of the original natural assemblage. The metaphor of domestication that so shaped Dubos's perspective on the microbial world applied equally to the macroscopic world. Critics complained that this approach to environmentalism was overly anthropocentric, but in Dubos's worldview, humans intimately interacted with other living organisms, both the microbiota that flourished internally and the external biota of broader ecosystems. Domestication was not so much a matter of control by humans as it was codetermination. For both humans and nature, balance and homeostatic regulation, tenuous as they might be, were the primary determinants of health.

Despite being both ecological and evolutionary, Dubos's views on symbiosis and the microbiota stood uneasily with the contemporary evolutionary ecology that was gaining ascendancy during the 1960s. Prominent historians, philosophers, and biologists have emphasized how competition and individual selection became the dominant foci of evolutionary debates during this period. Taking this individualistic perspective, orthodox evolutionary biologists often dismissed symbiosis, particularly as a creative evolutionary process.[36] Nevertheless, Dubos took a more Darwinian attitude toward symbiosis than might seem apparent. Mutualism did not develop in a linear fashion from competition or parasitism. The relationship between symbiotic partners was always a tenuous and shifting balance that could move back and forth among competition, parasitism, and mutualism depending upon exigencies of the environment or changing selective pressures. All of these processes occurred simultaneously within the digestive tract. Potentially harmful bacteria ("persisters") could barely get a toehold in the ecosystem of the gut because they were poor competitors, but this balance could shift in response to the overuse of antibiotics or other disturbances that eliminated mutualistic bacteria helpful in maintaining homeostasis. From this perspective health and disease were not clearly defined alternatives but formed broad areas on a continuum of states only partly determined by biology. At least in the case of humans, social conditions and culture, as well as individual choice and creativity, heavily mediated these broad existential categories.

Dubos's views on symbiosis, adaptation, and homeostasis were antiutopian but avoided the overt pessimism of Richards's medical physiology or Homer Smith's evolutionary physiology. Species, particularly humans, were not trapped by pathology or evolutionary dead ends but were resilient, adaptable, and creative in their responses to environmental changes. Species interactions, although never free from conflict, enriched the scope of these creative responses. For Dubos, the lesson learned from lichens or the microbiota of the digestive tract had less to do with shared nutrition than with the emergence of entirely new characteristics possessed by neither partner individually. In lichens this included the production of unique pigments, organic compounds, and fruiting bodies not found in either fungi or algae. In mammals symbiosis with the microbiota suggested the possibility of shared regulation of development and morphogenesis resulting from the intimate interaction of bacteria and host. In the short term, these examples of symbiosis involved adaptive homeostasis but also implied a coevolutionary process that led to the emergence of novelty rather than constancy. The intimate relationships among symbionts challenged the idea that organisms exist as autonomous individuals by claiming that symbiotic partners together formed entirely new forms of life. Such partnerships were not merely biological curiosities but a pervasive and fundamental characteristic of the natural world. Emphasizing metaphors of gardening and domestication rather than competition and warfare, Dubos drew far-ranging and optimistic lessons about the possibility of creative, symbiotic relationships between humans and the rest of nature.[37]

The Holistic Ecosystem: More than the Sum of the Parts

During the second half of the twentieth century, ecosystem ecology flourished around the core concepts of energy flow and biogeochemical cycling. Yet symbiosis and homeostasis, which so interested Dubos, were also central to the broader understanding of how ecosystems function. Eugene Odum, the dean of ecosystem ecologists in the United States, used the "lichen model" as a prominent feature of his thinking about ecosystems.[38] Lichens were the quintessential example of close interactions among species that characterized all ecosystems. They also seemed to exemplify a self-organizing tendency that characterized the development of mature ecosystems. The partnership between fungus and algae that resulted in a new type of composite organism served as a metaphor for Odum's holistic claim that in a mature ecosystem the whole is greater than the sum of its parts. In contrast to Dubos, who stressed indeterminacy in symbiosis and acknowledged the unstable balance struck between fungus and algae, Odum argued that lichens had evolved from a primitive parasitic condition to a more advanced state of mutualism. This progressive evolution stabilized the union, leading to greater homeostasis both in the combined organism and in the ecosystem of which it was a part. By liberating organic acids that helped to break down rocks to form soil, lichens often acted as pioneer species that prepared barren areas for the establishment of plant species during ecological succession. This developmental process led to increasing stability and self-regulation within maturing ecosystems. Indeed, for Odum, stability drove evolution in the sense that more stable systems tended to replace less-stable systems. Sometimes Odum invoked natural selection to explain this process, but in his more contentious moments, he denied that "chaotic" Darwinian selection could account for this progressive evolutionary trend. Like Dubos, Odum also used the mutually beneficial relationship found in lichens as a metaphor for a partnership between humans and the natural world. Dubos turned to domestication and gardening as ways to think about this partnership, and he avoided mechanistic explanations involving feedback or other cybernetic ideas. By contrast, mechanistic engineering perspectives held a great appeal for Odum. During the early years of spaceflight it became quite common to compare the life support systems of the biosphere with artificial life support systems in spacecraft. This idea of Spaceship Earth was a powerful metaphor that intrigued Odum but repelled Dubos. Although both scientists were attracted to James Lovelock's Gaia hypothesis, Odum emphasized the mechanistic, homeostatic, and cybernetic aspects of Lovelock's thinking about the biosphere, while Dubos applauded Gaian phenomena as examples of creative evolution.[39] From Odum's perspective humans had much to learn from nature, but a healthy partnership between humans and nature depended heavily upon the guiding hand and enlightened problem-solving skills of professional ecologists.

Despite his prominence in ecosystem ecology, Odum's formal training was in ornithology and traditional physiological ecology. Working under the direction of Charles Kendeigh in the Zoology Department at the University of Illinois, Odum measured the heart rates of birds under a variety of natural and seminatural conditions. This early interest in avian physiology continued throughout Odum's career, even as he moved decisively toward studying communities and ecosystems. In a broader sense, Kendeigh encouraged holistic and organic perspectives in ecology that were influenced by Frederic Clements's ideas about ecological succession as a developmental process analogous to ontogeny. Clements's views were controversial and had been attacked by Henry Allan Gleason, who drew a more atomistic conclusion about plant communities as collections of independently adapted species. The Clements–Gleason controversy was a lively part of Odum's graduate school experience, but he was strongly drawn to physiological perspectives. In retrospect he recalled that moving from the study of the heart rate of birds to the broader physiology of ecosystems was a natural transition.[40]

Odum later emphasized the importance of homeostasis as a bridging concept applicable at multiple levels of organization. It was natural, he claimed, to move from homeostasis in a traditional physiological context like the birds that he studied to self-regulation of populations, communities, and ecosystems. He credited his own epiphany to reading Cannon's *Wisdom of the Body* as a graduate student and assigning the book in the physiology course that he taught as a young assistant professor at the University of Georgia.[41] Despite that claim, this broader use of homeostasis developed gradually and somewhat inconsistently in Odum's work. Although he occasionally referred to homeostasis in his articles on the physiological ecology of birds, he developed and implemented the concept more fully in his writing about ecosystems, particularly in his influential textbook *Fundamentals of Ecology*.[42] Especially in later editions of the book, he embedded the idea of self-regulation within a broad perspective on nature and society that was holistic, interactive, progressive, and cooperative.

Eugene Odum readily acknowledged that this broader philosophical perspective owed much to his father, the sociologist Howard Washington Odum, and to his younger brother, the ecologist Howard Thomas (H. T.) Odum. The interactions among the Odums were complex and mutually reinforcing. For example, while his father was encouraging Eugene's early professional development and providing a strong social dimension to Eugene's thinking about ecosystems, he was also incorporating his son's ecological ideas into his own sociology.[43] The partnership between Eugene and his brother H. T. was episodic but multifaceted. At times the relationship was closely cooperative and at other times intensely competitive, but the brothers' combination of talents turned out to be crucially important for the development of ecosystem ecology during the second half of the twentieth century.

Eugene Odum credited his father with introducing him to the idea of the functional integration of parts into a larger whole, which he routinely applied to both ecosystems and human societies. As he later wrote in *Fundamentals of Ecology*, "the concept that uniquely different cultural units function together as wholes is, of course, parallel to the ecologist's concept of the 'ecosystem.'"[44] The elder Odum was a leading proponent of "regionalism." Unlike the older sociological theory of sectionalism, based on divisiveness and conflict, regionalism emphasized cooperation, interdependence, and integration among the various parts of the United States.[45] Each region contributed to the national whole by bringing a unique set of natural resources, economic opportunities, social structures, and cultural characteristics. Progressive social evolution led to greater equilibrium and harmony in the organic whole of the nation.[46] The Great Depression challenged this assumption, and H. W. Odum supported the growing role of the federal government in planning and coordinating social and economic progress. However, he took pains to place the growth of government within the context of traditional democratic ideals and to present big government as a necessary part of the continuing evolution of American society. His views reflected the progressive liberalism of the New Deal, to which he was deeply committed. For Eugene, his father's progressive ideas associated with the New Deal became important guides to his own broad social goal of creating a "new South," as well as his scientific goal of creating a "new ecology."[47]

Just as important for stability and social integration, was H. W. Odum's progressive commitment to expert planning and the role of academic scholars as practical problem solvers. This commitment later imbued Eugene's efforts to professionalize ecology at the University of Georgia, both as an academic practice and as a solution to environmental problems. His experience as a graduate student reinforced and extended the ecological implications of his father's progressive sociology. At the University of Illinois, Odum came under the sway of Victor Shelford, an ecologist who was deeply committed to holistic approaches and who stressed the social role of ecologists in promoting conservation and environmental protection. There were obvious parallels between Shelford's idea of ecological regions (biomes) and the sociological regions of Odum's father. Odum later claimed that Shelford's biome concept was a forerunner of the ecosystem concept.[48] Odum believed that managing nature for human benefit was the primary social responsibility of professional ecologists, and he sometimes went so far as to argue that all ecology was fundamentally "human ecology."[49]

Eugene Odum's interest in ecosystems coincided with his younger brother's dissertation research on biogeochemical cycles at Yale University under the direction of G. Evelyn Hutchinson. The resulting collaboration between the Odum brothers during the 1950s was multifaceted and had a profound effect on the development of ecosystem ecology. Not only did Eugene learn about the Yale ecology program from his brother, but as he wrote *Fundamentals of Ecology*, he also began a correspondence directly with Hutchinson. In Eugene's eyes, the prominent Yale ecologist was an ally in creating a "new ecology" that would focus on the movement of energy and matter through ecosystems, as well as optimism about the role of professional ecology in solving environmental problems.[50] His younger brother's dissertation on the biogeochemistry of strontium was a key development in Eugene's ambitious plans. H. T. contributed to the first edition of his brother's textbook, and he wrote the chapters on biogeochemical cycling and ecosystem energetics for the second edition. Perhaps more importantly, H. T. developed the field methods for measuring energy flow that the two brothers used in a collaborative study on a coral reef at Eniwetok Atoll that won the Mercer Award from the Ecological Society of America and did much to highlight the prospects for ecosystem ecology.[51]

Although the primary focus of the Eniwetok study was energy flow through the coral reef ecosystem, the Odums also emphasized the role of symbiosis in maintaining stability and balance in the system. The colonial coral animals that constructed the reef housed photosynthetic green algae within their bodies. By providing products of photosynthesis to the coral animals in exchange for protective shelter, the algae contributed importantly to the metabolism of the entire reef. By housing their photosynthetic partners, the corals had become "part plant." Within the unusual food web of the reef, the corals acted both as herbivores living off the productivity of their resident algae and as carnivores by capturing free-living microorganisms. This filter feeding not only supplemented the caloric needs of the coral animals but also provided scarce nutrients such as phosphorus for the endosymbiotic algae.[52] Once ingested, the phosphorus cycled back and forth between algae and coral animal, providing an important mechanism for nutrient conservation.

Various other non-endosymbiotic algae inhabited the skeletal remains of past corals beneath the living surface of the reef. Because these photosynthetic organisms bored into the rock, some biologists claimed they were destructive, but the Odums argued that these algae also contributed to the metabolism of the reef. In addition, free-living calcareous red algae growing on areas not inhabited by coral animals were both photosynthetic producers and reef builders. The reef, taken as a whole, presented a paradigmatic example of a system that maintained a steady state equilibrium through the intricate self-regulating mechanisms of its interacting constituents. It was highly dynamic in the constant flux of energy and cycling of chemicals but at the same time exhibited great stability and balance. Studying such a system, the Odums claimed in their report, provided important lessons for human civilizations, which were not in steady state and whose relationships with nature fluctuated "erratically and dangerously."[53]

Casting equilibrium in terms of a steady state harmonized stability and balance with dynamic, but orderly, change. It provided an intellectual bond between the physiological idea of homeostasis, particularly in its cybernetic form of negative feedback, and the broader systems approaches that H. T. Odum was pioneering both in his theoretical writing and in simulations using analog computers. The mutually beneficial symbiotic interactions among species served as important components of the information networks that H. T. dubbed the "invisible wires of nature." For Eugene, homeostasis became a principle that operated at all levels of biological and social organization. As a characteristic of all mature systems, homeostasis both developed and evolved in a progressive, linear fashion. This increased homeostasis went hand in hand with the progressive evolution of "reciprocal adaptation" in mutualistic partnerships.[54]

Simpler systems also illustrated Eugene Odum's general claim that mutualism was pervasive in nature and that it tended to replace "negative interactions" such as competition and parasitism. Lichens were a case in point. Odum claimed that in primitive lichens the fungal partner captured and parasitized algae. In these associations, the fungal filaments or hyphae penetrated the cell walls of the algae and extracted the products of photosynthesis. However, in more advanced lichens, Odum believed, a true mutualistic partnership evolved, with the fungus providing inorganic nutrients and protection from desiccation, while the algae provided energy-rich sugars and other organic compounds. In these advanced lichens, the fungal hyphae surrounded the photosynthetic partners without penetrating the cell walls. This progressive evolutionary claim was controversial and unsubstantiated by later phylogenetic analysis, but it became a core principle for Odum's discussions of symbiosis, adaptation, and homeostasis during the 1950s and 1960s.[55]

In sharp contrast to the indeterminacy that characterized René Dubos's perspective on lichens and other symbiotic associations, Odum presented various forms of symbiosis as well-defined categories involving positive, negative, or neutral effects of one species on another. Odum also emphasized a progressive evolutionary trend from negative to positive interactions that was absent in Dubos's writing. This trend toward mutualism, which Odum found throughout the living world, was a necessary characteristic of all stable ecosystems. Odum claimed that Darwinian "survival of the fittest" overemphasized negative interactions such as predation and competition, at the expense of mutually beneficial symbiotic interactions. In his later writings, Odum sometimes took a more critical view of Darwinism, equating it with a chaotic worldview that was at odds with the stability that one encountered in simple systems like lichens, as well as complex ecosystems like the coral reef.[56] For Odum, evolution was a developmental process that led to increased integration, stability, efficiency, and resilience at all levels of biological organization from populations to ecosystems.[57] From this developmental perspective, immature and mature ecosystems exhibited sharply contrasting properties involving species diversity, energy transformation, nutrient cycling, and homeostasis. Mature ecosystems were not simply collections of well-adapted populations but were themselves well-adapted entities of a higher order. Odum's early experience on the coral reef at Eniwetok provided an enduring example of such a well-adapted system characterized by a richly homeostatic control of the environment by the diverse mutualistic interactions of its constituent species.

The Cybernetic Nature of Ecosystems

Ideas of symbiosis and homeostasis were congenial to Eugene Odum's broad physiological and ecological perspective, but cybernetics and other forms of systems thinking increasingly marked his later thinking about ecosystems. Admitting that he was not highly proficient in mathematics, Odum acknowledged an intellectual debt to his younger brother and to other systems ecologists.[58] Despite the derivative nature of these ideas, systems ecology served a variety of purposes in Eugene Odum's ecology. He used cybernetics for visualizing energy flow, thinking about stability in a mechanistic but nonreductionist way, discussing teleology and goal-directedness in the development of ecosystems, and suggesting possibilities for ecological engineering. Because of his prominence as the president of the Ecological Society of America, director of a growing institute of ecology at the University of Georgia, and a leading textbook writer, Eugene Odum was able to disseminate the basics of systems ecology to a broad audience in a nontechnical style.

Although H. T. Odum shared his older brother's early enthusiasm for birds, he was drawn to the physical sciences even before becoming an ecologist.[59] A boyhood interest in electronics presaged his later enthusiasm for analog computer simulation and using circuit diagrams to visualize ecosystem processes. During World War II, he served as a meteorologist in the US Army Air Force, an experience that he later recalled introducing him to thinking about the behavior of large, complex physical systems. His dissertation research on the global biogeochemical cycle of strontium, completed under G. Evelyn Hutchinson at Yale University, was hailed as one of the top twenty scientific discoveries at the annual meeting of the American Association for the Advancement of Science in 1950. He also accompanied Hutchinson to a Macy Conference on Circular Causal Systems, where he discussed his research with major figures in the new field of cybernetics. During his graduate studies, Odum read Alfred Lotka's *Elements of Physical Biology*, which he claimed had a profound impact on his systems thinking. Notably, Odum credited Lotka with providing the intellectual foundation for his claim that the global strontium cycle represented a stable system characterized by self-regulating mechanisms.[60]

H. T. Odum's study of the strontium cycle had a lasting impact on biogeochemistry, but his interests quickly shifted to ecosystem energetics when he took a faculty position at the University of Florida shortly before defending his dissertation in 1950.[61] Here he documented the annual energy budget for an aquatic ecosystem at Silver Springs. The warm springs created a relatively constant physical environment that Odum analogized to thermostatic and chemostatic control mechanisms. Supporting diverse communities of organisms, the springs constituted "a giant constant temperature laboratory" for experimental study. The young ecologist enthusiastically added, "Thus there exists a marvelous opportunity to study community metabolism and productivity in the ready-made natural laboratory in which whole communities can be studied under controlled conditions."[62] The large-scale, five-year study was historically significant on several levels. Methodologically, it involved developing techniques for measuring photosynthetic productivity in flowing water that the Odum brothers later used on Eniwetok Atoll. H. T. Odum also created a visual representation of energy flow through the ecosystem, made up of boxes representing various trophic levels connected by arrows representing inputs and outputs of energy. The diagrams shared a family resemblance with earlier food chain diagrams, but the decreasing sizes of the boxes and arrows highlighted the importance of the second law of thermodynamics in the one-way energy flow. Eugene Odum later referred to his brother's innovative box-and-arrow diagram as an "Odum device," and the visual representations became a standard feature in biology textbooks.[63] Finally, Silver Springs provided an early model for how to study large ecosystems by reducing their complexity to energy flow and nutrient cycling. For the rest of his career, H. T. Odum pursued such studies experimentally in a wide variety of environmental settings but also theoretically through his development of a highly original approach to formal systems ecology.[64]

The younger Odum began simulating ecosystem energetics using the electronic circuits of analog computers soon after arriving at the University of Florida. The use of computer modeling was itself innovative, but so too was H. T. Odum's creation of symbolic representations of ecosystem components loosely based on the pictograms used in electronic circuitry.[65] This visual language, which he referred to as "energese," grew out of, and largely replaced, his earlier box-and-arrow diagrams—at least in Odum's own work. The circuits in his diagrams represented energy flow but also the "invisible wires of nature" comprising a complex array of self-regulatory feedback loops.

For Eugene Odum, the modeling approaches that his brother was exploring as part of the broader development of mathematical systems ecology were a central feature of the "new ecology." Indeed, he sometimes equated the new ecology and systems ecology as synonyms for approaches committed to a holistic understanding of ecosystems.[66] Although basic principles such as the laws of thermodynamics applied at all levels of organization, from cells to ecosystems, explanations at higher levels were not reducible to the operation of their constituent parts. Central to this holistic perspective was the concept of homeostasis. Self-regulation in organisms often involved centralized control by nerves and hormones, but according to Eugene Odum, ecosystems involved diffuse homeostatic regulation composed of a complex, interlocking, causal network. This self-regulation was cybernetic because it involved negative feedback, but it was not comparable to the thermostat with a set point, often used as a mechanical analogy for homeostasis. The more holistic "invisible wires of nature" symbolized by the looping arrows in his brother's circuit diagrams constituted a highly simplified representation of the complexly interlocking symbiotic interactions among organisms. Taken together, these processes comprised an "information network" that overlay the primary network of energy flow and material cycling. This diffuse information network brought stability and balance to the ecosystem analogous to homeostasis in organisms but required neither constancy nor centralized control.

Are Ecosystems Truly Cybernetic?

Eugene Odum's ideas about ecosystems and ecology were broadly representative of major themes in post–World War II ecology, but they were not universally accepted. The idea that ecosystems are cybernetic was criticized both by evolutionary ecologists who argued against group selection and by physiologists who denied a meaningful analogy between ecological feedback and homeostasis. Writing in the *American Naturalist*, the biophysicist Joseph Engelberg and physiologist Louis Boyarsky claimed that supposed examples of feedback control in ecological systems always involved "brute force" interactions such as predation.[67] In contrast, truly cybernetic systems, whether living or nonliving, always involved control of high-energy processes by low-energy signals. Homeostatic regulation of physiological processes by hormonal and neural control was the paradigmatic example of such signaling, but the basic principles of negative feedback also applied to automated machines and societies. From the perspective of Engelberg and Boyarsky, ecosystems were collections of cybernetic entities (organisms) but were not cybernetic themselves.

The attack by Engelberg and Boyarsky elicited a detailed response from Odum and his younger colleague Bernard C. Patten. Patten had earned a PhD from Rutgers University for a study of plankton diversity that used information theory to explain annual changes in the composition of species. Although acknowledging an early training in the "taxonomist-natural historian tradition," Patten drew inspiration from H. T. Odum and G. Evelyn Hutchinson.[68] As a graduate student, he led a seminar presentation on Hutchinson's research, and he corresponded with the older ecologist about how to develop his mathematical skills along the lines used by Odum and other members of the Yale ecology group. He earned a reputation as a leading systems ecologist during the decades after completing his dissertation in 1959.

In their response, Patten and Odum directly addressed two of the issues raised by Engelberg and Boyarsky. First, they argued that Engelberg and Boyarsky were confusing the "brute force" pathways of energy transfer and nutrient cycling with the information pathways that regulated these movements. Secondly, they argued that although ecosystem homeostasis was analogous to organismal homeostasis, the mechanisms involved were significantly different. In particular, self-regulation in ecosystems did not involve discrete "set points." Rather, the "invisible wires of nature" were "diffuse, decentralized, and anatomically indistinct."[69] Although systems ecologists described self-regulation using mechanistic terms and concepts, there were no thermostats in ecosystems. Engelberg and Boyarsky had barely mentioned evolution, but Patten and Odum also attacked what they viewed as a reductionist bias that restricted evolutionary explanation to the interactions among individual organisms. Stability itself was a cause of orderly change in ecosystems. More-stable ecosystems inevitably replaced less-stable collections of interacting species. The alternative evolutionary worldview, according to Patten and Odum, implied "chaos" that was at odds with the balance and stability of most ecosystems. Perhaps not surprisingly, the dueling articles stimulated several additional responses in the *American Naturalist* from opponents and proponents of cybernetics.

For Patten and Odum, "invisible wires of nature" was a metaphor for the totality of interactions among species. More concretely, they also suggested that information might be transmitted by pheromones, allelopaths, and other chemical secretions that acted as environmental hormones. This was an idea that Odum had elaborated earlier in *Fundamentals of Ecology*, and it was widely shared by other ecosystem ecologists. In responding to Engelberg and Boyarsky, several of these ecologists cited the destructive behavior of pine bark beetles as an example of information transfer among members of forest communities that had beneficial effects on overall forest productivity.[70] Female pine bark beetles laid eggs beneath the bark of trees, but usually the defensive resins produced by the tree overwhelmed a single invader. By recruiting other females, a beetle could tip the balance in favor of the parasites. The collective galleries excavated by the beetles to lay their eggs often killed the infested tree. However, beetles tended to attack older, less vigorous pine trees. As older trees died, the forest canopy opened for new growth by younger trees. The consequence for the ecosystem was increased productivity. The information network controlling this process involved substances produced by the trees that attracted beetles as well as a variety of pheromones produced by the beetles themselves—some of which were metabolites of plant compounds.

The behavioral ecologist John Alcock saw things very differently.[71] Alcock had little interest in ecosystems, but he was a firm opponent of group selection and was skeptical of cooperation that implied the evolution of behavior that benefited species or larger groups. Rather, Alcock argued, explanations of behavior rested on how individuals maximized fitness. For Alcock, the pine bark beetles provided a compelling case study for how individuals manipulated, deceived, and eavesdropped on members of their own species, as well as others. From the perspective of a female pine bark beetle, recruitment of other females was not a cooperative behavior, but rather a strategy for increasing individual reproductive success. Too few females meant death from tree defenses, but too many females meant crowding and reduced space for egg laying. Successful females produced aggregation pheromones for recruitment but also camouflaging pheromones to prevent further recruitment at high population densities. The timing and amount of pheromone production were important for maximizing reproductive success, but this optimum might be different for females looking for egg-laying sites than for resident females. Further complicating the system, beetle predators also tracked their prey by the pheromones intended for intraspecific signaling. Thus, although pheromones carried information, the use of these signals was open to both "eavesdropping" and "deception." The evolution of these chemical messengers was a complex process, but it involved natural selection acting on individuals and not populations, species, or ecosystems. Alcock had nothing to say about a regulatory effect of pine bark beetles on ecosystem productivity. Such ecosystem-wide effects, if they existed, were presumably fortuitous side effects of individual selection.

The richly cooperative ecosystems that Eugene Odum envisioned stood in stark contrast to the self-centered world of autonomous individuals that Alcock described. For many ecologists, this dichotomy formed a fundamental intellectual divide with no middle ground. For example, in the title of his contribution to the debate in the *American Naturalist*, the Swedish forest ecologist Lauri Oksanen posed the rhetorical question "Ecosystem organization:

mutualism and cybernetics or plain Darwinian struggle for existence?"[72] His argument left little doubt that the latter alternative provided an adequate foundation for ecology. Like Alcock, Oksanen argued that mutualism was simply mutual exploitation between selfish partners. The supposed cybernetic regulation of ecosystem processes was explicable in terms of competition, predation, and the exploitation of natural resources for individual fitness.

Mutualism was not necessarily contrary to a Darwinian worldview. Using game theory, Robert Axelrod and William Hamilton argued that under certain conditions, cooperation and mutualism could evolve even in distantly related species such as the algae and fungi in lichens.[73] Based on the Prisoner's Dilemma game, in which players choose to cooperate or act selfishly without knowing the action of the other player, mutualism might evolve if the benefits of cooperating outweighed the costs of selfishness. However, this tit-for-tat evolutionary mechanism was highly context dependent and did not lead to the type of progressive evolution that Odum favored. Mutualism was unstable and always held the potential for shifting toward exploitation of one member by the other. Starting from different premises, Dubos had also argued for a fluid and indeterminate interpretation of mutualism, but for Odum progress and cooperation were the bedrock for his optimistic and progressive views on natural ecosystems and human societies. Not surprisingly, although he acknowledged the importance of Axelrod and Hamilton's theoretical work, Odum continued to argue for a progressive evolution of mutualism, even when it was being widely rejected by other biologists.

In a more subtle but important way, Axelrod and Hamilton worked to undermine Odum's holistic perspective. Reciprocal altruism was thoroughly individualistic. Indeed, it rested not so much on the behavior of individual organisms as on the reproductive success of autonomous genes. Biologists, Axelrod and Hamilton argued, needed to shift to a "gene's-eye view of natural selection" to understand the dynamics of evolution. This evolutionary reductionism also appealed to George Williams, Robert Trivers, E. O. Wilson, and Richard Dawkins, whose theoretical studies of behavior formed an important background for Axelrod and Hamilton. This stance was far removed from the holistic, ecosystem-centered "new ecology" that Odum championed.

Ecosystem Engineers and Homeostatic Fortresses

Dubos clearly recognized the uneasy boundaries among mutualism, parasitism, and competition, while emphasizing creativity, adaptability, and resilience, both in organisms and in the superorganimal ecosystem of the digestive tract. Later researchers who studied the microbiomes of humans and other animals have not always acknowledged the earlier work of Dubos. Nonetheless, they share his broad ecological perspective that highlights the importance of symbiosis and homeostasis within an ecosystem composed of microbes and their multicellular hosts. Many of these researchers also share Dubos's critical view of the overuse of antibiotics and the need for avoiding military metaphors suggesting warfare against bacteria. Sometimes viewing their roles as analogous to wildlife biologists, microbiome researchers have often turned to ecology and evolutionary biology as important sources of theory for understanding how to manage, nurture, and even design microbiomes to enhance human health.[74] From this perspective, the metaphor of selfish genes is less compelling than intricate networks of host and bacterial genes that use "cross-talk" to promote adaptation and homeostasis.

Symbiosis and homeostasis have become well-established concepts in microbiome research, but what of ecological systems more generally? Eugene Odum's idea that homeostasis and mutualism increase progressively as ecosystems mature has not stood the test of time, but the broader idea that symbiotic interactions can lead to something like homeostatic self-regulation remains attractive to some biologists. Take for example the multispecies complexes in leafcutter ant colonies. The ants construct enormous underground nests made up of hundreds of chambers. In some of the chambers, the ants actively tend specific fungi that live on leaves that the ants bring to the nest. Some biologists who study this system characterize the relationship between ants and fungi as a form of domestication that rivals human agriculture.[75] Various other microbes inhabit both the nest and the bodies of the ants, forming relationships that may be mutualistic, parasitic, or competitive. The nature of these relationships may evolve in predictable ways as the nest grows and species coadapt. For example, parasites may become less virulent, and mutualists less cooperative, over time. According to this perspective, the entire subterranean ecosystem constitutes a kind of "homeostatic fortress" regulated by multiple symbiotic feedback loops.[76] Selfishness may exist, but shifting attention to a more holistic consideration of the nest as a system emphasizes a rich array of interspecies interactions with profound effects on both the living community and the external environment.

The idea that leafcutter ants act as fortress builders, as well as farmers, illustrates the conceptual flexibility of adaptation and homeostasis in recent ecology and evolutionary biology. Some evolutionary biologists have argued that the ideas of "ecosystem engineers" and "niche construction" allow one to discuss ecosystem homeostasis in cybernetic terms that avoid criticisms leveled at Odum and Patten.[77] Ecosystem engineers are not just adapted but also adaptable agents that actively build the worlds within which they live. To the extent that these modifications affect the fitness of later generations of the species or other members of the ecosystem, one can consider constructed niches as a form of *ecological inheritance* that may be as important as genetic inheritance in maintaining the resilience and stability of ecosystems. Metaphorically, ecological engineering and niche construction also provide ways of discussing symbiosis and homeostasis without reliance on notions of evolutionary progress and harmony that marked Odum's earlier characterization of ecosystem homeostasis. Perhaps more importantly, ecosystems exist as complex, cybernetic systems rich in feedback without falling back on overt analogies with organisms. The idea that communities or ecosystems are like organisms has haunted ecosystem ecology from its inception to the advent of James Lovelock's controversial Gaia hypothesis. From the perspective of ecological engineering, ecosystems are not superorganisms, but rather *superconstructions* that exhibit an intermediate status between living and nonliving systems.

Conclusion

Dubos's studies of the communities of microorganisms living in the mammalian digestive tract were an early precursor of current research on the gut microbiome. Biologists today continue to combine ecological and physiological terms and concepts to explain the microbiome as both an organ and an ecosystem. Although relying heavily upon techniques of molecular biology largely unknown to Dubos and his contemporaries, modern microbiome research retains a focus on the organism as the site of pervasive homeostasis and symbiotic interaction among microorganisms and host to maintain organic integrity.[1] From this perspective the boundary between internal and external environments is ambiguous. For microorganisms, the gut is the external environment, while for the host the collective microorganisms act as an organ within an organ, producing hormones and vitamins that help regulate body functions. The functioning or malfunctioning of this complex homeostatic system has been increasingly implicated in diverse phenomena related to health and disease. From an evolutionary or coevolutionary perspective, these interactions represent a shifting balance of symbiotic relationships of mutualism, commensalism, and parasitism sometimes influenced, for better or worse, by human interventions such as administering antibiotics. How various microorganisms respond to this shifting environment might be compared to a kind of arms race in which adaptive changes in one constituent are countered by evolutionary changes in others. But following Dubos's distaste for military analogies, microbiome researchers today often employ more peaceful metaphors drawn from agriculture, wildlife conservation, or engineering.[2] From these perspectives evolutionary fitness and the fit between organism and environment can be likened to harmony and cooperation, as well as struggle for existence.

Dubos's microbiological research illuminates the fluid boundaries separating medical research, ecology, and evolutionary biology after World War II. Despite his earliest scientific interests in soil microbes and decomposition published in the journal *Ecology*, and his later claims about the digestive tract as an ecosystem, Dubos remained largely within the orbit of medical research, and he never became deeply active in the professional discipline of ecology. Similarly, his broad interests in evolution diverged in important ways from the modern synthesis that was consolidating its influence during the post–World War II period in which Dubos worked. By his own admission, Dubos's thinking on the coevolution of symbiotic relationships was shaped more heavily by the philosophies of Henri Bergson and John Dewey than by more recent developments in population genetics.[3] This was not so much a rejection of Mendelian genetics or Darwinian selection, per se, as a reaction to what Dubos considered to be an overly deterministic perspective on adaptation, fitness, and homeostasis. Organisms were not simply adapted but were also adaptable agents capable of creatively responding to changes in environmental conditions. This was true even for the relatively simple algae and fungi making up the symbiotic partnerships of lichens that so fascinated Dubos. Although the nature of these partnerships might fluctuate between mutualism and parasitism, the lichen often exhibited characteristics that transcended those of its constituent members, and the interactions of algae and fungus often contributed to greater homeostasis for both.

Dubos's far-flung intellectual pursuits were unusually broad, but his combined interests in medicine, ecology, and evolution were not entirely idiosyncratic. During a period of increasing specialization, both institutional and disciplinary, many biologists after World War II called for a more unified and integrated science. *Life Out of Balance* focuses on a number of scientists originally trained in medicine who became deeply engaged in studying the interactions of organisms and natural environments. In quite different ways, Homer Smith, Edward Adolph, and Per Scholander laid the foundations for a physiological ecology that combined elements of both medical and comparative physiology, together with ecology, evolutionary biology, natural history, and behavior. For Smith, phylogenetic studies based on the structure and function of the nephron were seamless extensions of medically oriented research in renal physiology. By contrast, Scholander turned away from medicine after earning a medical degree to pursue a wide range of studies in natural history and comparative physiology, and even developed methods for measuring carbon dioxide in ice cores. Despite his prominence as a pioneer in physiological ecology, Scholander never completely abandoned his interest in medical physiology. That said, humans were only one species among many that interested Scholander. Unlike more traditional medical researchers, Scholander became fascinated by seals, sloths, and arctic foxes for understanding diverse strategies for adaptation, and not simply as model organisms for understanding human health and disease. By way of contrast, Edward Adolph's applied human research during World War II made a significant contribution to desert physiology and stimulated important later developments in physiological ecology of diverse desert animals from kangaroo rats to camels by the Schmidt-Nielsens after the war ended. What unified these eclectic studies was the recognition that the fit between organism and environment was both flexible and dynamic. Gaining this insight required researchers to leave the controlled environment of the laboratory and grapple with the complexity of natural environments. Here they came face-to-face with the multifaceted adaptive challenges that organisms encounter under constantly fluctuating environmental conditions but also the ways that organisms manipulate the biotic and abiotic environments in which they live to enhance survival and reproductive success. From this perspective, homeostasis came to mean resiliency and regulated change more than maintaining constancy of the internal environment.

The move from laboratory to field both encouraged and demanded broad and flexible interpretations of homeostasis that sometimes deviated significantly from Walter Cannon's original formulation of the concept. Cannon was sometimes criticized for being too focused on normality and for taking a mechanistic perspective on self-regulation that emphasized how organisms (or their parts) reacted to environmental disturbances to maintain constancy. He marveled that internal body temperature was so precisely regulated that thermometer makers could confidently mark normal body temperature of 98.6°F on their instruments. The analogy between temperature regulation and thermostatic control systems later became an iconic representation of Cannon's medically oriented interpretation of homeostasis that continues to be popularized in textbooks. For physiological ecologists and other biologists who took ecological or evolutionary perspectives on adaptation, this analogy seemed naive and misleading. Even among birds and mammals, body temperature was often a labile characteristic, sometimes only intermittently maintained at a constant level. This was not a deviation from normality or the symptom of pathological disturbance, but rather an adaptive strategy for conserving energy used especially by small birds and mammals to exist—and thrive—in periodically hot and cold environments. Similarly, the camel's ability to tolerate wide fluctuations in body temperature throughout the day was an important part of the repertoire of adaptations necessary for a large desert mammal incapable of escaping from solar radiation and extreme daily temperature oscillations in the Sahara. Faced with complex, varying external environments, organisms exhibited intricate combinations of avoidance, compensation, lability, and resiliency to maintain functional integrity—but not necessarily physiological constancy.

George Bartholomew used this expansive interpretation of homeostasis to justify an integrated approach to biology that focused on the organism, while at the same time incorporating lower and higher levels of biological organization from cells to ecosystems. For Bartholomew, life was defined as the homeostatic state that arose from the inseparably interactive system composed of the organism and its environment. Bartholomew's own comparative studies of physiological and behavioral adaptations of allopatric subspecies, closely related sympatric species, and distantly related taxa such as hummingbirds and sphinx moths, were exquisite examples of how natural history, physiology, ecology, behavioral biology, and evolutionary theory could be integrated to answer questions at multiple levels of organization. Understanding complex adaptations required carefully designed experiments conducted in the laboratory or field, mathematical modeling and analysis of data, careful consideration of physical constraints on biological function, and the phylogenetic histories of the species under study but also the naturalist's keen insights into the lives actually lived by animals in their natural environments. Organismal biology and the careful study of adaptation provided a fertile meeting place for traditional natural history and modern biological specialties oriented toward experimentation and reductionist thinking. To counter the "tunnel vision" of overly specialized disciplines, Bartholomew called for symbiotic partnerships among biologists: "We can form promising symbioses—the naturalist and the molecular biologist, the naturalist and the biophysicist, the naturalist and the cell biologist."[4] Such partnerships were necessary for building a truly integrated biology and also for the rigorous study of adaptation and homeostasis. The fit of organism and environment was not a "trivial facet of natural history, but a biological attribute so central as to be inseparable from life itself."[5] Calling for biological synthesis, Bartholomew remarked, "Biology is indivisible; biologists should be undivided."[6]

Bartholomew's deeply interdisciplinary approach to biological synthesis contrasts importantly with the well-known dichotomy between functional biology and evolutionary biology that Ernst Mayr formulated during the same period. Mayr applied this dichotomy to the work of individual biologists as well to biological disciplines. Although acknowledging that a complete biological explanation needed to include both proximate and ultimate causation, Mayr placed more emphasis on his claim that functional and evolutionary biology were "largely separate fields." Historians and philosophers have carefully analyzed Mayr's dichotomy as an attempt to privilege the modern synthesis and to defend organismal biology against both the encroachment of molecular biology and the reduction of biological explanation to physics. Without denying the divisions that marked post–World War II biology or dismissing the concerns that Mayr expressed for the future of organismal biology, Bartholomew's less defensive alternative approach to biological integration should alert historians to a richer diversity of perspectives on organismal biology. In quite different ways, Bartholomew, Laurence Irving, Per Scholander, and John Moore all pursued combinations of museum, field, and laboratory studies to understand the adaptive fit between organism and environment. Although the leaders of the modern synthesis have rightly attracted the attention of historians for their central roles in laying the evolutionary foundations for contemporary biology, other biologists also played significant roles in synthesizing and integrating biological disciplines. The transformation of the American Society of Zoologists into the newer Society for Integrative and Comparative Biology is one example of reinvigorating organismal biology by redirecting its focus to comparative physiology, ecology, behavioral biology, and evolutionary theory. Within this context, integration meant centering the organism within its environment, while embracing research perspectives from diverse levels of organization from molecules and cells to communities and ecosystems. This broad synthetic effort included prominent figures associated with the modern synthesis, but the intellectual direction of the new society was also importantly shaped by Bartholomew and other physiological ecologists. Indeed, after the SICB was formed, its members lionized Bartholomew for his holistic view of biology and his philosophical commitment to unifying biological disciplines.[7]

Ironically, at the same time that he was sharply distinguishing functional biology and evolutionary biology, Mayr was also applying a broad conception of homeostasis to both organisms and populations to explain functional adaptation, Darwinian adaptation, and speciation. In these discussions Mayr rarely distinguished between proximate and ultimate causation. Consciously expanding Cannon's physiological concept, Mayr and other evolutionary biologists used genetic homeostasis to explain how combinations of genes—perhaps including the entire genome—worked together to regulate development, maintain physiological balance, and also provide the raw material for the adaptation of both organisms and populations in heterogeneous, fluctuating environments. This homeostatic regulation might occasionally break down during "genetic revolutions," leading to speciation. To be successful, however, the resulting daughter species needed to reestablish homeostatic balance with new environmental conditions after these revolutionary episodes.

Applying homeostasis to populations, communities, and ecosystems has been controversial, particularly because both critics and some supporters identified this use of self-regulation with largely discredited ideas of group selection. Nonetheless, as current examples of interest in homeostatic microbiomes, homeostatic fortresses constructed by leafcutter ants and their many symbionts, and various ideas of homeostasis associated with niche construction by adaptable organisms and populations attest, broad concepts of homeostasis continue to hold attraction. For some biologists, homeostasis remains the "second law" of biology applicable to all levels of biological organization. Together with the "first law" of natural selection, homeostasis forms the bedrock for biological explanation.[8] Yet, unlike lawlike physical concepts of inertia or force that are precisely defined mathematically, homeostasis has taken a variety of meanings. This conceptual richness has itself been a source of controversy, but it has also provided a basis for fruitful attempts to combine functional and evolutionary approaches to studying diverse biological phenomena. Even critics who complain that homeostasis misleads biologists into thinking reactively, rather than proactively and adaptively, are unwilling to completely reject the concept, preferring instead to recast it as a more dynamic, encompassing "allostasis" (or as Dubos would have preferred "homeokinesis").[9] In an unbalanced world, life emerges from the often tenuous balance between organisms and the environments of which they are inseparable parts.

Notes

Introduction

1. J. Scott Turner, "Biology's Second Law: Homeostasis, Purpose, and Desire," in *Beyond Mechanism: Putting Life Back into Biology*, ed. Brian G. Henning and Adam C. Scarfe, 183–204 (Lanham, MD: Lexington Books, 2013).

2. Sean B. Carroll, *The Serengeti Rules: The Quest to Discover How Life Works and Why It Matters* (Princeton: Princeton University Press, 2016), 5–7.

3. Originated by the University of Chicago biologist Leigh Van Valen to explain extinction rates, the Red Queen hypothesis continues to stimulate investigation of a variety of evolutionary and ecological phenomena. An excellent general review of how the concept applies to the problem of adaptation is Richard C. Lewontin, "Adaptation," *Scientific American* 239, no. 3 (September 1978): 212–31.

4. C. Ladd Prosser, ed., *Comparative Animal Physiology* (Philadelphia: Saunders, 1950). Although five coauthors contributed to the book, Prosser wrote more than half of the chapters, as well as the introduction, which highlighted explicit connections among comparative physiology, ecology, and evolutionary biology.

5. Ernst Mayr, "Cause and Effect in Biology," *Science* 134 (1961): 1501–6.

6. Vassiliki Betty Smocovitis, "Unifying Biology: The Evolutionary Synthesis and Evolutionary Biology," *Journal of the History of Biology* 25 (1992): 1–65; John Beatty, "The Proximate–Ultimate Distinction in the Multiple Careers of Ernst Mayr," *Biology and Philosophy* 9 (1994): 333–56; Michael R. Dietrich, "Paradox and Persuasion: Negotiating the Place of Molecular Evolution within Evolutionary Biology," *Journal of the History of Biology* 31 (1998): 85–111; Joel B. Hagen, "Naturalists, Molecular Biologists, and the Challenges of Molecular Evolution," *Journal of the History of Biology* 32 (1999): 321–41; Erika Lorraine Milam, "The Equally Wonderful Field: Ernst Mayr and Organismic Biology," *Historical Studies in the Natural Sciences* 40 (2010): 279–317.

7. Beatty, "Proximate–Ultimate Distinction."

8. Carroll, *Serengeti Rules*, 5–7.

Chapter 1

1. Letter from Norton to Cannon, May 24, 1929, box 106, Walter Bradford Cannon Papers, Center for the History of Medicine, Francis A. Countway Library of Medicine, Harvard University; Eline L. Wolfe, A. Clifford Barger, and Saul Benison, *Walter B. Cannon, Science and Society* (Cambridge: Harvard University Press, 2000), 260–62.

2. Herbert Spencer Jennings, *Biological Basis of Human Nature* (New York: Norton, 1930). The book went through five printings during the time when Norton and Cannon were negotiating a contract. Norton clearly considered Jennings's successful book to be a model for *The Wisdom of the Body*. In both cases Norton wanted books firmly grounded in scientific practice that also made sweeping claims about the human condition—both biological and social. Norton took pains to pattern the appearance of Cannon's book after Jennings's. The dust jackets of the two books were based on the same plan, although Norton's designer consciously chose a color scheme highlighted with "Harvard crimson" for Cannon's book; see letters of Norton to Cannon, December 32, 1931, and February 12, 1932; and Cannon to Norton, February 10, 1932, box 106, Cannon Papers.

3. See letter from Norton to Cannon, May 11, 1931, and Cannon to Norton, May 12, 1931, box 106, Cannon Papers.

4. This criticism will be discussed in more detail later in this book (in chapters 2, 4, and 11). Although historians have generally credited Cannon's originality, Michael Ruse equated homeostasis with Herbert Spencer's discredited nineteenth-century ideas about organic equilibrium in *The Evolution Wars* (Santa Barbara, CA: ABC-CLIO, 2000), 245.

5. Walter B. Cannon, "Organization for Physiological Homeostasis," *Physiological Reviews* 9 (1929): 399–431.

6. Drawing close comparisons between the ideas of Bernard and Cannon has been widespread among biologists, although some historians have challenged equating the thinking of the two physiologists too closely; see Frederic Holmes, "Claude Bernard, the Milieu Intérieur, and Regulatory Physiology," *History and Philosophy of the Life Sciences* 8 (1986): 3–25; Stephen J. Cross and William R. Albury, "Walter B. Cannon, L. J. Henderson and the Organic Analogy," *Osiris* 3 (1987): 165–92; Mathieu Arminjon, "Birth of the Allostatic Model: From Cannon's Biocracy to Critical Physiology," *Journal of the History of Biology* 49 (2016): 397–423.

7. Garland Allen, *Life Sciences in the Twentieth Century* (New York: John Wiley, 1975), 103–6.

8. Cannon, "Organization for Physiological Homeostasis," 400, 427.

9. Cannon, "Organization for Physiological Homeostasis," 401.

10. Admitting that there was still a lot to be learned about self-regulation, Cannon claimed that the "chase" was intellectually more satisfying than the "quarry." For Cannon, homeostasis was the guiding principle in this chase for knowledge; see Walter

B. Cannon, "Stresses and Strains of Homeostasis," *American Journal of Medical Sciences* 189 (1935): 1–14, quotations on 14.

11. Cannon, "Stresses and Strains of Homeostasis," 13–14.

12. For example, see the critical essays in Jay Schulkin, ed., *Allostasis, Homeostasis, and the Costs of Physiological Adaptation* (Cambridge: Cambridge University Press, 2004); Arminjon, "Birth of the Allostatic Model." For a more positive presentation of Cannon's homeostasis as an exemplar of a general regulatory biology, see Sean B. Carroll, *The Serengeti Rules: The Quest to Discover How Life Works and Why It Matters* (Princeton: Princeton University Press, 2016), chap. 1.

13. The role of glucagon, another important hormone in the homeostatic regulation of blood glucose, was poorly understood at the time Cannon wrote his article and book. The unknown hormone contaminated most commercial preparations of insulin, and its existence was hypothesized on the basis of transient increases in blood glucose following insulin injection. The homeostatic function of glucagon in raising blood glucose was not widely recognized until the late 1950s; see Pierre Lefèbvre, "Early Milestones in Glucagon Research," *Diabetes, Obesity, and Metabolism* 13, Supplement 1 (2011): 1–4; Pierre Lefèbvre, "Glucagon's Golden Jubilee at the University of Liège," *British Journal of Diabetes and Vascular Disease* 12 (2012): 278–84.

14. Cannon, "Organization for Physiological Homeostasis," 422.

15. Robert Perlman, "The Concept of the Organism in Physiology," *Theory in Biosciences* 119 (2000): 174–86.

16. Letter from Cannon to Norton, May 12, 1931; letter from Norton to Cannon, August 12, 1931, box 106, Cannon Papers. The rhetorical question "How do we stay normal?" was prominently displayed on the dust jacket of *The Wisdom of the Body*; see copy in box 106, Cannon Papers.

17. Walter B. Cannon, *The Wisdom of the Body*, 2nd ed. (New York: Norton, 1939), 23–24.

18. Cannon, "Organization for Physiological Homeostasis," 427.

19. Cannon's reticence about discussing social homeostasis and Norton's insistence that it be included as a full chapter rather than as an epilogue were aired in an exchange of letters shortly before the book was published: Cannon to Norton, December 3, 1931, and December 16, 1931; Norton to Cannon, December 17, 1931; Norton to Cannon, January 5, 1932; Cannon to Norton, January 9, 1932, all in box 106, Cannon Papers. Cannon was mollified after the chapter was reviewed by the Harvard sociologist Pitirim Sorokin and economists Thomas Nixon Carver, Frank Taussig, and Edwin Gay. Cannon was particularly pleased that Carver, who he thought would be the harshest critic, gave the "bothersome" chapter the most "lenient" reading of the four reviewers (Cannon to Norton, January 9, 1932).

20. Walter B. Cannon, "The Body Physiologic and the Body Politic," *Science* 93 (1941): 1–10, 2.

21. For the broader socioeconomic context of Cannon's social homeostasis, see Cross and Albury, "Walter B. Cannon, L. J. Henderson and the Organic Analogy"; see also Arminjon, "Birth of the Allostatic Model."

22. Cannon, "Body Physiologic and the Body Politic," 6.

23. See the exchange of letters between Norton and Cannon, February 7 and 8, 1933, in box 106, Cannon Papers. Copies of the articles from the *New York Times* and *Publishers Weekly* are included in Norton's letter.

24. Norbert Wiener, *Cybernetics: Or Control and Communication in the Animal and the Machine* (Cambridge, MA: MIT Press, 1948), introduction; David A. Mindell, *Between Human and Machine: Feedback, Control, and Computing before Cybernetics* (Baltimore: Johns Hopkins University, 2002), 282–83; Debora Hammond, *The Science of Synthesis: Exploring the Social Implications of General Systems Theory* (Boulder: University Press of Colorado, 2003), chap. 4; Steven J. Cooper, "From Claude Bernard to Walter Cannon: Emergence of the Concept of Homeostasis," *Appetite* 51 (2008): 419–27; Wolfe, Barger, and Benison, *Walter B. Cannon*, 314, 476.

25. Wolfe, Barger, and Benison, *Walter B. Cannon*, 263.

26. William B. Provine, "The Origin of Dobzhansky's Genetics and the Origin of Species," in *The Evolution of Theodosius Dobzhansky*, ed. Mark B. Adams (Princeton: Princeton University Press, 1994), 99–114; Leah Ceccarelli, *Shaping Science with Rhetoric: The Cases of Dobzhansky, Schrödinger, and Wilson* (Chicago: University of Chicago, 2001). Some of the important historical details of the origin of Dobzhansky's book have been challenged and revised by Joe Cain, "Co-opting Colleagues: Appropriating Dobzhansky's 1936 Lectures at Columbia," *Journal of the History of Biology* 35 (2002): 207–19.

27. Provine, "Origin of Dobzhansky's Genetics and the Origin of Species," 112–13.

28. Ernst Mayr, *The Growth of Biological Thought: Diversity, Evolution, and Inheritance* (Cambridge: Belknap Press, 1982), 67, 558. For a brief history of Mayr's critique of beanbag genetics, along with responses from prominent defenders of the idea, see William B. Provine, "Ernst Mayr: Genetics and Speciation," *Genetics* 167 (2004): 1041–46.

29. Theodosius Dobzhansky, *Genetics and the Origin of Species* (New York: Columbia University Press, 1937), 126, 150, 170.

30. Dobzhansky, *Genetics and the Origin of Species*, 3.

31. Dobzhansky, *Genetics and the Origin of Species*, 8.

32. Garland E. Allen, "Theodosius Dobzhansky, the Morgan Lab, and the Breakdown of the Naturalist/Experimentalist Dichotomy, 1927–1947," in *The Evolution of Theodosius Dobzhansky*, ed. Mark B. Adams, 87–98 (Princeton: Princeton University Press, 1994).

33. David M. Steffes, "Panpsychic Organicism: Sewall Wright's Philosophy for Understanding Complex Genetic Systems," *Journal of the History of Biology* 40 (2007): 327–61. Steffes argues for strong indirect linkages between Wright's thinking about genetic systems and the physiological tradition of Claude Bernard, Lawrence J. Henderson,

and Walter B. Cannon via the writings of the biochemist and philosopher Benjamin Moore. For the background and development of Wright's interest in physiological genetics, see William B. Provine, *Sewall Wright and Evolutionary Biology* (Chicago: University of Chicago Press, 1986), 91–92, 202–6.

34. David J. Depew, "Adaptation as Process: The Future of Darwinism and the Legacy of Theodosius Dobzhansky," *Studies in History and Philosophy of Biological and Biomedical Sciences* 42 (2011): 89–98. Depew contrasts Dobzhansky's view with that of Julian Huxley, who emphasized that organisms were collections of more or less independent adaptations.

35. Dobzhansky, *Genetics and the Origin of Species*, 170. For a useful history of ideas associated with the norm of reaction, see Sahotra Sarkar, "From the *Reaktionsnorm* to the Adaptive Norm: The Norm of Reaction, 1909–1960, *Biology and Philosophy* 14 (1999): 235–52.

36. Dobzhansky, *Genetics and the Origin of Species*, 150.

37. Robson and Richards were enthusiastic followers of Joseph Henry Woodger, whose philosophy of biology has been widely discredited. For a recent attempt to reassess Woodger's contributions to biology and rehabilitate his philosophical legacy, see Daniel J. Nicholson and Richard G. Gawne, "Rethinking Woodger's Legacy in the Philosophy of Biology," *Journal of the History of Biology* 47 (2014): 243–92. The broader context of British debates over adaptation and a contrast with Dobzhansky's position is provided by Depew, "Adaptation as Process."

38. G. C. Robson and O. W. Richards, *The Variation of Animals in Nature* (New York: Longmans, 1936), 352. A somewhat similar critique of an "adaptationist programme" sometimes based on spurious "just-so stories" was later made by Stephen Jay Gould and R. C. Lewontin, "The Spandrels of San Marco and the Panglossian Paradigm: A Critique of the Adaptationist Programme," *Proceedings of the Royal Society of London*, Part B 205 (1979): 581–98. Stephen Jay Gould used the phrase "just so stories" as a pejorative against what he considered to be speculative claims about adaptation. There is no reason to think that Gould's critique was based on the earlier ideas of Robson and Richards, but the similarity highlights the long-standing historical debate over the nature and extent of adaptation.

39. The idea of plasticity was a recurring theme running through *The Variation of Animals in Nature*. At one point, Robson and Richards referred to the "seemingly infinite plasticity" of living organisms and their parts (7).

40. Robson and Richards, *Variation of Animals in Nature*, 378.

41. Robson and Richards, *Variation of Animals in Nature*, 358.

42. Robson and Richards, *Variation of Animals in Nature*, 347.

43. Ceccarelli, *Shaping Science with Rhetoric*, 168.

44. This potential cost to poorly adapted individuals was partly ameliorated by the "developmental plasticity" that broadened the norm of reaction and allowed genotypes to produce well-adapted phenotypes in a wide range of environments. The biological

and broader philosophical aspects of Dobzhansky's thinking about plasticity are discussed by John Beatty, "Dobzhansky and the Biology of Democracy: The Moral and Political Significance of Genetic Variation," in *The Evolution of Theodosius Dobzhansky*, ed. Mark B. Adams (Princeton: Princeton University Press, 1994), 195–218.

45. Dobzhansky, *Genetics and the Origin of Species*, 229.

46. Provine, *Sewall Wright*, chap. 9; Micahel R. Dietrich and Robert A. Skipper, "A Shifting Terrain: A Brief History of the Adaptive Landscape," in *The Adaptive Landscape in Evolutionary Biology*, ed. Erik I. Svensson and Ryan Calsbeek, (Oxford: Oxford University Press, 2012), 3–15. Other essays in this collection provide valuable insights into the historical influence and continuing importance of the adaptive landscape. Anya Plutynski, "The Rise and Fall of the Adaptive Landscape?" *Biology and Philosophy* 23 (2008): 605–23 provides another excellent historical overview of the reception of Wright's model. Other articles in the special issue of *Biology and Philosophy* highlight the continuing philosophical and theoretical disagreements about the landscape.

47. Although the peaks and valleys in the original model represented fitness, Ceccarelli, *Shaping Science with Rhetoric*, 169–70, emphasizes that a more ecological interpretation in terms of niches was crucial for popularizing the model for naturalists and nonmathematical biologists. The mixing of ecological and genetic interpretations of the landscape is particularly striking in Dobzhansky, *Genetics and the Origin of Species*, 3rd ed., 8–10, 276–82.

48. George M. Somero, "Clifford Ladd Prosser 1907–2002," *Biographical Memoirs of the National Academy of Sciences* (Washington, DC, 2008), 1–17.

49. C. Ladd Prosser, "The Making of a Comparative Physiologist," *Annual Reviews in Physiology* 48 (1986): 1–6.

50. C. Ladd Prosser, ed., *Comparative Animal Physiology* (Philadelphia: Saunders, 1950), 2.

51. C. Ladd Prosser, "Theory of Physiological Adaptation of Poikilotherms to Heat and Cold," Brody Memorial Lecture 5, *University of Missouri Agricultural Station Special Report* 59, 1965.

52. C. Ladd Prosser, "The 'Origin' after a Century: Prospects for the Future," *American Scientist* 47 (1959): 536–60, 545.

53. C. Ladd Prosser, "Physiological Variation in Animals," *Biological Reviews* 30 (1955): 229–61. In his review Prosser discussed the research of C. H. Waddington, I. I. Schmalhausen, Richard Goldschmidt, and Georgii Gause on these phenomena. For a historical discussion of the tenuous relationship between these developmental and evolutionary ideas and the modern synthesis, see Scott F. Gilbert, "Dobzhansky, Waddington, and Schmalhausen," in *The Evolution of Theodosius Dobzhansky*, ed. Mark B. Adams (Princeton: Princeton University Press, 1994), 143–54

54. Vassiliki Betty Smocovitis, "The 1959 Darwin Centennial Celebration in America," *Osiris* 14 (1999): 274–323.

55. Ernst Mayr, "Cause and Effect in Biology," *Science* 134 (1961): 1501–6.

56. Mayr, “Cause and Effect in Biology,” 1501.

57. Vassiliki Betty Smocovitis, “Unifying Biology: The Evolutionary Synthesis and Evolutionary Biology,” *Journal of the History of Biology* 25 (1992): 1–65; John Beatty, “The Proximate–Ultimate Distinction in the Multiple Careers of Ernst Mayr,” *Biology and Philosophy* 9 (1994): 333–56; Michael R. Dietrich, “Paradox and Persuasion: Negotiating the Place of Molecular Evolution within Evolutionary Biology,” *Journal of the History of Biology* 31 (1998): 85–111; Joel B. Hagen, “Naturalists, Molecular Biologists, and the Challenges of Molecular Evolution,” *Journal of the History of Biology* 32 (1999): 321–41; Erika Lorraine Milam, “The Equally Wonderful Field: Ernst Mayr and Organismic Biology,” *Historical Studies in the Natural Sciences* 40 (2010): 279–317.

58. William R. Dawson, “George A. Bartholomew’s Contributions to Integrative and Comparative Biology,” *Integrative and Comparative Biology* 45 (2005): 219–30; William R. Dawson, “George A. Bartholomew, 1916–2006,” *Biographical Memoirs of the National Academy of Sciences* (Washington, DC, 2011), 1–33.

59. George A. Bartholomew, “Interspecific Comparison as a Tool for Ecological Physiologists,” in *New Directions in Ecological Physiology*, ed. Martin E. Feder, Albert F. Bennett, Warren W. Burggren, and Raymond B. Huey (Cambridge: Cambridge University Press), 11–37; George A. Bartholomew, “The Role of Natural History in Contemporary Biology,” *Bioscience* 36 (1986): 324–29.

60. George A. Bartholomew, “The Roles of Physiology and Behaviour in the Maintenance of Homeostasis in the Desert Environment,” *Symposia of the Society for Experimental Biology* 18 (1964): 7–29.

61. Bartholomew, “Roles of Physiology and Behaviour,” 8.

62. Bartholomew, “Roles of Physiology and Behaviour,” 8; Joel B. Hagen, “Camels, Cormorants, and Kangaroo Rats: Integration and Synthesis in Organismal Biology after World War II,” *Journal of the History of Biology* 48 (2015): 169–99.

63. G. K. H. Zupanc and M. M. Zupanc, “Theodore H. Bullock: Pioneer of Integrative and Comparative Neurobiology,” *Journal of Comparative Physiology A* 194 (2008): 119–34.

64. Theodore Holmes Bullock, “In Praise of ‘Natural History,’” *Cellular and Molecular Neurobiology* 25 (2005): 217–21.

65. Theodore H. Bullock, “Homeostasis in Marine Organisms,” in *Perspectives in Marine Biology*, ed. A. A. Buzzati-Traverso, 199–210 (Berkeley: University of California Press, 1960); Theodore Holmes Bullock, “In Search of Principles of Integrative Biology,” *American Zoologist* 5 (1965): 745–55.

66. Milam, “Equally Wonderful Field.” E. O. Wilson, Mayr’s younger colleague at Harvard, used the proximate–ultimate distinction to defend organismal biology against what he considered aggressive attacks by the molecular biologist James Watson. Later, Wilson also used a version of the proximate–ultimate distinction to promote and defend sociobiology, which he considered to be a “new synthesis” of ethology, population biology, and evolutionary theory. For details, see Erika Lorraine Milam, *Creatures*

of Cain: The Hunt for Human Nature in Cold War America (Princeton: Princeton University Press, 2018), 241–45. Milam's account of the rancor surrounding sociobiology highlights the perhaps not-too-surprising observation that claims for synthesis often result in division as much as unification.

67. Bartholomew, "Interspecific Comparison."

68. Dawson, "George A. Bartholomew's Contributions."

69. Ernst Mayr, *Growth of Biological Thought*, 53, 59.

Chapter 2

1. D. W. Richards Jr. "The Circulation in Traumatic Shock in Man," *Bulletin of the New York Academy of Medicine* 20 (1944): 363–93.

2. Dickinson W. Richards, "Homeostasis versus Hyperexis: Or Saint George and the Dragon," *Scientific Monthly* 77 (1953): 289–94.

3. See the foreword to Dickinson W. Richards, *Medical Priesthoods and Other Essays*, Lakeville, CT: Connecticut Printers, 1970, x.

4. William F. Hamilton and Dickinson W. Richards, "The Output of the Heart," in *Circulation of the Blood: Men and Ideas*, ed. Alfred P. Fishman and Dickinson W. Richards (Bethesda: American Physiological Society, 1982), 71–126; Dickinson W. Richards, "Cardiovascular Physiology: Concepts and Development of Knowledge," in Richards, *Medical Priesthoods*, 99–119.

5. Dickinson W. Richards, "Right Heart Catheterization: Its Contributions to Physiology and Medicine," *Science* 125 (1957): 1181–85; Dickinson W. Richards, "Lawrence Joseph Henderson," in Richards, *Medical Priesthoods*, 75–84; André Cournand and Michael Meyer, *From Roots . . . to Late Budding: The Intellectual Adventures of a Medical Scientist* (New York: Gardner Press, 1986), chap. 4.

6. André Cournand, "Dickinson Woodruff Richards," *Biographical Memoirs of the National Academy of Sciences* (Washington DC, 1989), 458–87.

7. Cournand, "Dickinson Woodruff Richards."

8. Dickinson W. Richards Jr. and Alvan L. Barach, "Prolonged Residence in High Oxygen Atmospheres: Effects on Normal Individuals and on Patients with Chronic Cardiac and Pulmonary Insufficiency," *Quarterly Journal of Medicine* 3 (1934): 437–66.

9. Dickinson W. Richards Jr. and Marjorie L. Strauss, "Carbon Dioxide and Oxygen Tensions of the Mixed Venous Blood of Man at Rest," *Journal of Clinical Investigation* 9 (1930): 475–532.

10. Dickinson W. Richards Jr., André Cournand, and Natalie A. Bryan, "Applicability of Rebreathing Method for Determining Mixed Venous CO_2 in Cases of Chronic Pulmonary Disease," *Journal of Clinical Investigation* 14 (1935): 173–80.

11. M. G. Bourassa, "History of Cardiac Catheterization," *Canadian Journal of Cardiology* 21 (2005): 1011–14.

12. Henry D. Lauson, Stanley E. Bradley, and André Cournand, "The Renal Circulation in Shock," *Journal of Clinical Investigation* 23 (1944): 381–402.

13. Richards, "Circulation in Traumatic Shock." According to Cournand, by the end of the war the study involved over 125 subjects; see Cournand and Meyer, *From Roots*, 46.

14. Richards, "Circulation in Traumatic Shock."

15. Richards, "Circulation in Traumatic Shock," 370.

16. Both technical problems of preserving blood and widespread skepticism among doctors that stored blood was equivalent to freshly transfused blood delayed the widespread use of whole blood to treat shock prior to World War II; see Kara W. Swanson, *Banking on the Body: The Market in Blood, Milk, and Sperm in Modern America* (Cambridge: Harvard University Press, 2014), chap. 2.

17. Walter B. Cannon, *Traumatic Shock* (New York: Appleton, 1923), 4–5; Saul Benison, A. Clifford Barger, and Elin L. Wolfe, "Walter B. Cannon and the Mystery of Shock: A Study of Anglo-American Cooperation," *Medical History* 35 (1991): 217–49.

18. Richards, "Circulation in Traumatic Shock," 366.

19. Richards, "Circulation in Traumatic Shock," 393.

20. Richards, "The Effects of Hemorrhage on the Circulation," *Annals of the New York Academy of Sciences* 44 (1948): 534–41, 540; Richards, "Homeostasis versus Hyperexis," 291. See also André Cournand, "Some Aspects of the Pulmonary Circulation in Normal Man and in Chronic Cardiopulmonary Diseases," *Circulation* 2 (1950): 641–57.

21. Richards, "Homeostasis versus Hyperexis," 292; Dickinson W. Richards, "Homeostasis: Its Dislocations and Perturbations," *Perspectives in Biology and Medicine* 3 (1960): 238–51, 247–48.

22. Richards, "Homeostasis versus Hyperexis," 289.

23. Richards, "Homeostasis versus Hyperexis," 293.

24. Richards, "Homeostasis: Its Dislocations and Perturbations." Richards briefly criticized homeostasis in his Nobel Lecture, in which he lauded the important physiological contributions of Bernard and Henderson, while failing to recognize Cannon; see Richards, "Right Heart Catheterization."

25. Richards, "Homeostasis: Its Dislocations and Perturbations," 249.

26. Dickinson W. Richards, "The Right Heart and the Lung: With Some Observations on Teleology," *American Review of Respiratory Diseases* 94 (1966): 691–702; reprinted in Richards, *Medical Priesthoods and Other Essays*, 120–40.

27. Lawrence J. Henderson, *The Order of Nature* (Cambridge: Harvard University Press, 1917). Henderson developed many of these ideas on teleology and fitness earlier in *The Fitness of the Environment: An Inquiry into the Biological Significance of the Properties of Matter* (New York: Macmillan), 1913. For an analysis of Henderson's ideas within the broader historical context of teleological thinking, see Bruce H. Weber, "Lawrence Henderson's Natural Teleology," in *Water and Life: The Unique Properties of* H_2O, ed. Ruth M. Lynden-Bell, Simon Conway Morris, John D. Barlow, John L. Finney, and Charles L. Harper Jr. (Boca Raton, FL: CRC Press, 2010), 327–44.

28. Henderson, *Order of Nature*, 50.

29. See "Discussion and Summary," by Richard's colleague Alvan Barach in *Pulmonary Emphysema*, ed. Alvan L. Barach and Hylan A. Bickerman (Baltimore: Williams and Wilkins, 1956), 510.

30. Recent advocates of "allostasis" also criticize homeostasis as an oversimplified concept that places too much emphasis on maintaining balance through physiological reactions to disturbances, but unlike Richards they often advocate social activism in medicine. Rather than treating symptoms of disease, many of these critics call for a more holistic approach to medicine that emphasizes prevention, often involving social interventions; see Peter Sterling, "Allostasis: A Model of Predictive Regulation," *Physiology & Behavior* 106 (2011): 5–15; Jay Schulkin, *Rethinking Homeostasis: Allostatic Regulation in Physiology and Pathophysiology* (Cambridge, MA: MIT Press, 2003); Jay Schulkin, ed., *Homeostasis and the Costs of Physiological Adaptation* (Cambridge: Cambridge University Press, 2004); Mathieu Arminjon, "Birth of the Allostatic Model: From Cannon's Biocracy to Critical Physiology," *Journal of the History of Biology* 49 (2016): 397–423.

31. Richards, "Right Heart and Lung," 137–83. Although the subject continues to stimulate considerable interest, the basic idea that senescence is the inevitable result of natural selection acting primarily on prereproductive rather than postreproductive life stages was well established by the 1940s; for a concise historical survey, see Brian Charlesworth, "Fisher, Medawar, Hamilton and the Evolution of Aging," *Genetics* 156 (2000): 927–31.

32. Arturo Rosenblueth, "A Critique of Homeostasis," in *Perspectives in Biology: A Collection of Papers Dedicated to Bernardo A. Houssay on the Occasion of his 75th Birthday*, ed. Carl F. Cori, V. G. Foglia, L. F. Leloir, and S. Ochoa (New York: Elsevier, 1963), 323–31, 328.

33. David L. Drabkin, "Metabolism of the Hemin Chromoproteins," *Physiological Reviews* 31 (1951): 345–430, 423; see also Drabkin, "Imperfection: Biochemical Phobias and Metabolic Ambivalence," *Perspectives in Biology and Medicine* 2 (1959): 473–517.

34. Drabkin, "Metabolism," 423; see also Drabkin, "Imperfection," 477–78.

35. Drabkin, "Imperfection," 477.

36. Drabkin, "Imperfection," 509–10.

37. Drabkin, "Imperfection," 513.

38. Drabkin, "Imperfection," 500.

39. Drabkin, "Imperfection," 506.

40. Drabkin, "Imperfection," 512.

41. Rosenblueth, "Critique of Homeostasis," 326.

42. For a discussion of the collaboration, see the chapter "Arturo Rosenblueth and Wiener's Work in Physiology," by Pesi R. Masani, *Norbert Wiener, 1894–1964* (Basel: Birkhauser, 1990), 197–217.

43. Both his Mexican nationality and Jewish heritage worked against Rosenblueth, as mentioned in Cannon's letter to the pharmacologist Chauncey Leake, March 15, 1943. Leake was a vice president at the University of Texas in charge of the medical

school in Galveston. The tenuous situation at Harvard and the unlikelihood that Rosenblueth could become Cannon's successor are discussed in a letter from Cecil Drinker to Cannon, February 25, 1942. At the time, Drinker was a professor of physiology and the Dean of the Harvard School of Public Health. Both letters, and other correspondence about Rosenblueth's employment prospects, are in box 117, Walter Bradford Cannon Papers, Center for the History of Medicine, Francis A. Countway Library of Medicine, Harvard University.

44. Rosenblueth, "Critique of Homeostasis," 324.

45. Rosenblueth, "Critique of Homeostasis," 326.

46. Arthur C. Guyton, *Textbook of Medical Physiology* (Philadelphia: Saunders, 1956). Guyton continued to revise his textbook until his death in 2003. Later editions were coauthored by John Hall, who succeeded Guyton as chair of the Department of Physiology and Biophysics at the University of Mississippi School of Medicine. The book is currently in its twelfth edition (2011). For the origins of the textbook and Guyton's educational philosophy, see Arthur C. Guyton, "An Author's Philosophy of Physiology Textbook Writing," *Advances in Physiology Education* 19 (1998): S1–S5.

47. Guyton, *Textbook of Medical Physiology*, 24; see also Guyton, "Author's Philosophy."

48. Carroll Brinson and Janis Quinn, *Arthur C. Guyton: His Family, His Life, His Achievements* (Jackson, MS: Oakdale Press, 1989), chaps. 4 and 5.

49. Arthur C. Guyton, "Electronic Counting and Size Determination of Particles in Aerosols," *Journal of Industrial Hygiene and Toxicology* 28 (1946): 133–41.

50. Brinson and Quinn, *Arthur Guyton*, 51.

51. Brinson and Quinn, *Arthur Guyton*, 38, 81.

52. Guyton, *Textbook of Medical Physiology*, 2nd ed. (1961), 12.

53. Nerves and hormones constituted "backup control systems" that added stability to the more basic mechanisms of local self-regulation by tissues; see Arthur C. Guyton, Howard T. Milhorn, and Thomas G. Coleman, "Simulation of Physiological Mechanisms Part II," *Simulation* 9 (1967): 73–79.

54. Guyton, *Textbook of Medical Physiology*, 18.

55. Arthur C. Guyton, Carl E. Jones, and Thomas G. Coleman, *Circulatory Physiology: Cardiac Output and Its Regulation*, 2nd ed. (Philadelphia: Saunders, 1973), 140.

56. See Guyton's recollection in Brinson and Quinn, *Arthur Guyton*, 83.

57. Christian Crone, "The Autoregulation of the Microcirculation," *Acta Medica Scandinavica* 197 (1975): 15–18; Paul C. Johnson, "Autoregulation of Blood Flow," *Circulation Research* 59 (1989): 483–95.

58. Guyton, Milhorn, and Coleman, "Simulation of Physiological Mechanisms Part II"; see also Arthur C. Guyton, John E. Hall, Thomas E. Lohmeier, R. Davis Manning Jr., and Thomas E. Jackson, "Position Paper: The Concept of Whole Body Autoregulation and the Dominant Role of the Kidneys for Long-Term Blood Pressure Regulation," in *Frontiers in Hypertension Research* (1981): 125–34.

59. Harris J. Granger and Arthur C. Guyton, "Autoregulation of the Total Systemic

Circulation Following Destruction of the Central Nervous System in the Dog," *Circulation Research* 25 (1969): 379–88.

60. Arthur C. Guyton, Thomas G. Coleman, and Harris J. Granger, "Circulation: Overall Regulation," *Annual Review of Physiology* 34 (1972): 13–46.

61. Arthur C. Guyton, Howard T. Milhorn, and Thomas G. Coleman, "Simulation of Physiological Mechanisms Part I," *Simulation* 9 (1967): 15–20; Guyton, Milhorn, and Coleman, "Simulation Part II"; Howard T. Milhorn, *The Application of Control Theory in Physiology* (Philadelphia: Saunders, 1966).

62. Guyton, Milhorn, and Coleman, "Simulation of Physiological Mechanisms Part I"; Guyton, Milhorn, and Coleman, "Simulation of Physiological Mechanisms Part II."

63. Guyton, Milhorn, and Coleman, "Simulation of Physiological Mechanisms Part II."

64. Arthur C. Guyton, "Past-President's Address: Physiology: A Beauty and a Philosophy," *Physiologist* 18 (1975): 495–501.

65. Guyton, "Past-President's Address."

66. Guyton, "Past-President's Address."

67. For the broader cultural context of changes in higher education in the South after World War II, see Melissa Kean, *Desegregating Private Higher Education in the South* (Baton Rouge: Louisiana State University Press, 2008); see also Michael Vinson Williams, *Medgar Evers: Mississippi Martyr* (Fayetteville: University of Arkansas Press, 2011); Charles C. Bolton, *William F. Winter and the New Mississippi* (Jackson: University Press of Mississippi, 2013).

68. Richard Bode, "A Doctor Who's Dad to Seven Doctors—So Far," *Reader's Digest*, December 1982, 141–45; Nancy Shulins, "Doctor Overcomes Polio to Bring 8—or 10—Children into Medicine: For Arthur Guyton of Mississippi a Handicap Is Not Disabling," *Los Angeles Times*, August 3, 1988, http://articles.latimes.com/1986-08-03/news/mn-971_1_arthur-guyton/4; Stephanie Shapiro, "Is There a Doctor in the House?" *Hopkins Medicine Magazine*, Winter 2013, https://www.hopkinsmedicine.org/news/publications/hopkins_medicine_magazine/archives/winter_2013/is_there_a_doctor_in_the_house.

Chapter 3

1. Homer W. Smith, "The Retention and Physiological Role of Urea in the Elasmobranchii," *Biological Reviews* 49 (1936): 49–81, 53.

2. Homer W. Smith, *The Kidney: Structure and Function in Health and Disease* (New York: Oxford University Press, 1950), v–vi; Homer W. Smith, *From Fish to Philosopher* (Boston: Little, Brown, 1953), 214.

3. Robert F. Pitts, "Homer William Smith," *Biographical Memoirs of the National Academy of Sciences* (Washington, DC, 1967), 445–70; Stanley Bradley, "Homer William Smith: A Personal Memoir," in *A Laboratory by the Sea: The Mount Desert Island Biological Laboratory, 1898–1998*, ed. Franklin H. Epstein (Rhinebeck, NY: River Press,

1998), 97–101; Thomas H. Maren, "Eli Kennerly Marshall, Jr.," *Biographical Memoirs of the National Academy of Sciences* (Washington DC, 1986), 313–52.

4. Letter from Smith to Cannon, May 12, 1945, box 117, Walter Bradford Cannon Papers, Center for the History of Medicine, Francis A. Countway Library of Medicine, Harvard University.

5. David Evans, *Marine Physiology Down East: The History of the Mt. Desert Island Biological Laboratory* (New York: Springer, 2015), chap. 3; Franklin H. Epstein, ed., *A Laboratory by the Sea: The Mount Desert Island Biological Laboratory* (Rhinebeck, NY: River Press, 1998). For the broader historical context of biological research at American marine laboratories, see Jane Maienschein, "History of American Marine Laboratories: Why Do Research at the Seashore?," *American Zoologist* 28 (1988): 15–25; and Philip J. Pauly, "Summer Resort and Scientific Discipline: Woods Hole and the Structure of American Biology," in *The American Development of Biology*, ed. Ronald Rainger, Keith R. Benson, and Jane Maienschein (Philadelphia: University of Pennsylvania Press, 1988), 121–50.

6. Melanie P. Hoenig and Mark L. Zeidel, "Homeostasis, the Milieu Intérieur, and the Wisdom of the Nephron," *Renal Physiology* 9 (2014): 1–10.

7. For his historical perspective on developments in the field, see Homer W. Smith, "Renal Physiology," in *Circulation of the Blood: Men and Ideas*, ed. Alfred P. Fishman and Dickinson W. Richards (Bethesda, MD: American Physiological Society, 1982), 545–606. Originally published in 1964, the 1982 edition of this book was dedicated to Smith's memory. For a discussion of the "Smithian era" in renal physiology, see Pitts, "Homer William Smith."

8. Homer W. Smith, *Principles of Renal Physiology* (New York: Oxford University Press, 1956), 36.

9. Jeff M. Sands, "Micropuncture: Unlocking the Secrets of Renal Function," *American Journal of Physiology—Renal Physiology* 287 (2004): F866–87; Hoenig and Zeidel, "Wisdom of the Nephron."

10. Steven J. Peitzman, "The Flame Photometer as Engine of Nephrology: A Biography," *American Journal of Kidney Diseases* 56 (2010): 379–86.

11. Robert W. Berliner, "Remembering Homer Smith," *Kidney International* 43 (1993): 171–72.

12. Homer W. Smith, "Metabolism of the Lung-Fish, *Protopterus aethiopicus*," *Journal of Biological Chemistry* 88 (1930): 97–130.

13. Homer W. Smith, "Observations on the African Lung-Fish, *Protopterus aethiopicus*, and on Evolution from Water to Land Environments," *Ecology* 12 (1931): 164–81; Homer W. Smith, "Lung-Fish," *Scientific Monthly* 31 (1930): 467–70.

14. Smith, "Observations on African Lung-Fish," 179; Smith, "Metabolism of the Lung-Fish."

15. Homer W. Smith, *Kamongo* (New York: Viking Press, 1932), 31. For Smith's account of writing *Kamongo*, see "The Story of This Book" at the end of Homer W.

Smith, *Man and His Gods* (New York: Little, Brown, 1952), 447–48. Some critics complained that the two characters in *Kamongo*, a scientist and a priest, were caricatures; see John O'Hara Cosgrave, "With a Spice of Fiction," *Saturday Review of Literature*, April 30, 1932, 32; and Vincent Engels, "A Snob of Science," *Commonweal* 17 (November 16, 1932): 82–83. Other reviewers compared *Kamongo* favorably to a Joseph Conrad novel. For a thoroughly positive review, see R. L. Duffuss, "Taking the Lung-Fish as a Text for Philosophy," *New York Times*, April 17, 1932, 6.

16. Smith, *Kamongo*, 31.

17. For Smith's self-criticism, see "The Story of This Book."

18. Smith, *Kamongo*, 89–90.

19. Smith, *Kamongo*, 74.

20. Smith, *Kamongo*, 90.

21. Smith, *Kamongo*, 138. The unusual punctuation at the end of this quotation was Smith's way of emphasizing that Joel cuts Padre off in mid-sentence.

22. Smith, *Kamongo*, 138.

23. Smith, *Kamongo*, 146–47.

24. Smith, "Renal Physiology," 596.

25. Homer W. Smith, *From Fish to Philosopher*, 2 (master chemist); 143 (extravagance).

26. Smith, *Fish to Philosopher*, 47–48, 107–8; Smith, *Principles of Renal Physiology*, 21–22. Smith regarded the early studies of urine formation in aglomerular fish by his friend and collaborator Eli Marshall to be "theory-shaking" events in the history of renal physiology. See also Klaus W. Beyenbach, "Kidneys sans Glomeruli," *American Journal of Physiology—Renal Physiology* 286 (2004): F811–27.

27. Maren, "Eli Marshall." Although Marshall's earlier comparative studies provided an important foundation for the phylogenetic reconstruction, he was apparently much less interested in evolution than Smith. After the 1930 article, Marshall turned away from phylogenetic reconstruction, although he continued to do research on renal physiology of fish and amphibians at the Mount Desert Island Laboratory.

28. E. K. Marshall Jr. and Homer W. Smith, "The Glomerular Development of the Vertebrate Kidney in Relation to Habitat," *Biological Bulletin* 59 (1930): 135–53.

29. Beyenbach, "Kidneys sans Glomeruli."

30. E. K. Marshall, "The Comparative Physiology of the Kidney in Relation to Theories of Renal Secretion," *Physiological Reviews* 14 (1934): 133–59; Marshall and Smith, "Glomerular Development," 139–40.

31. Hoenig and Zeidel, "Homeostasis," suggest that linear thinking was reflected in the linear schematic diagrams of nephrons used by both Smith and Marshall.

32. Eldon J. Braun, "Osmotic and Ionic Regulation in Birds," in *Osmotic and Ionic Regulation: Cells and Animals*, ed. David H. Evans (Boca Raton, FL: CRC Press, 2009), 505–524; and William H. Dantzler and S. Donald Bradshaw, "Osmotic Regulation in Reptiles," in the same volume, 443–504.

33. Maren, "Eli Marshall," 324.

34. Homer W. Smith, "Water Regulation and Its Evolution in the Fishes," *Quarterly Review of Biology* 7 (1932): 1–26.

35. Homer W. Smith, "The Absorption and Excretion of Water and Salts by Marine Teleosts," *American Journal of Physiology* 93 (1930): 480–505.

36. Homer W. Smith, "The Excretion of Ammonia and Urea by the Gills of Fish," *Journal of Biological Chemistry* 81 (1929): 727–42.

37. Homer W. Smith, "The Composition of the Body Fluids of Elasmobranchs," *Journal of Biological Chemistry* 86 (1929): 407–19; Smith, "Physiological Role of Urea."

38. Smith, "Physiological Role of Urea," 65.

39. Smith, "Physiological Role of Urea," 70–71.

40. August Krogh, *Osmotic Regulation in Aquatic Animals* (Cambridge: Cambridge University Press, 1939); David H. Evans, "Teleost Fish Osmoregulation: What Have We Learned since August Krogh, Homer Smith, and Ancel Keys?" *American Journal of Physiology—Regulatory, Integrative and Comparative Physiology* 295 (2008): R704–13.

41. Ancel Keys, "The Heart-Gill Preparation of the Eel and Its Perfusion for the Study of a Natural Membrane in situ," *Zeitschrift für Vergleichende Physiologie* 15 (1931): 352–63; Ancel Keys, "Chloride and Water Secretion and Absorption by the Gills of the Eel," *Zeitschrift für Vergleichende Physiologie* 15 (1931): 364–89. Following this important contribution to comparative physiology, Keys would later become famous for his studies of the effects of starvation, the development of K-rations during World War II, and his promotion of the Mediterranean Diet after the war.

42. Homer W. Smith, "The Functional and Structural Evolution of the Vertebrate Kidney," *Sigma Xi Quarterly* 21 (1933): 141–51.

43. Homer W. Smith, "The Kidney," *Scientific American* 188, no. 1 (January 1953): 40–48; Smith, *From Fish to Philosopher*, 2.

44. Smith's "Evolution of the Kidney" was presented as part of the Porter Lecture Series at the University of Kansas Medical School in 1943 and was published as a chapter in Homer W. Smith, *Lectures on the Kidney* (Lawrence: University of Kansas, 1943), 4. It was later reprinted in the *Anatomical Record* 277A (2004): 345–54, along with Peter D. Vize, "A Homeric View of Kidney's Evolution: A Reprint of H. W. Smith's Classic Essay with a New Introduction," 344. The quotation is also found in *From Fish to Philosopher*, 4.

45. Waldemar Kaempffert, "The Ostracoderms Lead the Pageant," *New York Times*, November 29, 1953, 22.

46. Smith, *From Fish to Philosopher*, 20; see also Smith, "Evolution of the Kidney," 6.

47. Homer W. Smith, "Organism and Environment: Dynamic Oppositions," in *Adaptation*, ed. John Romano (Ithaca: Cornell University Press, 1949), 36–37.

48. Smith, "Organism and Environment," 40.

49. Smith, "Organism and Environment," 36; see also Smith, "Evolution of the Kidney," 21.

50. Smith employed these pessimistic characterizations in both his technical and

more popular writing. See Smith, "Role of Urea in the Elasmobranchii," 70–71; "Observations on the African Lung-Fish," 171, 173, 178; *Kamongo*, 92–93; *From Fish to Philosopher* 77, 133, 158.

51. Beyenbach, "Kidneys sans Glomeruli"; Braun, "Osmotic and Ionic Regulation in Birds."

52. Vassiliki Betty Smocovitis, *Unifying Biology: The Evolutionary Synthesis and Evolutionary Biology* (Princeton: Princeton University Press, 1996), chap. 5; Erika Lorraine Milam, *Creatures of Cain: The Hunt for Human Nature in Cold War America* (Princeton: Princeton University Press, 2017), 17–23.

53. George Gaylord Simpson, *The Meaning of Evolution* (New Haven: Yale University Press, 1949), 284.

54. For a detailed discussion of the correspondence between Simpson and Huxley on these issues, and on their disagreements, see Marc Swetlitz, "Julian Huxley and the End of Evolution," *Journal of the History of Biology* 28 (1995): 181–217. The broader intellectual context of the two biologists' views on progress is provided by Michael Ruse, *Monad to Man: The Concept of Progress in Evolutionary Biology* (Cambridge: Harvard University Press, 1996).

55. For a critical comparison of the views of Huxley and Simpson, see Timothy Shanahan, *The Evolution of Darwinism* (Cambridge: Cambridge University Press, 2004), chap. 8.

56. Simpson, *Meaning of Evolution*, 283–84.

57. Homer W. Smith, "Objectives and Objectivity in Science," in *Homer William Smith: His Scientific and Literary Achievements*, ed. Herbert Chasis and William Goldring (New York: New York University Press, 1965), 211. The essay was originally published together with contrasting articles on science and religion by John Courtney Murray and Theodore Greene in *Yale Scientific Magazine* 23, no. 5 (February 1949).

58. For Simpson's views on creativity and natural selection, as well as those of other leaders of the modern synthesis, see John Beatty, "The Creativity of Natural Selection? Part II: The Synthesis and Since," *Journal of the History of Biology* 52 (2019): 705–31.

59. Smith, *From Fish to Philosopher*, 22.

60. Smith, *From Fish to Philosopher*, 64.

61. Smith, *From Fish to Philosopher*, 141.

62. Smith, *From Fish to Philosopher*, 4.

63. Smith, "Evolution of the Kidney," 3; Smith, *From Fish to Philosopher*, 2.

64. Smith, *From Fish to Philosopher*, 133. Marshall and Smith, "Glomerular Development," described what they considered to be the "glomerular degeneration" in birds and remarked the lack of development of the loop of Henle in comparison to the mammalian nephron; see also Smith, "Evolution of the Kidney."

Chapter 4

1. Edward F. Adolph and Associates, *The Physiology of Man in the Desert* (New York: Interscience, 1947).

2. Edward F. Adolph, "Early Concepts of Physiological Regulations," *Physiological Reviews* 41 (1961): 737–70; Edward F. Adolph, "Seven Discoveries of Physiological Regulation," in *Neural Integration of Physiological Mechanisms and Behaviour*, ed. Gordon J. Mogenson and Franco R. Calaresu (Toronto: University of Toronto Press, 1975), 11–23; see also his presidential address to the American Physiological Society: Edward F. Adolph, "The Physiological Scholar," *American Journal of Physiology* 179 (1954): 607–12.

3. Adolph, "Early Concepts," 766.

4. Adolph, "Early Concepts," 739, 764.

5. Edward F. Adolph, *Physiological Regulations* (Lancaster, PA; Jacques Cattel Press, 1943), 302.

6. Edward F. Adolph, "Regulation of Water Metabolism in Stress," *Homeostatic Mechanisms, Brookhaven Symposia in Biology* 10 (1958): 147–61.

7. Adolph's idea of organisms as integrated systems was undoubtedly influenced by Henderson's systems thinking, which emphasized mutual interactions of parts rather than more reductive cause-and-effect relationships. For the development of Henderson's "pre-cybernetic" systems thinking in the biological and social sciences, see Mateo Jasmine Muñoz, "Lawrence Joseph Henderson: Bridging Laboratory and Social Life," PhD dissertation, Harvard University, 2014; see also Iris Fry, "On the Biological Significance of Properties of Matter: L. J. Henderson's Theory of the Fitness of the Environment," *Journal of the History of Biology* 29 (1996): 155–96; and John Parascandola, "Organismic and Holistic Concepts in the Thought of L. J. Henderson," *Journal of the History of Biology* 4 (1971): 63–113. For a detailed analysis of Haldane's organicist ideas and his critique of mechanistic thinking in biology, see Steve Sturdy, "Biology as Social Theory: John Scott Haldane and Physiological Regulation," *British Journal for the History of Science* 21 (1988): 315–40.

8. The early manuscript of "Seven Discoveries of Physiological Regulation," is in box 50, Edward F. Adolph Papers, Edward G. Miner Library, University of Rochester Medical Center.

9. Adolph, *Man in the Desert*, vii.

10. Adolph, *Physiological Regulations*, vii.

11. Edward F. Adolph, "The Regulation of the Water Content of the Human Organism," *Journal of Physiology* 55 (1921): 114–32.

12. Edward F. Adolph, "Living Water," *Quarterly Review of Biology* 5 (1930): 51–67.

13. Edward F. Adolph, "Excretion of Water by the Kidneys," *American Journal of Physiology* 65 (1923): 419–49.

14. Edward F. Adolph, "The Metabolism of the Water in Ameba as Measured in the Contractile Vacuole," *Journal of Experimental Zoology* 44 (1926): 355–81; Edward F. Adolph, "The Regulation of Volume and Concentration in the Body Fluids of Earthworms," *Journal of Experimental Zoology* 47 (1927): 31–62; Edward F. Adolph, "The Skin and the Kidneys as Regulators of the Body Volume of Frogs," *Journal of Experimental Zoology* 47 (1927): 1–30.

15. Adolph, "Skin and Kidneys."

16. Adolph, "Living Water," 57.

17. Letter from E. F. Adolph to Charles Thomas, March 8, 1940, and letter from Thomas to Adolph April 26, 1940, in box 30, Adolph Papers.

18. Correspondence between Adolph and Cattell in box 30, Adolph Papers.

19. Carl L. Hubbs, "Physiological Regulations," *American Naturalist* 78 (1944): 82–83.

20. August Krogh, *Osmotic Regulation in Aquatic Animals* (Cambridge: Cambridge University Press, 1939).

21. Krogh, *Osmotic Regulation*, 7.

22. Bodil Schmidt-Nielsen, *August and Marie Krogh: Lives in Science* (New York: Oxford University Press, 1995); Heinz-Gerd Zimmer, "August Krogh," *Clinical Cardiology* 29 (2006): 231–33.

23. John B. West, *High Life: A History of High-Altitude Physiology and Medicine* (New York: Oxford University Press, 1998), 107–13.

24. Edward F. Adolph and William B. Fulton, "The Effects of Exposure to High Temperatures upon the Circulation in Man," *American Journal of Physiology* 67 (1924): 573–88.

25. Extensive correspondence between the two men before and after the one-month expedition is in box 2, Adolph Papers; see also Edward F. Adolph and David Bruce Dill, "Observations on Water Metabolism in the Desert," *American Journal of Physiology* 123 (1938): 369–78. For the broader historical context of heat-related death and disability among construction workers at the dam, see Michael Hiltzik, *Colossus: Hoover Dam and the Making of the American Century* (New York: Free Press, 2010).

26. David Bruce Dill, *Life, Heat and Altitude: Physiological Effects of Hot Climates and Great Heights* (Cambridge: Harvard University Press, 1938). For the development of the Harvard Fatigue Laboratory, see Sarah W. Tracy, "The Physiology of Extremes: Ancel Keys and the International High Altitude Expedition of 1935," *Bulletin of the History of Medicine* 86 (2012): 627–60; Andi Johnson, "'They Sweat for Science': The Harvard Fatigue Laboratory and Self-Experimentation in American Exercise Physiology," *Journal of the History of Biology* 48 (2015): 425–54; and Robin Wolfe Scheffler, "The Power of Exercise and the Exercise of Power: The Harvard Fatigue Laboratory, Distance Running, and the Disappearance of Work," *Journal of the History of Biology* 48 (2015): 391–423.

27. For the broader context of self-experimentation conducted by Dill and other members of the Harvard Fatigue Laboratory, see Johnson, "They Sweat for Science."

28. Edward F. Adolph, "Heat Exchanges of Man in the Desert," *American Journal of Physiology* 123 (1938): 486–99.

29. Dill, *Life, Heat, Altitude*, 44.

30. Dill, *Life, Heat, Altitude*, 41; see also David Bruce Dill, F. G. Hall, and H. T. Edwards, "Changes in Composition of Sweat during Acclimatization to Heat," *American Journal of Physiology* 123 (1938): 412–19.

31. Based on this research, the use of salt tablets was widely promoted to prevent fatigue in workers in hot conditions; see Scheffler, "Power of Exercise."

32. Adolph and Dill, "Observations on Water Metabolism in the Desert," 377.

33. Contract letter from A. N. Richards to E. F. Adolph, July 31, 1942, box 20, Adolph Papers. Richards was the chair of the Committee on Medical Research of O.S.R.D.

34. See the postwar summary and analysis for the Office of the Quartermaster General by Elizabeth Schickele, "Environment and Fatal Heat Stroke: An Analysis of 157 Cases Occurring in the Army in the U.S. during World War II," *Military Surgeon* 100 (1947): 235–56.

35. Adolph, *Man in the Desert*, 194.

36. Adolph, *Man in the Desert*, 344. Unlike the soldiers, but like the donkey used in Boulder City, Adolph's experimental dogs quickly replenished lost water after being dehydrated (194).

37. Adolph, *Man in the Desert*, 333, 350.

38. Edward F. Adolph, "Desert," in *Physiology of Heat Regulation and the Science of Clothing*, ed. L. H. Newburgh (Philadelphia: W. B. Saunders, 1949), 330–338.

39. Adolph, *Man in the Desert*, 334.

40. Adolph, *Man in the Desert*, 339.

Chapter 5

1. Edward F. Adolph, "Do Rats Thrive When Drinking Seawater?" *American Journal of Physiology* 140 (1943): 25–32; Bodil Schmidt-Nielsen and Knut Schmidt-Nielsen, "Do Kangaroo Rats Thrive When Drinking Seawater?" *American Journal of Physiology* 160 (1950): 291–94.

2. Bodil Schmidt-Nielsen, "The Resourcefulness of Nature in Physiological Adaptation to the Environment," *Physiologist* 1 (1958): 4–20. For critiques of the Schmidt-Nielsens' early research, see Glenn E. Walsberg, "Small Mammals in Hot Deserts: Some Generalizations Revisited," *BioScience* 50 (2000): 109–20; Randall L. Tracy and Glen E. Walsberg, "Kangaroo Rats Revisited: Re-Evaluating a Classic Case of Desert Survival," *Oecologia* 133 (2002): 449–57; Douglas A. Kelt, "Comparative Ecology of Desert Small Mammals: A Selective Review of the Past 30 Years," *Journal of Mammalogy* 92 (2011): 1158–78.

3. Knut Schmidt-Nielsen, *The Camel's Nose: Memoirs of a Curious Scientist* (Washington, DC: Island Press, 1998), chap. 3; see also Steven Vogel, "Knut Schmidt-Nielsen. 24 September 1915–25 January 2007," *Biographical Memoirs of the Fellows of the Royal Society* 54 (2008): 319–31; and Joel B. Hagen, "Camels, Cormorants, and Kangaroo Rats: Integration and Synthesis in Organismal Biology after World War II," *Journal of the History of Biology* 48 (2015): 169–99.

4. K. Schmidt-Nielsen, *Camel's Nose*, chap. 6; Joel B. Hagen, "Bergmann's Rule, Adaptation, and Thermoregulation in Arctic Animals: Conflicting Perspectives from Physiology, Evolutionary Biology, and Physical Anthropology after World War II," *Journal of the History of Biology* 50 (2016): 235–65.

5. Irving apparently gave the Schmidt-Nielsens wide latitude in selecting a research project. He himself was not particularly interested in desert mammals, and at the

time he was much more deeply involved in establishing an arctic research program in Alaska; see Robert Elsner, "The Irving–Scholander Legacy in Polar Physiology," *Comparative Biochemistry and Physiology—Part A* 126 (2000): 137–42. The military contract and the problem of continued funding for the research is briefly discussed in a letter from Irving to Knut Schmidt-Nielsen, July 12, 1947, and Knut Schmidt-Nielsen to Irving, September 3, 1947, box 2, Knut Schmidt-Nielsen Papers, Rubenstein Library, Duke University.

6. See the undated letter from Knut Schmidt-Nielsen to Irving and the letter from Irving's daughter Susan to Bodil and Knut, August 16, 1947, box 2, Schmidt-Nielsen Papers.

7. Irving to Knut Schmidt-Nielsen, October 12, 1947, box 2, Schmidt-Nielsen Papers. The report was also sent to the Army Air Forces to justify extending the contract that paid for the desert research. The initial results were published nearly a year later: Knut Schmidt-Nielsen, Bodil Schmidt-Nielsen, and Adelaide Brokaw, "Urea Excretion in Desert Rodents Exposed to High Protein Diets," *Journal of Cellular and Comparative Physiology* 32 (1948): 361–80; and Bodil Schmidt-Nielsen, Knut Schmidt-Nielsen, Adelaide Brokaw, and Howard Scheiderman, "Water Conservation in Desert Rodents," *Journal of Cellular and Comparative Physiology* 32 (1948): 331–60. See also Bodil Schmidt-Nielsen and Knut Schmidt-Nielsen, "The Water Economy of Desert Mammals," *Scientific Monthly* 69 (1949): 180–85.

8. Although many of their articles were published in physiological journals, the Schmidt-Nielsens placed their research within a context that combined elements of laboratory physiology, field studies in ecology, and traditional natural history. Of particular importance was the earlier study by A. Brazier Howell and I. Gersh, "Conservation of Water by the Rodent *Dipodomys*," *Journal of Mammalogy* 16 (1935): 1–9.

9. Knut Schmidt-Nielsen, *How Animals Work* (Cambridge: Cambridge University Press, 1972), 1.

10. Schmidt-Nielsen et al., "Urea Excretion." Surprisingly, white rats survived longer on average than kangaroo rats, although they suffered greater dehydration. According to the authors, part of the explanation may have been that because white rats were bred for lives in the laboratory, they were under less stress than the wild animals subjected to captivity.

11. Bodil Schmidt-Nielsen and Roberta O'Dell, "Structure and Concentrating Mechanism in the Mammalian Kidney," *American Journal of Physiology* 200 (1961): 1119–24.

12. Bodil Schmidt-Nielsen and Knut Schmidt-Nielsen, "Water Loss in Desert Rodents in Their Natural Habitat," *Ecology* 31 (1950): 75–85; August Krogh, "A Microhabitat Recorder," *Ecology* 21 (1940): 275–78.

13. Knut Schmidt-Nielsen, "Countercurrent Systems in Animals," *Scientific American* 244, no. 5 (May 1981): 118–28.

14. Walsberg, "Small Mammals in Hot Deserts"; Tracy and Walsberg, "Kangaroo Rats Revisited."

15. Schmidt-Nielsen and Schmidt-Nielsen, "Water Economy of Desert Mammals."

16. Homer W. Smith, *From Fish to Philosopher* (Boston: Little, Brown, 1953), 156–62.

17. See the letter from Knut Schmidt-Nielsen to Smith, April 20, 1954, box 4, Schmidt-Nielsen Papers. The Schmidt-Nielsens were thrilled by Smith's public comments about the importance of their work; see K. Schmidt-Nielsen, *Camel's Nose*, 95.

18. K. Schmidt-Nielsen, *Camel's Nose*, 105. Bodil worked in Smith's lab during the summer of 1952. Smith also invited the couple to join the lab after they completed research on camels in Algeria; see Smith to Knut Schmidt-Nielsen, April 26, 1954, box 4, Schmidt-Nielsen Papers.

19. See the correspondence between Flanagan and Schmidt-Nielsen beginning in April 1951 and continuing through May 1952, box 3, Schmidt-Nielsen Papers. Although Flanagan originally planned to publish the article in 1951, it did not appear until July 1953: Knut Schmidt-Nielsen and Bodil Schmidt-Nielsen, "The Desert Rat," *Scientific American* 189, no. 1 (July 1953): 173–78.

20. Knut Schmidt-Nielsen to Adolph, June 5, 1951, box 2, Schmidt-Nielsen Papers; see also Schmidt-Nielsen to Adolph, August 3, 1949, in the same box.

21. Knut Schmidt-Nielsen and Bodil Schmidt-Nielsen, "Problems in the Physiology of the Camel," undated memo to Laurence Irving, series 2, box 12, Laurence Irving Papers, University of Alaska Archives.

22. See the letter from Irving to Detlev Bronk, December 7, 1948, series 2, box 12, Laurence Irving Papers. At the time, Bronk was chair of the National Research Council and president of the Johnson Research Foundation. See also Knut Schmidt-Nielsen, "An Outline of Plans for a Physiological Expedition to Sahara," undated memo to Irving in the same box.

23. After three years of planning, the Arid Zone Research Programme began in 1951, shortly before the Schmidt-Nielsens began their camel research project. Although the AZRP focused primarily on geography, hydrology, and agriculture, physiology (both human and animal) was also an important consideration; see Maurice Goldsmith, "A New Deal for the World's Arid Lands," *UNESCO Courier* 4, no. 6 (1951): 13; Michel Batisse, "A Long Look at the World's Arid Lands," *UNESCO Courier* 47, no. 1 (1994): 34–39; Malcolm Hadley, "Nature to the Fore: The Early Years of UNESCO's Environmental Program, 1945–1965," in *Sixty Years of Science at UNESCO, 1945–2005*, ed. P. Petitjean, V. Zharov, G. Glaser, J. Richardson, B. de Padirac, and G. Archibald, 201–32 (Paris: UNESCO Publishing, 2006).

24. Bodil Schmidt-Nielsen, "Water Conservation in Small Desert Rodents," and Knut Schmidt-Nielsen, "Heat Regulation in Small and Large Desert Mammals," both in *Biology of Deserts*, ed. J. L. Cloudsley Thompson (London: Institute of Biology, 1954), 173–81, 182–87; Bodil Schmidt-Nielsen and Knut Schmidt-Nielsen "The Camel: Facts and Fables," *UNESCO Courier* 8, no. 8 (1955): 70.

25. K. Schmidt-Nielsen, *Camel's Nose*, 101, 108.

26. The Schmidt-Nielsens were assisted in their research by Richard Houpt, a

veterinarian from the University of Pennsylvania, and Stig Jarnum, a Danish physician.

27. Knut Schmidt-Nielsen, "The Physiology of the Camel," *Scientific American* 201, no. 6 (December 1959): 140–51.

28. Bodil Schmidt-Nielsen, Knut Schmidt-Nielsen, T. Richard Houpt, and Stig A. Jarnum, "Water Balance of the Camel," *American Journal of Physiology* 185 (1956): 185–94; Knut Schmidt-Nielsen, Bodil Schmidt-Nielsen, T. Richard Houpt, and Stig A. Jarnum, "The Question of Water Storage in the Stomach of the Camel," *Mammalia* 20 (1956): 1–15; Knut Schmidt-Nielsen, Bodil Schmidt-Nielsen, Stig A. Jarnum, and T. Richard Houpt, "Body Temperature of the Camel and Its Relation to Water Economy," *American Journal of Physiology* 188 (1956): 103–12; Knut Schmidt-Nielsen, *Desert Animals: Physiological Problems of Heat and Water* (Oxford: Oxford University Press, 1964).

29. B. Schmidt-Nielsen, "Resourcefulness of Nature."

30. Knut Schmidt-Nielsen, "Comparative Physiology of Desert Mammals," Brody Memorial Lecture 2, *University of Missouri Agricultural Experiment Station Special Report #21*, 1962.

31. B. Schmidt-Nielsen and O'Dell, "Structure and Concentrating Mechanism."

32. K. Schmidt-Nielsen, "Countercurrent Systems"; Donald C. Jackson and Knut Schmidt-Nielsen, "Countercurrent Heat Exchange in the Respiratory Passages," *Proceedings of the National Academy of Sciences* 51 (1964): 1192–97.

33. P. F. Scholander, "The Wonderful Net," *Scientific American* 196, no. 4 (April 1957): 97–107.

34. K. Schmidt-Nielsen, *How Animals Work.*

35. K. Schmidt-Nielsen, *Desert Animals*, 27.

Chapter 6

1. George A. Bartholomew, "The Roles of Physiology and Behaviour in the Maintenance of Homeostasis in the Desert Environment," *Symposia of the Society for Experimental Biology* 18 (1964): 7–29.

2. George A. Bartholomew, "Interspecific Comparison as a Tool for Ecological Physiologists," in Martin E. Feder, Albert F. Bennett, Warren W. Burggren, and Raymond B. Huey, *New Directions in Ecological Physiology* (Cambridge: Cambridge University Press, 1987), 13; Joel B. Hagen, "Camels, Cormorants, and Kangaroo Rats: Integration and Synthesis in Organismal Biology after World War II," *Journal of the History of Biology* 48 (2015): 169–99.

3. At the time Miller was modernizing the MVZ by strengthening ties with the Zoology Department at Berkeley, and by encouraging experimental and behavioral studies by curators; see Mary E. Sunderland, "Modernizing Natural History: Berkeley's Museum of Vertebrate Zoology in Transition," *Journal of the History of Biology* 46 (2013): 369–400.

4. George A Bartholomew, "Interspecific Comparison"; William R. Dawson, "George A. Bartholomew's Contributions to Integrative and Comparative Biology,"

Integrative and Comparative Biology 45 (2005): 219–30; William R. Dawson, "George A. Bartholomew, 1916–2006," *Biographical Memoirs of the National Academy of Sciences* (Washington, DC, 2011), 1–33.

5. George A. Bartholomew, "The Role of Natural History in Contemporary Biology," *Bioscience* 36 (1986): 324–29.

6. George A. Bartholomew, "The Role of Physiology in the Distribution of Terrestrial Vertebrates," in *Zoogeography*, ed. Carl L. Hubbs (Washington, DC: American Association for the Advancement of Science, 1958), 81–95; Bartholomew, "Roles of Physiology and Behaviour."

7. Bartholomew, "Roles of Physiology and Behaviour," 8–9.

8. Bartholomew, "Roles of Physiology and Behaviour," 9–10.

9. George A. Bartholomew, "Interaction of Physiology and Behavior under Natural Conditions," in *The Galapagos*, ed. Robert I. Bowman (Berkeley: University of California Press, 1966), 39–45.

10. Bartholomew, "Interspecific Comparison." According to Bartholomew, "environments are classified as 'extreme' by us physiologists, not by the animals that live in them" (18).

11. Bartholomew, "Roles of Physiology and Behaviour," 11.

12. William R. Dawson, "The Relation of Oxygen Consumption to Temperature in Desert Rodents," *Journal of Mammalogy* 36 (1955): 543–53; George A. Bartholomew and Jack W. Hudson, "Effects of Sodium Chloride on Weight and Drinking in the Antelope Ground Squirrel," *Journal of Mammalogy* 40 (1959): 354–60; George A. Bartholomew and Jack W. Hudson, "Desert Ground Squirrels," *Scientific American* 205, no. 5 (November 1961): 107–116; Jack W. Hudson, "The Role of Water in the Biology of the Antelope Ground Squirrel, *Citellus leucurus*," *University of California Publications in Zoology* 64 (1962): 1–56.

13. Bartholomew, "Roles of Physiology and Behaviour," 12.

14. Bartholomew, "Roles of Physiology and Behaviour," 11.

15. Bartholomew, "Interaction of Physiology and Behavior."

16. Charles Darwin, *The Voyage of the Beagle*, Project Gutenberg e-book (posted June 24, 2013), http://www.gutenberg.org/files/944/944-h/944-h.htm, chap. 17.

17. George A. Bartholomew, "A Field Study of Temperature Relations in the Galápagos Marine Iguana," *Copeia* 1966 (1966): 241–50.

18. Knut Schmidt-Nielsen, *The Camel's Nose* (Washington, DC: Island Press, 1998), 150–51.

19. Bartholomew asked to use unpublished manuscripts of Schmidt-Nielsen's early studies on cormorants in a graduate seminar that he was teaching and offered to send his manuscripts on salt regulation in passerine birds. March 21, 1957, box 5, Schmidt-Nielsen Papers.

20. Knut Schmidt-Nielsen, C. Barker Jörgensen, and Humio Osaki, "Extrarenal Salt Excretion in Birds," *American Journal of Physiology* 193 (1958): 101–7; Knut

Schmidt-Nielsen and Ragnär Fange, "Salt Glands in Marine Reptiles," *Nature* 182 (1958): 783–85; Knut Schmidt-Nielsen, "The Salt-Secreting Gland of Marine Birds," *Circulation* 21 (1960): 955–67.

21. Schmidt-Nielsen, "Salt-Secreting Gland," speculated that chloride was actively transported, although more recent evidence implicates a sodium ion pump.

22. Schmidt-Nielsen and Fange, "Salt Glands," 785.

23. K. Schmidt-Nielsen to Bartholomew, November 10, 1958, box 5, Schmidt-Nielsen Papers.

24. George A. Bartholomew and Tom J. Cade, "The Water Economy of Land Birds," *Auk* 80 (1963): 504–39; see also Tom J. Cade and George A. Bartholomew, "Sea-Water and Salt Utilization by Savannah Sparrows," *Physiological Zoology* 32 (1959): 230–38; Thomas L. Poulson and George A. Bartholomew, "Salt Balance in the Savannah Sparrow," *Physiological Zoology* 35 (1962): 109–99.

25. Cade and Bartholomew, "Sea-Water and Salt Utilization," 236.

26. As Bartholomew and Cade, "Water Economy of Land Birds," admitted, the term "urine" required some qualification. Birds lack a urinary bladder, and all wastes are eliminated through the cloaca. In his experiments Bartholomew found clear liquid that was distinct from feces. However, it was not possible to determine whether this "urine" was formed completely by the kidneys or partly by the gut.

27. John S. Cooke, "Sectionalization," in *History of the American Physiological Society: The First Century, 1887–1987*, ed. John R. Brobeck, Orr E. Reynolds, and Tobey A. Appel (Bethesda: American Physiological Society, 1987), 435–61.

28. To join the APS, candidates needed to be nominated by two members and elected by the society at one of its biannual meetings; see Joseph F. Saunders and Aubrey E. Taylor, "Membership," in John R. Brobeck, Orr E. Reynolds, and Tobey A. Appel, *History of the American Physiological Society: The First Century, 1887–1987* (Bethesda, MD: American Physiological Society, 1987), 301–14. Schmidt-Nielsen's letter of nomination written to APS secretary Ray Daggs, March 6, 1963, is in the Schmidt-Nielsen Papers, box 6. For more on the nomination, see Schmidt-Nielsen to Bartholomew, March 7, 1963, and Bartholomew to Schmidt-Nielsen, March 13, 1963, in the same box.

29. The tensions within American zoology that resulted from specialization and the institutional problems of integrating disparate groups of zoologists during this period are explored in Kristin Johnson, "The Return of the Phoenix: The 1963 International Congress on Zoology and the American Zoologists in the Twentieth Century," *Journal of the History of Biology* 42 (2009): 417–56.

30. Bartholomew, "Role of Natural History," 327.

31. His first PhD student, William Dawson, described Bartholomew's arguments for an integrated biology as a "voice of reason" in "George A. Bartholomew's Contributions," 228; see Johnson, "Return of the Phoenix," for a discussion of the disciplinary fragmentation in the American Society of Zoologists.

32. Vassiliki Betty Smocovitis, "Unifying Biology: The Evolutionary Synthesis and

Evolutionary Biology," *Journal of the History of Biology* 25 (1992): 1–65; Erika Lorraine Milam, "The Equally Wonderful Field: Ernst Mayr and Organismic Biology," *Historical Studies in the Natural Sciences* 40 (2010): 279–317; see also John Beatty, "The Proximate–Ultimate Distinction in the Multiple Careers of Ernst Mayr," *Biology and Philosophy* 9 (1994): 333–56; Michael R. Dietrich, "Paradox and Persuasion: Negotiating the Place of Molecular Evolution within Evolutionary Biology, *Journal of the History of Biology* 31 (1998): 85–111; Joel B. Hagen, "Naturalists, Molecular Biologists, and the Challenges of Molecular Evolution, *Journal of the History of Biology* 32 (1999): 321–41.

33. Bartholomew, "Interspecific Comparison."

34. Ernst Mayr, *The Growth of Biological Thought: Diversity, Evolution, and Inheritance* (Cambridge: Belknap Press, 1982), 455.

35. Schmidt-Nielsen's short book *How Animals Work* (1972) summarized his early research in physiological ecology and epitomized his "engineering" perspective. The approach was broadly comparative and explicitly focused on adaptation, but without mention of natural selection or other evolutionary mechanisms.

36. Bartholomew, "Distribution of Terrestrial Vertebrates."

37. Milam, "Equally Wonderful Field."

38. Bartholomew, "Role of Natural History," 329.

Chapter 7

1. Walter B. Cannon, *The Wisdom of the Body*, 2nd ed. (New York: Norton, 1939), 177. References to the thermostat analogy were scattered throughout the book; see 185, 199, 289 for examples.

2. Cannon acknowledged comparable thermoregulation in birds, but he had little to say about them. He justified this omission because his own work focused on mammals but also because he claimed that little was known about thermoregulation in birds. *Wisdom of the Body*, 2nd ed., 300–301.

3. P. F. Scholander, Walter Flagg, Vladimir Walters, and Laurence Irving, "Climatic Adaptation in Arctic and Tropical Poikilotherms," *Physiological Zoology* 26 (1953): 67–92, 84.

4. C. Ladd Prosser, "Theory of Physiological Adaptation of Poikilotherms to Heat and Cold," Brody Memorial Lecture 5, *University of Missouri Agricultural Station Special Report #59*, 1965.

5. Brian K. McNab, "Energy Expenditure: A Brief History," in *Mammalian Energetics: Interdisciplinary Views of Metabolism and Reproduction*, ed. Thomas Edward Tomasi and Teresa Helen Horton (Ithaca: Cornell University Press, 1992), 1–15.

6. George A. Bartholomew and Tom J. Cade, "Temperature Regulation, Hibernation, and Aestivation in the Little Pocket Mouse, *Perognathus longimembris*," *Journal of Mammalogy* 38 (1957): 60–72.

7. George A. Bartholomew, "Aspects of Timing and Periodicity of Heterothermy," in *Hibernation and Hypothermia: Perspectives and Challenges*, ed. Frank E. South, John P.

Hannon, John R. Willis, Eric T. Pengelley, and Norman R. Alpert, 663–80 (New York: Elsevier, 1972); George A. Bartholomew, "A Matter of Size: An Examination of Endothermy in Insects and Terrestrial Vertebrates," in *Insect Thermoregulation*, ed. Bernd Heinrich (New York: John Wiley, 1981), 45–78.

8. Bartholomew, "Matter of Size," 46.

9. Bernd Heinrich, "A Brief Historical Survey," in *Insect Thermoregulation*, ed. Bernd Heinrich (New York: John Wiley, 1981), 7–18.

10. Bernd Heinrich, "Thoracic Temperature Stabilization by Blood Circulation in a Free-Flying Moth," *Science* 168: (1970): 580–82; Bernd Heinrich and George A. Bartholomew, "An Analysis of Pre-flight Warm-Up in the Sphinx Moth, *Manduca sexta*," *Journal of Experimental Biology* 55 (1971): 223–39; Bernd Heinrich and George A. Bartholomew, "Temperature Control in Flying Moths," *Scientific American* 226, no. 6 (June 1972): 70–77; Bartholomew, "Matter of Size."

11. Bartholomew, "Matter of Size," 72–74.

12. Bartholomew, "Matter of Size," 46.

13. Bartholomew, "Matter of Size," 71.

14. George A. Bartholomew, "Interspecific Comparison as a Tool for Ecological Physiologists," in *New Directions in Ecological Physiology*, ed. Martin E. Feder, Albert F. Bennett, Warren W. Burggren, and Raymond B. Huey (Cambridge: Cambridge University Press, 1987), 22.

15. P. F. Scholander, "Evolution of Climatic Adaptation in Homeotherms," *Evolution* 9 (1955): 15–26.

16. Brian K. McNab, *Extreme Measures: The Ecological Energetics of Birds and Mammals* (Chicago: University of Chicago Press), 2012.

17. Robert Elsner, "The Irving–Scholander Legacy in Polar Physiology," *Comparative Biochemistry and Physiology—Part A* 126 (2000): 137–42; William R. Dawson, "Laurence Irving: An Appreciation," *Physiological and Biochemical Zoology* 80 (2007): 9–24; Joel B. Hagen, "Camels, Cormorants, and Kangaroo Rats: Integration and Synthesis in Organismal Biology after World War II," *Journal of the History of Biology* 48 (2015): 169–99; Joel B. Hagen, "Bergmann's Rule, Adaptation, and Thermoregulation in Arctic Animals: Conflicting Perspectives from Physiology, Evolutionary Biology, and Physical Anthropology after World War II," *Journal of the History of Biology* 50 (2017): 235–65.

18. P. F. Scholander, *Enjoying a Life in Science: The Autobiography of P. F. Scholander* (Fairbanks: University of Alaska Press, 1990); Knut Schmidt-Nielsen, "Per Scholander 1905–1980," *Biographical Memoirs of the National Academy of Sciences*, Washington DC, 1987, 387–412; Joel B. Hagen, "The Diving Reflex and Asphyxia: Working across Species in Physiological Ecology," *History and Philosophy of the Life Sciences* 40, no. 1 (2018): 18.

19. P. F. Scholander, Vladimir Walters, Raymond Hock, and Laurence Irving, "Body Insulation of Some Arctic and Tropical Mammals and Birds," *Biological Bulletin* 99 (1950): 225–36; Laurence Irving, "Heterothermous Operation of Warm-Blooded

Animals," *Physiologist* 2 (1959): 18–32; Laurence Irving, "Heterothermy in the Cold Adaptations of Warm Blooded Animals," in *Comparative Physiology of Temperature Regulation*, ed. John P. Hannon (Fort Wainwright, AK: Arctic Aeromedical Laboratory, 1962), 133–74.

20. P. F. Scholander, Raymond Hock, Vladimir Walters, Fred Johnson, and Laurence Irving, "Heat Regulation in Some Arctic Mammals and Birds," *Biological Bulletin* 99 (1950): 237–58; P. F. Scholander, Raymond Hock, Vladimir Walters, and Laurence Irving, "Adaptation to Cold in Arctic and Tropical Mammals and Birds in Relation to Body Temperature, Insulation, and Basal Metabolic Rate," *Biological Bulletin* 99 (1950): 259–71.

21. Irving, "Heterothermous Operation," 25.

22. P. F. Scholander, Flagg, Walters, and Irving, "Adaptation in Arctic and Tropical Poikilotherms."

23. Scholander, "Evolution of Climatic Adaptation," 15; Laurence Irving, "The Usefulness of Scholander's Views on Adaptive Insulation of Animals," *Evolution* 11 (1957): 257–59, 259.

24. Scholander, "Evolution of Climatic Adaptation."

25. The correspondence between the two scientists includes letters from Mayr to Scholander, August 24, 1955, and September 14, 1955; Scholander to Mayr, September 22, 1955, Correspondence boxes 4 and 17, Ernst Mayr Papers, Pusey Library, Harvard University.

26. Scholander to Mayr, September 22, 1955.

27. Ernst Mayr, "Geographical Character Gradients and Climatic Adaptation," *Evolution* 10 (1956): 105–8; P. F. Scholander, "Climatic Rules," *Evolution* 10 (1956): 339–40; Marshall T. Newman, "Adaptation of Man to Cold Climates," *Evolution* 10 (1956): 101–5; Irving, "Usefulness of Scholander's Views"; Ernst Mayr, *Animal Species and Evolution* (Cambridge: Belknap Press, 1963).

28. Ernst Mayr, *Systematics and the Origin of Species* (New York: Columbia University Press, 1942), 94.

29. In his meticulous survey of physiological ecology, Brian McNab highlights the failed reconciliation of physiological and evolutionary perspectives. Like other physiological ecologists, McNab paid tribute to the pioneering work of Scholander and Irving on thermoregulation and agreed that surface area and volume provide an inadequate explanation for Bergmann's rule. At the same time, he criticized the early physiological ecologists for making broad evolutionary generalizations that largely ignored the importance of phylogeny and natural selection; see Brian K. McNab, *The Physiological Ecology of Vertebrates: A View from Energetics* (Cornell University Press, 2002), 92; and Brian K. McNab, *Extreme Measures*, 26, 93–100.

30. Scholander, "Evolution of Climatic Adaptation," 23; Scholander, "Climatic Rules."

31. Irving, "Heterothermous Operation," 20.

32. Matthew Farish, "Creating Cold War Climates: The Laboratories of American

Globalism," in *Environmental Histories of the Cold War*, ed. J. R. McNeill and Corinna R. Unger (Cambridge: Cambridge University Press, 2010), 51–84; Matthew Farish, "The Lab and the Land: Overcoming the Arctic in Cold War Alaska," *Isis* 104 (2013): 1–29.

33. Karen Brewster, "Native Contributions to Arctic Science at Barrow, Alaska," *Arctic* 50 (1997): 277–88; Dawson, "Laurence Irving: An Appreciation"; Hagen, "Bergmann's Rule."

34. Correspondence concerning the planning of these expeditions can be found in the Per Fredrik Scholander Papers, University of California at San Diego Archives, boxes 7 and 8.

35. Carleton S. Coon, *The Origin of Races* (New York: Knopf, 1962), 60–68.

36. P. F. Scholander, H. T. Hammel, K. Lange Andersen, and Y. Løyning, "Metabolic Acclimatization to Cold in Man," *Journal of Applied Physiology* 12 (1958): 1–8; H. T. Hammel, "Terrestrial Animals in Cold: Recent Studies of Primitive Man," in David Bruce Dill, Edward F. Adolph, and C. G. Wilber, eds., *Handbook of Physiology. Section 4: Adaptation to the Environment* (Washington, DC: American Physiological Society, 1964), 413–34; Scholander, *Enjoying a Life in Science*, chaps. 17 and 18. Extensive correspondence related to planning the expedition is in box 8, Scholander Papers.

37. P. F. Scholander, H. T. Hammel, J. S. Hart, D. H. LeMessurier, and J. Steen, "Cold Adaptation in Australian Aborigines," *Journal of Applied Physiology* 13 (1958): 211–18. See correspondence between Scholander and Australian biologists, including Cedric Stanton Hicks, about the expedition, box 7, Scholander Papers. Hicks's earlier research during the 1930s on thermoregulation in Australian Aborigines is described by Warwick Anderson, *The Cultivation of Whiteness: Science, Health, and Racial Destiny in Australia* (New York: Basic Books, 2003), 211–14. Anderson provides a detailed study of shifting views on genetics, environment, and race in Australian anthropology and medical research prior to World War II.

38. Arlene Pastorius, "'Ice-Box' Treatment Tests Native Resistance to Cold," *University of Washington Daily*, April 17, 1958. A copy of this news article is in box 17, Scholander Papers.

39. H. T. Hammel, "Terrestrial Animals in Cold."

40. Coon, *Origin of Races*, 40.

41. Laurence Irving, "Human Adaptation to Cold," *Nature* 185 (1960): 572–74. Though hardly conducting a definitive study, Irving wanted to make his point in a striking and provocative way; see the letter from Irving to Scholander, May 27, 1959, in series 2, box 12, Laurence Irving Papers, University of Alaska Archives. Irving later placed this experiment within the broader context of heterothermy in human and arctic animals in his longer review article "Heterothermy in the Cold Adaptations."

42. David H. Price, *Threatening Anthropology: McCarthyism and the FBI's Surveillance of Activist Anthropologists* (Durham, NC: Duke University Press, 2004), 177–84. According to Price, the FBI concluded that Newman was not a Communist but that his "poor judgement" might lead him to be duped into subversive activities. Newman's

supervisor at the Smithsonian Institution criticized his "championing the underdog," particularly in racial matters. This caused particular concern when Newman sought permission to act as an expert witness for the NAACP in the case of a couple accused of violating Virginia's miscegenation statute.

43. Marshall T. Newman, "Review of *Races: A Study of the Problems of Race Formation in Man*," *Boletín Bibliográfico de Antropología Americana* 13 (1950): 188–92.

44. Marshall T. Newman, "The Application of Ecological Rules to the Racial Anthropology of the Aboriginal New World," *American Anthropologist* 55 (1953): 311–27; Newman, "Adaptation of Man."

45. Mayr, *Animal Species and Evolution*, 644–48, 657–58.

46. John Jackson, "In Ways Unacademical: The Reception of Carleton S. Coon's *The Origin of Races*," *Journal of the History of Biology* 34 (2001): 247–85; Paul Lawrence Farber, *Mixing Races: From Scientific Racism to Modern Evolutionary Ideas* (Baltimore: Johns Hopkins University Press, 2011); Paul Lawrence Farber, "Dobzhansky and Montagu's Debate on Race: The Aftermath," *Journal of the History of Biology* 49 (2016): 625–39; see also Peter Sachs Collopy, "Race Relationships: Collegiality and Demarcation in Physical Anthropology," *Journal of the History of the Behavioral Sciences* 51 (2015): 237–60; Rachel Caspari, "From Types to Populations: A Century of Race, Physical Anthropology, and the American Anthropological Association," *American Anthropologist* 105 (2003): 65–76.

47. Mayr, *Animal Species and Evolution*, 643–44, 47.

48. Mayr wrote a long and rather glowing review of Coon's *The Origin of Human Races*: Ernst Mayr, "Origin of Human Races," *Science* 138 (1962): 420–22.

49. Coon thanked Mayr for sending a reprint of his response to Scholander in *Evolution* and urged him to send a copy to Ted Hammel, the University of Pennsylvania physiologist who worked with Scholander on human thermoregulation; letter from Coon to Mayr, January 27, 1958, box 5, Mayr Papers.

50. William Provine describes the influence of I. Michael Lerner and Theodosius Dobzhansky on Mayr's ideas about genetic homeostasis at the population level; see William B. Provine, "Ernst Mayr: Genetics and Speciation," *Genetics* 167 (2004): 1041–46. However, Mayr's interest in a physiological explanation for Bergmann's rule in terms of thermoregulation suggests that his interests in homeostasis were broad and provided a promising basis for unifying organismal and population biology.

51. Elsner, "Scholander–Irving Legacy"; Dawson, "Laurence Irving."

Chapter 8

1. Ernst Mayr, *The Growth of Biological Thought: Diversity, Evolution, and Inheritance* (Cambridge: Belknap Press, 1982), 73–74.

2. John Beatty, "The Proximate–Ultimate Distinction in the Multiple Careers of Ernst Mayr," *Biology and Philosophy* 9 (1994): 333–56; Vassiliki Betty Smocovitis, *Unifying Biology: The Evolutionary Synthesis and Evolutionary Biology* (Princeton: Princeton

University Press, 1996), 176–78; Michael R. Dietrich, "Paradox and Persuasion: Negotiating the Place of Molecular Evolution within Evolutionary Biology," *Journal of the History of Biology* 31 (1998): 85–111; Joel B. Hagen, "Naturalists, Molecular Biologists, and the Challenges of Molecular Evolution," *Journal of the History of Biology* 32 (1999): 321–41; Erika Lorraine Milam, "The Equally Wonderful Field: Ernst Mayr and Organismic Biology," *Historical Studies in the Natural Sciences* 40 (2010): 279–317. Some critics contend that this philosophical distinction has a pernicious effect on evolutionary biology; see Kevin N. Laland, Kim Sterelny, John Odling-Smee, William Hoppitt, and Tobias Uller, "Cause and Effect in Biology Revisited: Is Mayr's Proximate–Ultimate Dichotomy Still Useful?" *Science* 334 (2011): 1512–16.

3. Mayr, "Cause and Effect," 1505.

4. For three letters and Mayr's response, see William H. Kane, Silvio Fiala, C. H. Waddington, and Ernst Mayr, "On Cause and Effect in Biology," *Science* 135 (1962): 972–81.

5. Jonathan M. W. Slack, "Conrad Hal Waddington: The Last Renaissance Biologist?" *Nature Reviews Genetics* 3 (2002): 889–94.

6. The meeting was held at Villa Serbelloni in Bellagio, Italy, in 1966, and the proceedings were published in C. H. Waddington, ed., *Towards a Theoretical Biology* (Chicago: Aldine, 1968). Although Mayr's paper was apparently the point of departure for planning the meeting, he later acknowledged that the participants were more interested in mathematical and theoretical biology than in the more philosophical issues raised by "Cause and Effect in Biology." See the correspondence between Waddington and Mayr about planning the meeting beginning in January 1962 (box 9) and Mayr's letter to Waddington on November 9, 1966 (box 14) in the Ernst Mayr Papers, Pusey Library, Harvard University.

7. Mayr later revised his historical account of the rise of philosophy of biology so substantially that Waddington was completely ignored in *The Growth of Biological Thought.*

8. Waddington's letter to the editor of *Science* was reprinted as a commentary to Mayr's paper in Waddington, *Towards a Theoretical Biology*, 55–56.

9. Letter from Mayr to Waddington, January 29, 1962, box 9, Mayr Papers.

10. See the letter from Mayr to Waddington, January 19, 1954, box 2, Mayr Papers.

11. Ernst Mayr, *Animal Species and Evolution* (Cambridge: Belknap Press, 1963), preface.

12. Mayr, *Growth of Biological Thought*, 66–67; William B. Provine, "Ernst Mayr: Genetics and Speciation," *Genetics* 167 (2004): 1041–46.

13. Mayr, *Animal Species and Evolution*, 60–66.

14. Mayr, *Animal Species and Evolution*, 277.

15. Mayr, *Animal Species and Evolution*, 275–78.

16. Mayr, *Animal Species and Evolution*, 264.

17. Mayr, *Animal Species and Evolution*, 555.

18. Provine, "Ernst Mayr."

19. Mayr, *Animal Species and Evolution*, 263, 542 (homeostasis as conservative force), and 333 (ecotypic variation as a centrifugal force).

20. I. Michael Lerner, *Genetic Homeostasis* (New York: John Wiley, 1954); Brian K. Hall, "Fifty Years Later: I. Michael Lerner's *Genetic Homeostasis* (1954)—A Valiant Attempt to Integrate Genes, Organisms, and Environment," *Journal of Experimental Zoology* 304B (2005): 187–97; R. W. Allard, "Israel Michael Lerner, 1910–1977," *Biographical Memoirs of the National Academy of Sciences*, Washington DC, 1996: 165–75.

21. Lerner, *Genetic Homeostasis*, 1–7.

22. Lerner, *Genetic Homeostasis*, 10.

23. Lerner, *Genetic Homeostasis*, 64.

24. Lerner, *Genetic Homeostasis*, 12–13.

25. Lerner, *Genetic Homeostasis*, 5.

26. Lerner, *Genetic Homeostasis*, 64.

27. Richard C. Lewontin, *The Genetic Basis of Evolutionary Change* (New York: Columbia University Press, 1974); John Beatty, "Weighing the Risks: Stalemate in the Classical/Balance Controversy," *Journal of the History of Biology* 20 (1987): 289–319; Michael R. Dietrich, "The Origins of the Neutral Theory of Molecular Evolution," *Journal of the History of Biology* 27 (1994): 21–59.

28. Theodosius Dobzhansky, *Genetics and the Origin of Species* (New York: Columbia University Press, 1937), 124. The passage in slightly modified forms was repeated in all of the later editions of Dobzhansky's book, as well as in his later *Genetics of the Evolutionary Process* (New York: Columbia University Press, 1970), 100.

29. Theodosius Dobzhanksy, "Genetic Homeostasis," *Evolution* 9 (1955): 100–101; Bruce Wallace, "Genetic Homeostasis," *Science* 121 (1955): 558. Dobzhansky and Wallace acknowledged their intellectual debt to Lerner in Theodosius Dobzhansky and Bruce Wallace, "The Genetics of Homeostasis in *Drosophila*," *Proceedings of the National Academy of Sciences* 39 (1953): 162–71.

30. Dobzhansky and Wallace, "Genetics of Homeostasis in *Drosophila*."

31. Dobzhansky and Wallace, "Genetics of Homeostasis in *Drosophila*."

32. R. C. Lewontin, "Studies on Homeostasis and Heterozygosity I. General Considerations. Abdominal Bristle Number in Second Chromosome Homozygotes of *Drosophila melanogaster*," *American Naturalist* 90 (1956): 237–55.

33. Lewontin, "Studies on Homeostasis and Heterozygosity I," 238.

34. R. C. Lewontin, "Studies on Heterozygosity and Homeostasis II: Loss of Heterosis in a Constant Environment," *Evolution* 12 (1958): 494–503.

35. C. H. Waddington, "The Resistance to Evolutionary Change," *Nature* 175 (1955): 51–52.

36. C. H. Waddington, *The Strategy of the Genes: A Discussion of Some Aspects of Theoretical Biology* (London: George Allen and Unwin, 1957).

37. Brian K. Hall, "Waddington's Legacy in Development and Evolution," *American*

Zoologist 32 (1992): 113–22; Mark Siegal and Aviv Bergman, "Waddington's Canalization Revisited: Developmental Stability and Evolution," *Proceedings of the National Academy of Sciences* 99 (2002): 10528–32; Erika Crispo, "The Baldwin Effect and Genetic Assimilation: Revisiting Two Mechanisms of Evolutionary Change Mediated by Phenotypic Plasticity," *Evolution* 61 (2007): 2469–79; Jan Baedke, "The Epigenetic Landscape in the Course of Time: Conrad Hal Waddington's Methodological Impact on the Life Sciences," *Studies in History and Philosophy of the Biological and Biomedical Sciences* 44 (2013): 756–73.

38. C. H. Waddington, "Canalization of Development and the Inheritance of an Acquired Trait," *Nature* 150 (1942): 563–65. These ideas were much more fully developed in Waddington, *Strategy of the Genes.*

39. George Gaylord Simpson, "The Baldwin Effect," *Evolution* 7 (1953): 110–17; Mayr, *Animal Species and Evolution,* 190, 610–12. Although later critical of the idea, Mayr was interested in genetic assimilation, and he discussed it with Waddington during the writing of *The Strategy of the Genes*; see Waddington's letters of February 27, 1956, and March 29, 1956, and Mayr's letter of March 12, 1956, all of which are in box 4, Mayr Papers. For a historical and philosophical analysis of the genetic assimilation and related evolutionary concepts, see Brian K. Hall, "Organic Selection: Proximate Environmental Effects on the Evolution of Morphology and Behavior," *Biology and Philosophy* 16 (2001): 215–37.

40. Simpson, "Baldwin Effect," in particular, placed genetic assimilation within a historical context that emphasized the neo-Lamarckian ties.

41. Scott F. Gilbert and David Epel, *Ecological Developmental Biology: Integrating Epigenetics, Medicine, and Evolution* (Sunderland, MA: Sinnauer, 2009), 441–46.

42. Theodosius Dobzhansky, *The Biological Basis of Human Freedom* (New York: Columbia University Press, 1956). Dobzhansky's lectures were reported in two articles in the *Charlottesville Daily Progress* (March 2 and 3, 1954) and three articles in the university newspaper, the *Cavalier Daily* (March 2–4, 1954). Although summarizing the highlights of Dobzhansky's discussion of human evolution, none of the articles mentioned race.

43. One of five cases combined in the Supreme Court decision involved a lawsuit against the school board of Prince Edward County, less than one hundred miles away from the University of Virginia. The case drew attention because it originated as a walkout by 450 students from the segregated (and decidedly unequal) Moton High School in the town of Farmville.

44. For the broader context of Dozhansky's participation in the nature–nurture debate, see Diane B. Paul, "Dobzhansky in the Nature–Nurture Debate," in *The Evolution of Theodosius Dobzhansky*, ed. Mark B. Adams (Princeton: Princeton University Press, 1994), 219–31; Diane B. Paul, *The Politics of Heredity: Essays on Eugenics, Biomedicine, and the Nature–Nurture Debate* (Albany: State University of New York Press,

1998); Erika Lorraine Milam, *Creatures of Cain: The Hunt for Human Nature in Cold War America* (Princeton: Princeton University Press, 2018), chaps. 1 and 2.

45. Dobzhansky, *Biological Basis*, 74.

46. Dobzhansky, *Biological Basis*, 78.

47. For a detailed examination of Dobzhansky's role in these controversies, see John Beatty, "Dobzhansky and the Politics of Democracy: The Moral and Political Significance of Genetic Variation," in *The Evolution of Theodosius Dobzhansky*, ed. Mark B. Adams (Princeton: Princeton University Press, 1994), 195–218.

48. Dobzhansky was particularly critical of arguments for human degeneration made by Robert C. Cook, *Human Fertility: The Modern Dilemma* (New York: Sloane, 1951); and Garrett Hardin, *Nature and Man's Fate* (New York: Rhinehart), 1959.

49. Curt R. Richter, "Rats, Man, and the Welfare State," *American Psychologist* 14 (1959): 18–28; Jay Schulkin, *Curt Richter: A Life in the Laboratory* (Baltimore: Johns Hopkins University Press, 2005).

50. Ernst Mayr, *Animal Species and Evolution*, 661–62. For his more egalitarian views on teamwork in the coadapted genotype, see 264, 277, 280, and 295.

51. Dobzhansky, *Biological Basis*, 80–85, 132–35; Theodosius Dobzhansky, *Mankind Evolving: The Evolution of the Human Species* (New Haven: Yale University Press, 1962), 315–18, 325–27.

52. Dobzhansky, *Mankind Evolving*, 327–30.

53. Dobzhansky, *Biological Basis*, 131.

54. C. H. Waddington, *The Ethical Animal* (London: George Allen and Unwin, 1960), chaps. 2 and 8. For the broader historical context of Waddington's and Dobzhansky's ideas on ethics, see Paul Lawrence Farber, *The Temptations of Evolutionary Ethics* (Berkeley: University of California Press, 1994).

55. Dobzhansky, *Mankind Evolving*, 344.

56. Dobzhansky, *Mankind Evolving*, 322–25.

57. I. Michael Lerner, *Heredity, Evolution and Society* (San Francisco: Freeman, 1968).

58. Lerner, *Heredity, Evolution and Society*, 273.

59. Lerner, *Heredity, Evolution and Society*, x.

60. Lerner, *Heredity, Evolution and Society*, 292.

61. Walter Landauer, "The Genetics of Dynamic Equilibria," *American Naturalist* 89 (1955): 183–85.

62. R. C. Lewontin, "The Adaptations of Populations to Varying Environments," *Cold Spring Harbor Symposia on Quantitative Biology* 22 (1956): 395–408.

63. Richard Dawkins, *The Selfish Gene* (Oxford: Oxford University Press, 1976). Although Dawkins's book was a critical and commercial success, it popularized ideas already developed by Dawkins and likeminded evolutionary biologists over the preceding decade.

64. Dawkins, *Selfish Gene*, 92.

65. Mayr, *Growth of Biological Thought*, 66–67, 588–91.

66. Dawkins, *Selfish Gene*, 12. As Dawkins acknowledged, his ideas on the units of selection were heavily influenced by George C. Williams. For a useful analysis of Dawkins's thinking about genes and organisms, see Kim Sterelny, *Dawkins vs. Gould: Survival of the Fittest* (Cambridge, UK: Icon Books, 2007), 9–11, 24–27.

Chapter 9

1. The idea of species as evolving populations was widely shared by proponents of the modern synthesis, although they differed on important details such as the origins of reproductive isolation; see Ernst Mayr, *The Growth of Biological Thought: Diversity, Evolution, and Inheritance* (Cambridge,: Belknap Press, 1982), 273–75.

2. C. Ladd Prosser, ed., *Comparative Animal Physiology* (Philadelphia: Saunders, 1950), 2.

3. Prosser defined homeostasis broadly to include not only maintaining constancy but also resiliency and conforming to environmental change in an orderly manner; see C. Ladd Prosser, "Theory of Physiological Adaptation of Poikilotherms to Heat and Cold," Brody Memorial Lecture 5, *University of Missouri Agricultural Station Special Report #59*, 1965.

4. Prosser, *Comparative Animal Physiology*, 5.

5. C. Ladd Prosser, "The 'Origin' after a Century: Prospects for the Future," *American Scientist* 47 (1959): 536–60; C. Ladd Prosser, "The Species Problem from the Viewpoint of a Physiologist," in *The Species Problem*, ed. Ernst Mayr, 339–69 (Washington, DC: American Association for the Advancement of Science, 1957); C. Ladd Prosser, "Comparative Physiology in Relation to Evolutionary Theory," in *Evolution after Darwin*, ed. Sol Tax (Chicago: University of Chicago Press, 1960), 569–94. For the broader historical context of the Darwin centennial celebrations in the United States, see Vassiliki Betty Smocovitis, "The 1959 Darwin Centennial Celebration in America," *Osiris* 14 (1999): 274–323.

6. See the letters from Mayr to Prosser, February 6 and March 12, 1956, in box 4, Ernst Mayr Papers, Pusey Library, Harvard University.

7. Bruce Alberts, "John Alexander Moore, 1915–2002," in *Biographical Memoirs of the National Academy of Sciences* (Washington DC, 2011), 1–24; Rudolfo Ruibel, Vaughn H. Shoemaker, and Margaret M. Stewart, "Historical Perspective: John A. Moore," *Copeia* 4 (2001): 1155–57.

8. Rony Armon, "Between Biochemists and Embryologists—The Biochemical Study of Embryonic Induction in the 1930s," *Journal of the History of Biology* 45 (2012): 65–108.

9. John A. Moore, "Temperature Tolerances and Rates of Development in the Eggs of Amphibia," *Ecology* 20 (1939): 459–78. Although the main focus was on frogs, Moore also reported results for two species of toads (*Bufo*) and four species of salamanders (*Ambystoma*).

10. Theodosius Dobzhansky, *Genetics and the Origin of Species*, 2nd ed. (New York: Columbia University Press, 1941), 206.

11. Joel B. Hagen, "Experimentalists and Naturalists in Twentieth-Century Botany: Experimental Taxonomy, 1920–1950," *Journal of the History of Biology* 17 (1984): 249–70; Robert E. Kohler, *Landscapes and Labscapes: Exploring the Lab-Field Border in Biology* (Chicago: University of Chicago Press, 2002), chap. 5.

12. John A. Moore, "The Role of Temperature in the Speciation of Frogs," *Biological Symposia* 6 (1942): 189–213, 211–12.

13. John A. Moore, "Geographic Variation of Adaptive Characters in *Rana pipiens* Schreber," *Evolution* 3 (1949): 1–24, 22; Moore, "Role of Temperature in the Speciation of Frogs."

14. Alfred C. Weed, "New Frogs from Minnesota," *Proceedings of the Biological Society of Washington* 35 (1922): 107–10.

15. John A. Moore, "An Embryological and Genetical Study of *Rana burnsi* Weed," *Genetics* 27 (1942): 408–16.

16. Later studies by population geneticists more convincingly demonstrated that the color patterns described by Weed resulted from the interaction of at least two sets of alleles at loci on two different chromosomes in the frog genome; see David J. Merrell, "Migration and Gene Dispersal in *Rana pipiens*," *American Zoologist* 10 (1970): 47–52.

17. John Alexander Moore, "Geographic Variation in *Rana pipiens* Schreber of the Eastern United States," *Bulletin of the American Museum of Natural History* 82 (1944): 345–70.

18. John A. Moore, "*R. pipiens*: The Changing Paradigm," *American Zoologist* 15 (1975): 837–49.

19. Moore, "Geographic Variation in *R. pipiens*," 351.

20. Joel B. Hagen, "The Statistical Frame of Mind in Systematic Biology from *Quantitative Zoology* to *Biometry*," *Journal of the History of Biology* 36 (2003): 353–84.

21. Moore, "Geographic Variation in *R. pipiens*," 368.

22. This is not to claim that Mayr and Dobzhansky shared exactly the same species concept. Dobzhansky argued that species were a *stage* in the process of speciation, while Mayr claimed that species were the *result* of the process. Mayr and Dobzhansky differed more substantively on the evolution of isolating mechanisms. Dobzhansky had argued that natural selection against hybrids was the final step in speciation, when two previously separate populations once again became sympatric. Mayr consistently argued that isolating mechanisms were by-products of divergence between populations resulting from adaptation to different environments. This disagreement provided an important context for Moore's embryological experiments, and he used his results to support Mayr's claim; see John A. Moore, "An Embryologist's View of the Species Problem," *The Species Problem*, ed. Ernst Mayr (Washington, DC: American Association for the Advancement of Science, 1957), 325–38; see also Mayr, *Growth of Biological Thought*, 274.

23. Frogs were so ubiquitous in experimental biology that together with its close relatives, *R. pipiens* sometimes seemed to serve as the "representative vertebrate" or even "the animal" for physiologists; see John A. Moore, "Preface," in *The Physiology of the Amphibia*, ed. John A. Moore, (New York: Academic Press, 1964); Frederic L. Holmes, "The Old Martyr of Science: The Frog in Experimental Physiology," *Journal of the History of Biology* 26 (1993): 311–28; Rachel A. Ankeny and Sabina Leonelli, "What's So Special about Model Organisms?" *Studies in History and Philosophy of Science* 42 (2011): 313–23.

24. John A. Moore, "Studies on the Development of Frog Hybrids. II. Competence of the Gastrula Ectoderm of *Rana pipiens* x *Rana sylvatica* Hybrids," *Journal of Experimental Zoology* 105 (1947): 349–70.

25. John A. Moore, "Incipient Intraspecific Isolation Mechanisms in *R. pipiens*," *Genetics* 31 (1946): 304–26; John A. Moore, "Further Studies on *R. pipiens* Racial Hybrids," *American Naturalist* 84 (1950): 247–54.

26. Hagen, "Experimentalists and Naturalists in Twentieth-Century Botany."

27. Moore, "*R. pipiens*: The Changing Paradigm."

28. M. J. Littlejohn and R. S. Oldham, "*Rana pipiens* Complex: Mating Call Structure and Taxonomy," *Science* 162 (1968): 1003–5. For informative histories of the development and use of sound spectrographs by ornithologists and behavioral biologists, see Don Stap, *Birdsong: A Natural History* (New York: Scribner, 2005); and Joeri Bruyninckx, *Listening in the Field: Recording and the Science of Birdsong* (Cambridge: MIT Press, 2018).

29. Although the development of electrophoresis dates back to the 1930s, its use in evolutionary and systematic research was an outgrowth of pioneering studies by Richard Lewontin, Jack Hubby, and Harry Harris during the mid-1960s. For the history of this development, see Howard Hsueh-Hao Chiang, "The Laboratory Technology of Discrete Molecule Separation: The Historical Development of Gel Electrophoresis and the Material Epistemology of Biomolecular Science, 1945–1970," *Journal of the History of Biology* 42 (2009): 495–527; R. C. Lewontin, "Twenty-Five Years in Genetics: Electrophoresis in the Development of Evolutionary Genetics: Milestone or Millstone?," *Genetics* 128 (1991): 657–62; Michael Dietrich, "The Electrophoresis Revolution," *Perspectives on Molecular Evolution* (2004), http://authors.library.caltech.edu/5456/1/hrst.mit.edu/hrs/evolution/public/techniques/electrophoresis.html.

30. Stanley N. Salthe, "Geographic Variation of the Lactate Dehydrogenases of *Rana pipiens* and *Rana palustris*," *Biochemical Genetics* 2 (1969): 271–303.

31. David M. Hillis, "Systematics of the *R. pipiens* Complex: Puzzle and Paradigm," *Annual Review of Ecology and Systematics* 19 (1988): 39–63.

32. Theodosius Dobzhansky, *Genetics of the Evolutionary Process* (New York: Columbia University Press, 1970), 332–33.

33. Richard D. Sage and Robert K. Selander, "Hybridization between Species of the *Rana pipiens* Complex in Central Texas," *Evolution* 33 (1979): 1069–88; see also Lauren

Brown, "Speciation in the *Rana pipiens* Complex," *American Zoologist* 13 (1973): 73–79; and Ann E. Pace, *Systematic and Biological Studies of the Leopard Frogs (*Rana pipiens *Complex) of the United States,* Miscellaneous Publications of the Museum of Zoology No. 148. Lansing: University of Michigan, 1974.

34. For the importance of research schools in shaping the development of scientific traditions, see Gerald L. Geison, "Research Schools and New Directions in the Historiography of Science," *Osiris* 8 (1993): 227–38.

35. Moore, "*R. pipiens*: The Changing Paradigm."

36. Ernst Mayr, *Animal Species and Evolution* (Cambridge: Belknap Press, 1963), 32–34, 38. Moore would also certainly have been aware of Dobzhansky's important studies of sibling species in *Drosophila*. It is significant that Moore himself identified sibling species that were morphologically indistinguishable but genetically isolated—although he conducted this research on Australian amphibians rather than North American leopard frogs.

37. The existence of these evolutionary "puzzles" is described by Hillis, "Systematics of the *R. pipiens* Complex"; David M. Hillis and Thomas P. Wilcox, "Phylogeny of the New World True Frogs (*Rana*)," *Molecular Phylogeny and Evolution* 34 (2005): 299–314.

38. Moore, "*R. pipiens*: The Changing Paradigm," 849.

Chapter 10

1. Introduction in Norbert Wiener, *Cybernetics: Or Control and Communication in the Animal and the Machine* (Cambridge: MIT Press, 1948); David A. Mindell, *Between Human and Machine: Feedback, Control, and Computing before Cybernetics* (Baltimore: Johns Hopkins University Press, 2002), 282–83; Debora Hammond, *The Science of Synthesis: Exploring the Social Implications of General Systems Theory* (Boulder: University Press of Colorado, 2003), chap. 4; Steven J. Cooper, "From Claude Bernard to Walter Cannon: Emergence of the Concept of Homeostasis," *Appetite* 51 (2008): 419–27; Eline L. Wolfe, A. Clifford Barger, and Saul Benison, *Walter B. Cannon, Science and Society* (Cambridge: Harvard University Press, 2000), 314, 476.

2. Ranulph Glanville, "Cybernetics: Thinking through Technology," 45–77; and John Bruni, "Expanding the Self-Referential Paradox: The Macy Conferences and the Second Wave of Cybernetic Thinking," in *Traditions of Systems Theory: Major Figures and Contemporary Developments,* ed. Darrell P. Arnold (New York: Routledge, 2014), 78–83.

3. Lawrence K. Frank, "Foreword," *Annals of the New York Academy of Sciences* 50 (1948): 189–96. Frank acknowledged that the title of the conference, "Teleological Mechanisms," might seem "perplexing" and "difficult to accept," but he argued that systems approaches allowed discussion of holism, purpose, and self-regulation without recourse to outmoded commitments to vitalism.

4. G. Evelyn Hutchinson, "Circular Causal Systems in Ecology," *Annals of the New York Academy of Sciences* 50 (1948): 221–46.

5. Sharon E. Kingsland, *Modeling Nature: Episodes in the History of Population*

Ecology, (Chicago: University of Chicago Press, 1985), chap. 2; Peter J. Taylor, "Technocratic Optimism, H. T. Odum, and the Partial Transformation of Ecological Metaphor after World War II," *Journal of the History of Biology* 21 (1988): 213–44.

6. For a broad overview of the importance of Hutchinson and his students on the development of ecology, see Nancy G. Slack, *G. Evelyn Hutchinson and the Invention of Modern Ecology* (New Haven: Yale University Press, 2011).

7. Hutchinson, "Circular Causal Systems," 237.

8. Sharon E. Kingsland, "Mathematical Figments, Biological Facts: Population Ecology in the Thirties," *Journal of the History of Biology* 19 (1986): 235–56; James P. Collins, "'Evolutionary Ecology' and the Use of Natural Selection in Ecological Theory," *Journal of the History of Biology* 19 (1986): 257–88.

9. Even critics of density dependence accepted self-regulation of population size in certain cases. The Australian ecologists H. G. Andrewartha and L. C. Birch forcefully argued for density independence in their book *The Distribution and Abundance of Animals* (Chicago: University of Chicago Press, 1954). However, Andrewartha conceded that self-regulation occurred in at least some species of birds and mammals with well-developed social behavior; see H. G. Andrewartha, "Self-Regulatory Mechanisms in Animal Populations," *Australian Journal of Sciences* 22 (1959): 200–205.

10. Hutchinson, "Circular Causal Systems," 242.

11. R. C. Lewontin, "The Units of Selection," *Annual Review of Ecology and Systematics* 1 (1970): 1–18, 14. The earlier and more detailed critique of the aesthetic bases for theories of population regulation is found in George C. Williams, *Adaptation and Natural Selection* (Princeton: Princeton University Press, 1966), 232–34. Shortly after Williams's and Lewontin's critiques, Frank Egerton traced the history of balance of nature ideas from antiquity to the controversies in modern ecology in "Changing Concepts of the Balance of Nature," *Quarterly Review of Biology* 48 (1973): 322–50. Updated analyses of these controversies are presented by Gregory J. Cooper, *The Science of the Struggle for Existence: On the Foundations of Ecology* (Cambridge: Cambridge University Press, 2003), chap. 3; and John Kricher, *The Balance of Nature: Ecology's Enduring Myth* (Princeton: Princeton University Press, 2009). A careful analysis of the complex relationship between balance of nature and equilibrium concepts (both metaphorical and mathematical) in population ecology is provided by Kim Cuddington, "The 'Balance of Nature' Metaphor and Equilibrium in Population Ecology, *Biology and Philosophy* 16 (2001): 463–79.

12. In attacking Allee, Williams referred to an obscure theoretical article that the older ecologist had written on the evolution of sociality. Soil formation and the activity of earthworms, which was the basis for Williams's critique, was not actually discussed by Allee in the article cited. See Williams, *Adaptation and Natural Selection*, 18; and W. C. Allee, "Concerning the Origin of Sociality in Animals," *Scientia* 67 (1940): 154–60.

13. W. C. Allee, *The Social Life of Animals* (New York: Norton, 1938), 49; W. C. Allee, Alfred E. Emerson, Orlando Park, Thomas Park, and Karl P. Schmidt, *Principles of Animal Ecology* (Chicago: University of Chicago Press, 1949), 484–85.

14. Allee et al., *Principles of Animal Ecology*, 672, 695, 728–29; W. C. Allee, *Animal Aggregations* (Chicago: University of Chicago Press, 1931), 81–83.

15. Gregg Mitman, *The State of Nature: Ecology, Community, and American Social Thought, 1900–1950* (Chicago: University of Chicago Press, 1992), chap. 2.

16. Allee, *Social Life of Animals*, 45–46.

17. Alfred E. Emerson, "Social Coordination and the Superorganism," *American Midland Naturalist* 21 (1939): 182–209; Alfred E. Emerson, "Regenerate Behavior and Social Homeostasis of Termites," *Ecology* 37 (1956): 248–58.

18. Allee et al., *Principles of Animal Ecology*, 664, 695.

19. Williams, *Adaptation and Natural Selection*, 4–5.

20. Mark E. Borello, *Evolutionary Restraints: The Contentious History of Group Selection* (Chicago: University of Chicago Press, 2010); see also William C. Kimler, "Advantage, Adaptiveness, and Evolutionary Ecology," *Journal of the History of Biology* 19 (1986): 215–33.

21. V. C. Wynne-Edwards, *Animal Dispersion in Relation to Social Behaviour* (London: Oliver and Boyd, 1962), 9, 13.

22. Wynne-Edwards, *Animal Dispersion*, 10–11.

23. Wynne-Edwards, *Animal Dispersion*, 485.

24. Collins, "Evolutionary Ecology."

25. David E. Davis, "Early Behavioral Research on Populations," *American Zoologist* 27 (1987): 825–37, 828; see also Dennis Chitty, *Do Lemmings Commit Suicide? Beautiful Hypotheses and Ugly Facts* (Oxford: Oxford University Press, 1996), viii, 148.

26. See the announcement of the Mercer Award in the *Bulletin of the Ecological Society of America* 39, no. 1 (1958): 2–3. The award-winning article was John J. Christian, "Adrenal and Reproductive Responses to Population Size in Mice from Freely Growing Populations," *Ecology* 37 (1956): 258–73. Biographical details are from Stuart O. Landry, "John Jermyn Christian: 1917–1997," *Journal of Mammalogy* 79 (1998): 1432–38.

27. Jay Schulkin, *Curt Richter: A Life in the Laboratory* (Baltimore: Johns Hopkins University Press, 2005); Christine Keiner, "Wartime Rat Control, Rodent Ecology, and the Rise of Chemical Rodenticides," *Endeavor* 29 (2005): 119–25; Edmund Ramsden and Jon Adams, "Escaping the Laboratory: The Rodent Experiments of John B. Calhoun & Their Cultural Influence," *Journal of Social History* 42 (2009): 761–92; Edmund Ramsden, "From Rodent Utopia to Urban Hell: Population, Pathology, and the Crowded Rats of NIMH," *Isis* 102 (2011): 659–88; Edmund Ramsden, "Rats, Stress, and the Built Environment," *History of the Human Sciences* 25 (2012): 123–47; Dawn Day Biehler, *Pests in the City: Flies, Bedbugs, Cockroaches, and Rats* (Seattle: University of Washington Press, 2013), chap. 4.

28. John J. Christian, "The Adreno-Pituitary System and Population Cycles in Mammals," *Journal of Mammalogy* 31 (1950): 247–59; John J. Christian and H. L. Ratcliffe, "Shock Disease in Captive Wild Mammals," *American Journal of Pathology* 28 (1952): 725–37.

29. R. G. Green and C. L. Larson, "Shock Disease and the Snowshoe Hare Cycle,"

Science 87 (1938): 298–99; R. G. Green and C. L. Larson, "A Description of Shock Disease in the Snowshoe Hare," *American Journal of Hygiene* 28 (1938): 190–212. Green's studies of shock disease and experimental treatment with glucose injections were reported earlier in *Minnesota Wildlife Disease Investigation* 2 (1935–1936) to the Minnesota Department of Conservation, Robert Gladding Green Papers, University of Minnesota Archives.

30. Susan D. Jones, "Population Cycles, Disease, and Networks of Ecological Knowledge," *Journal of the History of Biology* 50 (2017): 357–91.

31. Letter from Elton to Green, April 8, 1938, box 1, Green Papers.

32. Hans Selye, "A Syndrome Produced by Diverse Nocuous Agents," *Nature* 138 (1936): 32; Hans Selye, "The General Adaptation Syndrome and the Diseases of Adaptation," *Journal of Clinical Endocrinology* 6 (1946): 117–230; see also Hans Selye, *The Stress of Life* (New York: McGraw-Hill, 1956), 60.

33. Mark Jackson, *The Age of Stress: Science and the Search for Stability* (Oxford: Oxford University Press, 2013); Mark Jackson, "Evaluating the Role of Hans Selye in the Modern History of Stress," in *Stress, Shock, and Adaptation in the Twentieth Century*, ed. David Cantor and Edmund Ramsden (Rochester, NY: University of Rochester Press, 2014), 21–48.

34. Christian and Ratcliffe, "Shock Disease in Captive Wild Mammals"; Selye, "General Adaptation Syndrome."

35. Christian, "Adreno-Pituitary System."

36. Although Christian discussed a variety of mammals in his early paper, he focused considerable attention on voles, snowshoe hares, and other species that exhibit population cycles. After working with rats and mice, which are noncyclic, Christian de-emphasized population cycles, as he developed a theory to explain population regulation in general. Christian's colleague Davis's later "Early Behavioral Research" claimed that the "mystical attraction" of population cycles had a disastrous effect on population studies.

37. Schulkin, *Curt Richter*, 83–84; Keiner, "Wartime Rat Control"; Biehler, *Pests in the City*, 128–34.

38. Wesley E. Lanyon, Stephen T. Emlen, and Gordon H. Orians, "In Memoriam: John Thompson Emlen, Jr.," *Auk* 117 (2000): 222–27; see also Sumner W. Matteson, "John T. Emlen, Jr.: A Naturalist for All Seasons, Part 2: Of Adventure, Innovation, and Conscience," *Passenger Pigeon* 60 (1998): 203–50.

39. John J. Christian, "In Memoriam: David E. Davis, 1913–1994," *Auk* 112 (1995): 491–92.

40. After the war Emlen returned to his primary interest in studying the natural history and ecology of birds at the University of Wisconsin; see Lanyon, Emlen, and Orians, "In Memoriam"; Matteson, "Naturalist for All Seasons."

41. John T. Emlen, Allen W. Stokes, and David E. Davis, "Methods for Estimating Populations of Brown Rats in Urban Habitats," *Ecology* 30 (1949): 430–42.

42. David E. Davis, "The Characteristics of Rat Populations," *Quarterly Review of*

Biology 28 (1953): 373–401; John T. Emlen, Allen W. Stokes, and Charles P. Winsor, "The Rate of Recovery of Decimated Populations of Brown Rats in Nature," *Ecology* 29 (1948): 133–45; David E. Davis, "A Comparison of Reproductive Potential of Two Rat Populations," *Ecology* 32 (1951): 469–75.

43. Davis, "Characteristics of Rat Populations," 375.

44. Davis, "Characteristics of Rat Populations," estimated that the rat population in Baltimore declined from 400,000 in 1943 to 65,000 in 1949. For the social and political context for the rat control campaigns, see Keiner, "Wartime Rat Control," and Biehler, *Pests in the City*. According to Biehler, both Emlen and Davis were deeply skeptical of Richter's program of poisoning rats to control populations.

45. John B. Calhoun, "Mortality and Movement of Brown Rats (*Rattus norvegicus*) in Artificially Supersaturated Populations," *Journal of Wildlife Management* 12 (1948): 167–72.

46. David E. Davis and John J. Christian, "Changes in Norway Rat Populations Induced by Introduction of Rats," *Journal of Wildlife Management* 20 (1956): 378–83.

47. Schulkin, *Curt Richter*.

48. John B. Calhoun, "The Social Aspects of Population Dynamics," *Journal of Mammalogy* 33 (1952): 139–59.

49. John B. Calhoun, "A Method for Self-Control of Population Growth among Mammals Living in the Wild," *Science* 109 (1949): 333–35.

50. Ramsden and Adams, "Escaping the Laboratory"; Ramsden, "From Rodent Utopia to Urban Hell"; Ramsden, "Rats, Stress, and the Built Environment."

51. Davis, "Early Behavioral Research," 828.

52. John J. Christian, "Phenomena Associated with Population Density," *Proceedings of the National Academy of Sciences* 47 (1961): 428–49.

53. Christian, "Phenomena Associated with Population Density," 446.

54. Carleton S. Coon, "Chairman's Opening Remarks," *Proceedings of the National Academy of Sciences* 47 (1961): 427.

55. Davis, "Early Behavioral Research," 828.

56. Ian McHarg and John Christian, *The House We Live In*, Philadelphia, WCAU–TV, aired February 26, 1961. Both the full transcript of the hour-and-a-half program and a highly edited version titled "Ecology and Environment," are in the Ian McHarg Papers, Architectural Archives, University of Pennsylvania.

57. David E. Davis, *Integral Animal Behavior* (New York: Macmillan, 1966).

58. Davis, *Integral Animal Behavior*, 48.

59. Davis, *Integral Animal Behavior*, 105.

60. John J. Christian and David E. Davis, "Endocrines, Behavior, and Population," *Science* 146 (1964): 1550–60; see also J. J. Christian, "Endocrine Adaptive Mechanisms and the Physiologic Regulation of Population Growth," in *Physiological Mammalogy*, Vol. 1, ed. William V. Mayer and Richard G. Van Gelder (New York: Academic Press, 1963), 189–353.

61. Dennis Chitty, *Do Lemmings Commit Suicide?*, 99; Peter Crowcroft, *Elton's Ecologists: A History of the Bureau of Animal Population* (Chicago: University of Chicago Press, 1991), 83.

62. Charles J. Krebs, *Population Fluctuations in Rodents* (Chicago: University of Chicago Press, 2013), 177.

Chapter 11

1. René Dubos and Russell W. Schaedler, "The Digestive Tract as an Ecosystem," *American Journal of Medical Sciences* 248 (1964): 267–71.

2. René J. Dubos, *Man Adapting* (New Haven: Yale University Press, 1965), chap. 5.

3. Dwayne C. Savage, "Microbial Biota of the Human Intestine: A Tribute to Some Pioneering Scientists," *Current Issues in Intestinal Microbiology* 2 (2001): 1–15.

4. Dubos, *Man Adapting*, 110. See also René Dubos, "Environmental Biology," *Bioscience* 14 (1964): 11–14.

5. Joel B. Hagen, *An Entangled Bank: The Origins of Ecosystem Ecology* (New Brunswick, NJ: Rutgers University Press, 1992); Peder Anker, "The Context of Ecosystem Theory," *Ecosystems* 5 (2005): 611–13; Peter Ayres, *Shaping Ecology: The Life of Arthur Tansley* (Chichester, UK: John Wiley and Sons, 2012).

6. Carol L. Moberg, *René Dubos: Friend of the Good Earth* (Washington, DC: ASM Press, 2005), 122; see also James G. Hirsch and Carol L. Moberg, "René J. Dubos, February 20, 1901–February 20, 1982," *Biographical Memoirs of the National Academy of Sciences* (Washington DC, 1989), 133–61.

7. Moberg, *René Dubos.*

8. For Winogradsky's important influence on ecology, see Lloyd T. Akert Jr., "The Role of Microbes in Agriculture: Sergei Vinogradskii's Discovery and Investigation of Chemosynthesis 1880–1910," *Journal of the History of Biology* 39 (2006): 373–406; and Akert, "The 'Cycle of Life" in Ecology: Sergei Vinogradskii's Soil Microbiology, 1885–1940," *Journal of the History of Biology* 40 (2007): 109–45.

9. René J. Dubos, "Influence of Environmental Conditions on the Activities of Cellulose Decomposing Organisms in the Soil," *Ecology* 9 (1928): 12–27.

10. Hirsch and Moberg, "René J. Dubos," 140.

11. René J. Dubos, "Infection into Disease," *Perspectives in Biology and Medicine* 1 (1958): 425–35.

12. Mark Honigsbaum, "René Dubos, Tuberculosis, and the 'Ecological Facets' of Virulence," *History and Philosophy of the Life Sciences* 39 (2017): 15. For the broader context of Dubos's contributions to "disease ecology," see Warwick Anderson, "Natural Histories of Infectious Disease: Ecological Vision in Twentieth-Century Biomedical Science," *Osiris* 19 (2004): 39–61.

13. René J. Dubos, *The Bacterial Cell in Its Relation to Problems of Virulence, Immunity, and Chemotherapy* (Cambridge: Harvard University Press, 1945).

14. For the broader historical context of these competing perspectives on bacterial

life, see Mathias Grote, “Petri Dish versus Winogradsky Column: A Long Durée Perspective on Purity and Diversity in Microbiology, 1880s–1980s,” *History and Philosophy of the Life Sciences* 40, no. 1 (2018): 11.

15. Dubos, *Bacterial Cell*, 157.

16. Dubos, *Bacterial Cell*, chap. 6, especially 189.

17. Dubos was limited in his hands-on experimentation by arthritis and poor eyesight, but he excelled in designing experiments that were conducted by other members of his laboratory; see Moberg, *René Dubos*, 27–28.

18. Russell W. Schaedler, René Dubos, and Richard Costello, “The Development of the Bacterial Flora in the Gastrointestinal Tract of Mice,” *Journal of Experimental Medicine* 122 (1965): 59–66.

19. Savage, “Microbial Biota.”

20. Russell W. Schaedler, René J. Dubos, and Richard Costello, “Association of Germfree Mice with Bacteria Isolated from Normal Mice,” *Journal of Experimental Medicine* 122 (1965): 77–82; Savage, “Microbial Biota”; Trenton R. Schoeb and Roger P. Orcutt, “Historical Overview,” in *Gnotobiotics*, ed. Trenton R. Schoeb and Kathryn A. Eaton (San Diego: Academic Press, 2017), 13–14. For the broader context of the development of laboratory mice and a brief discussion of the origin of the Swiss strain, see Karen A. Rader, *Making Mice: Standardizing Animals for American Biomedical Research, 1900–1955* (Princeton: Princeton University Press, 2004), especially 183–84.

21. René J. Dubos, Russell W. Schaedler, and Mallory Stephens, “The Effect of Antibacterial Drugs on the Fecal Flora of Mice,” *Journal of Experimental Medicine* 117 (1963): 231–43; René J. Dubos, Russell W. Schaedler, and Richard Costello, “The Effect of Antibacterial Drugs on the Weight of Mice,” *Journal of Experimental Medicine* 117 (1963): 245–57.

22. Schoeb and Orcutt, “Historical Overview.”

23. Robert G. W. Kirk, “Life in a Germ-Free World: Isolating Life from the Laboratory Animal to the Bubble Boy,” *Bulletin of the History of Medicine* 86 (2012): 237–75.

24. Moberg, *René Dubos*, 49.

25. The experiments conducted in Dubos’s lab were set within a broader context of controversies over the importance of bacteria for synthesizing vitamins and other nutrients in humans and other animals; see Theodor Rosebury, *Microorganisms Indigenous to Man* (New York: McGraw-Hill, 1962), 351–55. Although encyclopedic in his review of the literature, Rosebury concluded that “the ecology of the indigenous microbiota of man is in its earliest infancy,” 369.

26. René Dubos, *The Mirage of Health: Utopias, Progress, and Biological Change* (New York: Harper and Row, 1959), 26.

27. Dubos, *Mirage of Health*, 151–65.

28. Dubos, *Mirage of Health*, 28.

29. Dubos, *Mirage of Health*, 25–29, 117–24.

30. Dubos, *Mirage of Health*, 34.

31. Dubos, *Mirage of Health*, 118.

32. Dubos, *Man Adapting*, chap. 10; René Dubos, "Homeostasis, Illness, and Biological Creativity," *Lahey Clinic Foundation Bulletin* 23 (1974): 94–100.

33. Moberg, *René Dubos*, 159.

34. Dubos, *Mirage of Health*, 88–94; René Dubos and Alex Kessler, "Integrative and Disintegrative Factors in Symbiotic Associations," in *Symbiotic Associations*, ed. P. S. Nutman and Barbara Moss (Cambridge: Cambridge University Press, 1963), 1–11. The ideas in this essay were later elaborated in René Dubos, "Symbiosis of the Earth and Humankind," *Science* 193 (1976): 459–62; and René Dubos, "The Lichen Sermon," in *The World of René Dubos*, ed. Gerard Piel and Osborn Segerberg Jr. (New York: Henry Holt, 1990), 409–15. For the broader historical context of Dubos's ideas about symbiosis, see Jan Sapp, *Evolution by Association: A History of Symbiosis* (New York: Oxford University Press, 1994).

35. René Dubos, *The Resilience of Ecosystems: An Ecological View of Environmental Restoration* (Boulder: Colorado Associated University Press, 1978).

36. Scott F. Gilbert, Jan Sapp, and Alfred Tauber, "A Symbiotic View of Life: We Have Never Been Individuals," *Quarterly Review of Biology* 87 (2012): 325–41; Les Dethlefsen, Margaret McFall-Ngai, and David A. Relman, "An Ecological and Evolutionary Perspective on Human–Microbe Mutualism and Disease," *Nature* 449 (2007): 811–18. Both Honigsbaum, "René Dubos," and Moberg, *René Dubos*, 171, contrast Dubos's perspective on symbiosis with the more competitive views of prominent academic ecologists and evolutionary biologists during the 1960s.

37. Dubos described himself as a "despairing optimist" and contributed columns under that title to the *American Scholar* during the 1970s.

38. Eugene P. Odum, *Fundamentals of Ecology*, 3rd ed. (Philadelphia: Saunders, 1971), 233.

39. Eugene P. Odum, *Basic Ecology* (Philadelphia: Saunders, 1983), 24–29; René Dubos, "Gaia and Creative Evolution," *Nature* 282 (1979): 154–55.

40. Hagen, *Entangled Bank*, 123.

41. Hagen, *Entangled Bank*, 128–29. Odum made the comment during a 1988 interview with the author.

42. Joel B. Hagen, "Teaching Ecology during the Environmental Age, 1965–1980," *Environmental History* 13 (2008): 704–23; Joel B. Hagen, "Eugene Odum and the Homeostatic Ecosystem: The Resilience of an Idea," in Darrell P. Arnold, ed., *Traditions of Systems Theory: Major Figures and Contemporary Developments* (New York: Routledge, 2014), 179–93. A survey of biologists found *Fundamentals of Ecology* one of the most influential books of the twentieth century; see Gary W. Barrett and Karen E. Mabry, "Twentieth Century Classic Books and Benchmark Publications in Biology," *Bioscience* 52 (2002): 282–86.

43. Howard W. Odum and Harry Estill Moore, *American Regionalism: A Cultural-Historical Approach to National Integration* (New York: Henry Holt, 1938). Much of the

chapter on human ecology and ecological approaches to regionalism was based on an unpublished manuscript written by Eugene when he was still a graduate student at the University of Illinois. For a detailed account of Eugene's relationship with his father and brother, see Betty Jean Craige, *Eugene Odum: Ecosystem Ecologist & Environmentalist* (Athens: University of Georgia Press, 2001).

44. Odum, *Fundamentals of Ecology*, 512; see also Eugene P. Odum, "The Emergence of Ecology as a New Integrative Discipline," *Science* 195 (1977): 1289–93.

45. Odum, *Fundamentals of Ecology*, chap. 2; see also Howard W. Odum, Harold D. Meyer, B. S. Holden, and Fred M. Alexander, *American Democracy Anew* (New York: Henry Holt, 1940), 51–52. For the broader cultural context of Howard W. Odum's progressive regionalism, see Dewey W. Grantham, *Southern Progressivism: The Reconciliation of Progress and Tradition* (Knoxville: University of Tennessee Press, 1983).

46. Odum and Moore, *American Regionalism*, 43.

47. See the letter from Eugene Odum to G. Evelyn Hutchinson, May 23, 1950, box 39, G. Evelyn Hutchinson Papers, Yale University Archives.

48. Robert A. Croker, *Pioneer Ecologist: The Life and Works of Victor Ernest Shelford, 1877–1968* (Washington, DC: Smithsonian Institution Press, 1991), 101. Shelford's perspective reflected older ideas on cooperation in nature and society developed by ecologists at the University of Chicago earlier in the twentieth century. For this episode in the history of ecology, see Gregg Mitman, *The State of Nature: Ecology, Community, and American Social Thought, 1900–1950* (Chicago: University of Chicago Press, 1992).

49. Odum, *Fundamentals of Ecology*, 406. The relationship between professional ecology and applied ecology (including environmental preservation) has been a perennial issue among ecologists; see Sharon Kingsland, *The Evolution of American Ecology, 1890–2000* (Baltimore: Johns Hopkins University Press, 2005); Dorothy Nelkin, "Scientists and Professional Responsibility: The Experience of American Ecologists," *Social Studies of Science* 7 (1977): 75–95; Stephen J. Bocking, *Ecologists and Environmental Politics: A History of Contemporary Ecology* (New Haven: Yale University Press, 1997); Robert McIntosh, *The Background of Ecology: Concept and Theory* (Cambridge: Cambridge University Press, 1985), chap. 8; Abby J. Kinchy, "On the Borders of Post-War Ecology: Struggles over the Ecological Society of America's Preservation Committee, 1917–1945," *Science as Culture* 15 (2006): 23–44.

50. See the letters from Eugene Odum to Hutchinson, February 1, 1949, and October 12, 1953, box 39, Hutchinson Papers; see also Eugene P. Odum, "The New Ecology," *Bioscience* 14 (1964): 14–16. Odum referred to comments made by Hutchinson about ecology and environmental problem solving in a popular article: G. Evelyn Hutchinson, "On Living in the Biosphere," *Scientific Monthly* 67 (1948): 393–97.

51. Howard T. Odum and Eugene P. Odum, "Trophic Structure and Productivity of a Windward Coral Reef Community on Eniwetok Atoll," *Ecological Monographs* 25 (1955): 291–320.

52. Odum and Odum, "Trophic Structure"; Odum, *Fundamentals of Ecology*, 344–49.

53. Odum and Odum, "Trophic Structure," 291. The idea was further elaborated in Odum, *Fundamentals of Ecology*, 349.

54. Odum, *Fundamentals of Ecology*, 221–25.

55. Hagen, "Eugene Odum and the Homeostatic Ecosystem."

56. Bernard C. Patten and Eugene P. Odum, "The Cybernetic Nature of Ecosystems," *American Naturalist* 118 (1981): 886–95.

57. Odum made this comparison explicit in his 1966 presidential address to the Ecological Society of America; see Eugene P. Odum, "The Strategy of Ecosystem Development," *Science* 164 (1969): 262–70.

58. Eugene P. Odum, "New Ecology."

59. Peter J. Taylor, "Technocratic Optimism: H. T. Odum and the Partial Transformation of Ecological Metaphor after World War II," *Journal of the History of Biology* 21 (1988): 213–44. Additional information on Odum's career is provided by Charles A. S. Hall, ed., *Maximum Power: The Ideas and Applications of H. T. Odum* (Niwot: University Press of Colorado, 1995); John J. Ewel, "Resolution of Respect: Howard Thomas Odum (1924–2002)," *Bulletin of the Ecological Society of America* 84 (2003): 13–15.

60. Howard T. Odum, "The Stability of the World Strontium Cycle," *Science* 114 (1951): 407–11; Taylor, "Technocratic Optimism." For the broader influence of Lotka on Hutchinson and his students, see Sharon E. Kingsland, *Modeling Nature: Episodes in the History of Population Ecology* (Chicago: University of Chicago Press, 1985).

61. Karin E. Limburg, "The Biogeochemistry of Strontium: A Review of H. T. Odum's Contributions," *Ecological Modeling* 178 (2004): 31–33. Odum's original plan to publish his dissertation as a monograph through the American Museum of Natural History never materialized, partly because of his new interests in studying Silver Springs; see the correspondence among Odum, Hutchinson, and the publisher in box 39, Hutchinson Papers.

62. Howard T. Odum, "Trophic Structure and Productivity of Silver Springs, Florida," *Ecological Monographs* 27 (1957): 55–112.

63. Eugene P. Odum, "Energy Flow in Ecosystems: A Historical Review," *American Zoologist* 8 (1968): 11–18.

64. For a discussion of some important differences between Odum's theoretical approach and those of other systems ecologists, see Taylor, "Technocratic Optimism"; see also Debora Hammond, "Ecology and Ideology in the General Systems Community," *Environment and History* 3 (1997): 197–207.

65. Howard T. Odum, "Ecological Potential and Analogue Circuits for the Ecosystem," *American Scientist* 48 (1960): 1–8; Howard T. Odum, *Environment, Power, and Society* (New York: John Wiley, 1971); Taylor, "Technocratic Optimism"; Hall, *Maximum Power.*

66. Odum, "New Ecology."

67. Joseph Engelberg and Louis L. Boyarsky, "The Noncybernetic Nature of Ecosystems," *American Naturalist* 114 (1979): 317–24.

68. Bernard C. Patten, "This Week's Citation Classic," *Current Contents* #5 (February 4, 1985): 22; letters from Patten to Hutchinson, December 16, 1957, and January 28, 1958; and letter from Hutchinson to Patten, December 16, 1957, box 41, Hutchinson Papers.

69. Patten and Odum, "Cybernetic Nature of Ecosystems," 894.

70. S. J. McNaughton and Michael B. Coughenour, "The Cybernetic Nature of Ecosystems," *American Naturalist* 117 (1981): 985–90; Robert L. Knight and Dennis P. Swaney, "In Defense of Ecosystems," *American Naturalist* 117 (1981): 991–92.

71. John Alcock, "Natural Selection and Communication among Bark Beetles," *Florida Entomologist* 65 (1981): 17–32.

72. Lauri Oksanen, "Ecosystem Organization: Mutualism and Cybernetics or Plain Darwinian Struggle for Existence?" *American Naturalist* 131 (1988): 424–44.

73. Robert Axelrod and William D. Hamilton, "The Evolution of Cooperation," *Science* 211 (1981): 1390–96.

74. Dethlefsen, McFall-Ngai, and Relman, "Human–Microbe Mutualism"; Antonio Gonzalez, Jose C. Clemente, Ashley Shade, Jessica L. Metcalfe, Sejin Song, Bharath Prithiviraj, Brent E. Palmer, and Rob Knight, "Our Microbial Selves: What Ecology Can Teach Us," *EMBO Reports* 12 (2011): 775–84; John W. Pepper and Simon Rosenfeld, "The Emerging Medical Ecology of the Human Gut Microbiome," *Trends in Ecology and Evolution* 27, no. 7 (2012): 381–84.

75. David P. Hughes, Naomi E. Pierce, and Jacobus J. Boomsma, "Social Insect Symbionts: Evolution in Homeostatic Fortresses," *Trends in Ecology and Evolution* 23 (2008): 672–77; F. John Odling-Smee, Kevin N. Laland, and Marcus W. Feldman, *Niche Construction: The Neglected Process in Evolution* (Princeton: Princeton University Press, 2003), 3–7.

76. Hughes, et al., "Social Insect Symbionts: Evolution in Homeostatic Fortresses."

77. Odling-Smee, Laland, and Feldman, *Niche Construction*, 330–36; Clive G. Jones, John H. Lawton, and Moshe Shachak, "Organisms as Ecosystem Engineers," *Oikos* 69 (1994): 373–86.

Conclusion

1. The literature linking homeostasis to the gut microbiome is vast. A good example of analogizing the microbiome to both an ecosystem and an organ is found in Nishat Tasnim, Nijiati Abulizi, Jason Pither, Miranda M. Hart, and Deanna L. Gibson, "Linking the Gut Microbial Ecosystem with the Environment: Does Gut Health Depend on Where We Live?" *Frontiers in Microbiology* 8 (2017): 1–8.

2. Carl Zimmer, "Tending the Body's Microbial Garden," *New York Times*, June 18, 2012, http://www.nytimes.com/2012/06/19/science/studies-of-human-microbiome-yield-new-insights.html?pagewanted=all; Kent H. Redford, Julia A. Segre, Nick Salafsky, Carlos Martinez del Rio, and Denise McAloose, "Conservation and the Microbiome," *Conservation Biology* 26 (2012): 195–97.

3. René Dubos, "Homeostasis, Illness, and Biological Creativity," *Lahey Clinic Foundation Bulletin* 23 (1974): 94–100.

4. George A. Bartholomew, "The Role of Natural History in Contemporary Biology," *BioScience* 36 (1986): 324–29.

5. George A. Bartholomew, "Integrative Biology: An Organismic Biologist's Point of View," *Integrative and Comparative Biology* 45 (2005): 330–32. Bartholomew never directly addressed the well-known criticisms raised by S. J. Gould and R. C. Lewontin, "The Spandrels of San Marco and the Panglossian Paradigm: A Critique of the Adaptationist Programme," *Proceedings of the Royal Society of London*, Part B 205 (1979): 581–98. Although sensitive to the uncritical use of adaptation, Bartholomew argued that natural selection and adaptation formed the foundation of organismal biology and undergirded the integration of biology more generally.

6. Bartholomew, "Role of Natural History in Contemporary Biology."

7. William R. Dawson, "George A. Bartholomew's Contributions to Integrative and Comparative Biology," *Integrative and Comparative Biology* 45 (2005): 219–30; Raymond B. Huey and Albert F. Bennett, "Bart's Familiar Quotations: The Enduring Wisdom of George A. Bartholomew," *Physiological and Biochemical Zoology* 8 (2008): 519–25; William R. Dawson, "A Profound Loss to SICB and to Integrative and Comparative Biology Generally: The Passing of Professor George A. Bartholomew (1919–2006)," *SICB Newsletter* 11 (Fall 2006): 1–2.

8. J. Scott Turner, "Biology's Second Law: Homeostasis, Purpose, and Desire," in *Beyond Mechanism: Putting Life Back into Biology*, ed. Brian G. Henning and Adam C. Scarfe (Lanham, MD: Lexington Books, 2013), 183–204; Sean B. Carroll, *The Serengeti Rules: The Quest to Discover How Life Works and Why It Matters* (Princeton: Princeton University Press, 2016), 28.

9. Jay Schulkin, ed., *Allostasis, Homeostasis, and the Costs of Physiological Adaptation* (Cambridge: Cambridge University Press, 2004); Carol L. Moberg, *René Dubos: Friend of the Good Earth* (Washington, DC: ASM Press, 2005), 159.

Bibliography

Archival Sources

Edward F. Adolph Papers. Edward G. Miner Library. University of Rochester Medical Center. Rochester, New York.

Walter Bradford Cannon Papers. Center for the History of Medicine. Francis A. Countway Library of Medicine. Boston, Massachusetts.

Robert Gladding Green Papers. University of Minnesota Archives. Minneapolis, Minnesota.

G. Evelyn Hutchinson Papers. Yale University Archives. Sterling Library. New Haven, Connecticut.

Laurence Irving Papers. University of Alaska Archives. Elmer E. Rasmussen Library. Fairbanks, Alaska.

Ernst Mayr Papers. Harvard University Archives. Pusey Library. Cambridge, Massachusetts.

Ian McHarg Papers. Architectural Archives. University of Pennsylvania. Philadelphia, Pennsylvania.

Eugene P. Odum Papers. University of Georgia Archives. Athens, Georgia.

Howard T. Odum Papers. University of Florida Archives. Gainesville, Florida.

Knut Schmidt-Nielsen Papers. David M. Rubenstein Library. Duke University. Durham, North Carolina.

Per Fredrik Scholander Papers. University of California at San Diego Archives. San Diego, California.

Published Sources

Adolph, Edward F. "Desert." In *Physiology of Heat Regulation and the Science of Clothing,* edited by L. H. Newburgh, 330–38. Philadelphia: W. B. Saunders, 1949.

———. "Do Rats Thrive When Drinking Seawater?" *American Journal of Physiology* 140 (1943): 25–32.

———. "Early Concepts of Physiological Regulations." *Physiological Reviews* 41 (1961): 737–70.

———. "Excretion of Water by the Kidneys." *American Journal of Physiology* 65 (1923): 419–49.

———. "Heat Exchanges of Man in the Desert." *American Journal of Physiology* 123 (1938): 486–99.

———. "Living Water." *Quarterly Review of Biology* 5 (1930): 51–67.

———. "The Metabolism of the Water in Ameba as Measured in the Contractile Vacuole." *Journal of Experimental Zoology* 44 (1926): 355–81.

———. *Origins of Physiological Regulations.* New York: Academic Press, 1968.

———. *Physiological Regulations.* Lancaster, PA; Jacques Cattel Press, 1943.

———. "The Physiological Scholar." *American Journal of Physiology* 179 (1954): 607–12.

———. "The Regulation of the Water Content of the Human Organism." *Journal of Physiology* 55 (1921): 114–32.

———. "The Regulation of Volume and Concentration in the Body Fluids of Earthworms." *Journal of Experimental Zoology* 47 (1927): 31–62.

———. "Regulation of Water Metabolism in Stress." *Homeostatic Mechanisms. Brookhaven Symposia in Biology* 10 (1958): 147–61.

———. "Seven Discoveries of Physiological Regulation." In *Neural Integration of Physiological Mechanisms and Behaviour,* ed. Gordon J. Mogenson and Franco R. Calaresu, 11–23. Toronto: University of Toronto Press, 1975.

———. "The Skin and the Kidneys as Regulators of the Body Volume of Frogs." *Journal of Experimental Zoology* 47 (1927): 1–30.

Adolph, Edward F., and Associates. *The Physiology of Man in the Desert.* New York: Interscience, 1947.

Adolph, Edward F., and David Bruce Dill. "Observations on Water Metabolism in the Desert." *American Journal of Physiology* 123 (1938): 369–78.

Adolph, Edward F., and William B. Fulton. "The Effects of Exposure to High Temperatures upon the Circulation in Man." *American Journal of Physiology* 67 (1924): 573–88.

Akert, Lloyd T. "The 'Cycle of Life' in Ecology: Sergei Vinogradskii's Soil Microbiology, 1885–1940." *Journal of the History of Biology* 40 (2007): 109–45.

———. "The Role of Microbes in Agriculture: Sergei Vinogradskii's Discovery and Investigation of Chemosynthesis 1880–1910." *Journal of the History of Biology* 39 (2006): 373–406.

Alberts, Bruce. "John Alexander Moore, 1915–2002." *Biographical Memoirs of the National Academy of Sciences,* 1–24. Washington, DC, 2011.

Alcock, John. "Natural Selection and Communication among Bark Beetles." *Florida Entomologist* 65 (1981): 17–32.

Allard, R. W. "Israel Michael Lerner 1910–1977." *Biographical Memoirs of the National Academy of Sciences*, 165–75. Washington, DC, 1996.

Allee, W. C. *Animal Aggregations*. Chicago: University of Chicago Press, 1931.

———. "Concerning the Origin of Sociality in Animals." *Scientia* 67 (1940): 154–60.

———. *The Social Life of Animals*. New York: Norton, 1938.

Allee, W. C., Alfred E. Emerson, Orlando Park, Thomas Park, and Karl P. Schmidt. *Principles of Animal Ecology*. Chicago: University of Chicago Press, 1949.

Allen, Garland. *Life Sciences in the Twentieth Century*. New York: John Wiley, 1975.

———. "Theodosius Dobzhansky, the Morgan Lab, and the Breakdown of the Naturalist/Experimentalist Dichotomy, 1927–1947." In *The Evolution of Theodosius Dobzhansky*, edited by Mark B. Adams, 87–98. Princeton: Princeton University Press, 1994.

Anderson, Warwick. *The Cultivation of Whiteness: Science, Health, and Racial Destiny in Australia*. New York: Basic Books, 2003.

———. "Natural Histories of Infectious Disease: Ecological Vision in Twentieth-Century Biomedical Science." *Osiris* 19 (2004): 39–61.

Andrewartha, H. G. "Self-Regulatory Mechanisms in Animal Populations." *Australian Journal of Sciences* 22 (1959): 200–205.

Andrewartha, H. G., and L. C. Birch. *The Distribution and Abundance of Animals*. Chicago: University of Chicago Press, 1954.

Ankeny, Rachel A., and Sabina Leonelli. "What's So Special about Model Organisms?" *Studies in History and Philosophy of Science* 42 (2011): 313–23.

Anker, Peder. "The Context of Ecosystem Theory." *Ecosystems* 5 (2005): 611–13.

Arminjon, Mathieu. "Birth of the Allostatic Model: From Cannon's Biocracy to Critical Physiology." *Journal of the History of Biology* 49 (2016): 397–423.

Armon, Rony. "Between Biochemists and Embryologists—The Biochemical Study of Embryonic Induction in the 1930s." *Journal of the History of Biology* 45 (2012): 65–108.

Axelrod, Robert, and William D. Hamilton. "The Evolution of Cooperation." *Science* 211 (1981): 1390–96.

Ayres, Peter. *Shaping Ecology: The Life of Arthur Tansley*. Chichester: John Wiley and Sons, 2012.

Baedke, Jan. "The Epigenetic Landscape in the Course of Time: Conrad Hal Waddington's Methodological Impact on the Life Sciences." *Studies in History and Philosophy of the Biological and Biomedical Sciences* 44 (2013): 756–73.

Barach, Alvan L., and Hylan A. Bickerman, eds. *Pulmonary Emphysema*. Baltimore: Williams and Wilkins, 1956.

Barrett, Gary W., and Karen E. Mabry. "Twentieth Century Classic Books and Benchmark Publications in Biology." *Bioscience* 52 (2002): 282–86.

Bartholomew, George A. "Aspects of Timing and Periodicity of Heterothermy." In

Hibernation and Hypothermia: Perspectives and Challenges, edited by Frank E. South, John P. Hannon, John R. Willis, Eric T. Pengelley, and Norman R. Alpert, 663–80. New York: Elsevier, 1972.

———. "A Field Study of Temperature Relations in the Galápagos Marine Iguana." *Copeia* 1966 (1966): 241–50.

———. "Integrative Biology: An Organismic Biologist's Point of View." *Integrative and Comparative Biology* 45 (2005): 330–32.

———. "Interaction of Physiology and Behavior under Natural Conditions." In *The Galapagos*, edited by Robert I. Bowman, 39–45. Berkeley: University of California Press, 1966.

———. "Interspecific Comparison as a Tool for Ecological Physiologists." In *New Directions in Ecological Physiology*, edited by Martin E. Feder, Albert F. Bennett, Warren W. Burggren, and Raymond B. Huey, 11–37. Cambridge: Cambridge University Press, 1987.

———. "A Matter of Size: An Examination of Endothermy in Insects and Terrestrial Vertebrates." In *Insect Thermoregulation*, edited by Bernd Heinrich, 45–78. New York: John Wiley, 1981.

———. "The Role of Natural History in Contemporary Biology." *Bioscience* 36 (1986): 324–329.

———. "The Role of Physiology in the Distribution of Terrestrial Vertebrates." In *Zoogeography*, edited by Carl L. Hubbs, 81–95. Washington, DC: American Association for the Advancement of Science, 1958.

———. "The Roles of Physiology and Behaviour in the Maintenance of Homeostasis in the Desert Environment." *Symposia of the Society for Experimental Biology* 18 (1964): 7–29.

Bartholomew, George A., and Tom J. Cade. "Temperature Regulation, Hibernation, and Aestivation in the Little Pocket Mouse, *Perognathus longimembris*." *Journal of Mammalogy* 38 (1957): 60–72.

———. "The Water Economy of Land Birds." *Auk* 80 (1963): 504–39.

Bartholomew, George A., and Jack W. Hudson. "Desert Ground Squirrels." *Scientific American* 205, no. 5 (November 1961): 107–16.

———. "Effects of Sodium Chloride on Weight and Drinking in the Antelope Ground Squirrel." *Journal of Mammalogy* 40 (1959): 354–60.

Batisse, Michel. "A Long Look at the World's Arid Lands." *UNESCO Courier* 47, no. 1 (1994): 34–39.

Beatty, John. "The Creativity of Natural Selection? Part II: The Synthesis and Since." *Journal of the History of Biology* 52 (2019): 705–31.

———. "Dobzhansky and the Biology of Democracy: The Moral and Political Significance of Genetic Variation." In *The Evolution of Theodosius Dobzhansky*, edited by Mark B. Adams, 195–218. Princeton: Princeton University Press, 1994.

———. "The Proximate–Ultimate Distinction in the Multiple Careers of Ernst

Mayr." *Biology and Philosophy* 9 (1994): 333–56.

———. "Weighing the Risks: Stalemate in the Classical/Balance Controversy." *Journal of the History of Biology* 20 (1987): 289–319.

Bennison, Saul, A. Clifford Barger, and Elin L. Wolfe. "Walter B. Cannon and the Mystery of Shock: A Study of Anglo-American Cooperation." *Medical History* 35 (1991): 217–49.

Berliner, Robert W. "Remembering Homer Smith." *Kidney International* 43 (1993): 171–72.

Beyenbach, Klaus W. "Kidneys sans Glomeruli." *American Journal of Physiology—Renal Physiology* 286 (2004): F811–27.

Biehler, Dawn Day. *Pests in the City: Flies, Bedbugs, Cockroaches, and Rats.* Seattle: University of Washington Press, 2013.

Bocking, Stephen J. *Ecologists and Environmental Politics: A History of Contemporary Ecology.* New Haven: Yale University Press, 1997.

Bode, Richard. "A Doctor Who's Dad to Seven Doctors—So Far." *Reader's Digest,* December 1982, 141–45.

Bolton, Charles C. *William F. Winter and the New Mississippi.* Jackson: University Press of Mississippi, 2013.

Borello, Mark E. *Evolutionary Restraints: The Contentious History of Group Selection.* Chicago: University of Chicago Press, 2010.

Bourassa, M. G. "History of Cardiac Catheterization." *Canadian Journal of Cardiology* 21 (2005): 1011–14.

Bradley, Stanley. "Homer William Smith: A Personal Memoir." In *A Laboratory by the Sea: The Mount Desert Island Biological Laboratory, 1898–1998,* edited by Franklin H. Epstein, 97–101. Rhinebeck, NY: River Press, 1998.

Braun, Eldon J. "Osmotic and Ionic Regulation in Birds." In *Osmotic and Ionic Regulation: Cells and Animals,* edited by David H. Evans, 505–24. Boca Raton, FL: CRC Press, 2009.

Brewster, Karen. "Native Contributions to Arctic Science at Barrow, Alaska." *Arctic* 50 (1997): 277–88.

Brinson, Carroll, and Janis Quinn. *Arthur C. Guyton: His Family, His Life, His Achievements.* Jackson, MS: Oakdale Press, 1989.

Brown, Lauren. "Speciation in the *Rana pipiens* Complex." *American Zoologist* 13 (1973): 73–79.

Bruni, John. "Expanding the Self-Referential Paradox: The Macy Conferences and the Second Wave of Cybernetic Thinking." In *Traditions of Systems Theory: Major Figures and Contemporary Developments,* edited by Darrell P. Arnold, 78–83. New York: Routledge, 2014.

Bruyninckx, Joeri. *Listening in the Field: Recording and the Science of Birdsong.* Cambridge, MA: MIT Press, 2018.

Bullock, Theodore H. "Homeostasis in Marine Organisms." In *Perspectives in Marine*

Biology, edited by Adriano A. Buzzati-Traverso, 199–210. Berkeley: University of California Press, 1960.

———. "In Praise of 'Natural History.'" *Cellular and Molecular Neurobiology* 25 (2005): 217–21.

———. "In Search of Principles of Integrative Biology." *American Zoologist* 5 (1965): 745–55.

Cade, Tom J., and George A. Bartholomew. "Sea-Water and Salt Utilization by Savannah Sparrows." *Physiological Zoology* 32 (1959): 230–38.

Cain, Joseph. "Co-opting Colleagues: Appropriating Dobzhansky's 1936 Lectures at Columbia." *Journal of the History of Biology* 35 (2002): 207–19.

———. "Ernst Mayr as Community Architect: Launching the Society for the Study of Evolution and the Journal *Evolution*." *Biology and Philosophy* 9 (1994): 387–427.

Calhoun, John B. "A Method for Self-Control of Population Growth among Mammals Living in the Wild." *Science* 109 (1949): 333–35.

———. "Mortality and Movement of Brown Rats (*Rattus norvegicus*) in Artificially Supersaturated Populations." *Journal of Wildlife Management* 12 (1948): 167–72.

———. "The Social Aspects of Population Dynamics." *Journal of Mammalogy* 33 (1952): 139–59.

Cannon, Walter B. "The Body Physiologic and the Body Politic." *Science* 93 (1941): 1–10.

———. "Organization for Physiological Homeostasis." *Physiological Reviews* 9 (1929): 399–431.

———. "Stresses and Strains of Homeostasis." *American Journal of Medical Sciences* 189 (1935): 1–14.

———. *Traumatic Shock*. New York: Appleton, 1923.

———. *The Wisdom of the Body*. 2nd ed. 1939. New York: Norton, 1932.

Carroll, Sean B. *The Serengeti Rules: The Quest to Discover How Life Works and Why It Matters*. Princeton: Princeton University Press, 2016.

Caspari, Rachel. "From Types to Populations: A Century of Race, Physical Anthropology, and the American Anthropological Association." *American Anthropologist* 105 (2003): 65–76.

Ceccarelli, Leah. *Shaping Science with Rhetoric: The Cases of Dobzhansky, Schrödinger, and Wilson*. Chicago: University of Chicago Press, 2001.

Charlesworth, Brian. "Fisher, Medawar, Hamilton and the Evolution of Aging." *Genetics* 156 (2000): 927–31.

Chiang, Howard Hsueh-Hao. "The Laboratory Technology of Discrete Molecule Separation: The Historical Development of Gel Electrophoresis and the Material Epistemology of Biomolecular Science, 1945–1970." *Journal of the History of Biology* 42 (2009): 495–527.

Chitty, Dennis. *Do Lemmings Commit Suicide? Beautiful Hypotheses and Ugly Facts*. Oxford: Oxford University Press, 1996.

Christian, John J. "Adrenal and Reproductive Responses to Population Size in Mice from Freely Growing Populations." *Ecology* 37 (1956): 258–73.

———. "The Adreno-Pituitary System and Population Cycles in Mammals." *Journal of Mammalogy* 31 (1950): 247–59.

———. "Control of Population Growth in Rodents by Interplay between Population Density and Endocrine Physiology." *Wildlife Disease* 1 (January 1959): 1–38.

———. "Endocrine Adaptive Mechanisms and the Physiologic Regulation of Population Growth." In *Physiological Mammalogy*, Vol. 1, edited by William V. Mayer and Richard G. Van Gelder, 189–353. New York: Academic Press, 1964.

———. "In Memoriam: David E. Davis, 1913–1994." *Auk* 112 (1995): 491–92.

———. "Phenomena Associated with Population Density." *Proceedings of the National Academy of Sciences* 47 (1961): 428–49.

Christian, John J., and David E. Davis. "Endocrines, Behavior, and Population." *Science* 146 (1964): 1550–60.

Christian, John J., and H. L. Ratcliffe. "Shock Disease in Captive Wild Mammals." *American Journal of Pathology* 28 (1952): 725–37.

Collins, James P. "'Evolutionary Ecology' and the Use of Natural Selection in Ecological Theory." *Journal of the History of Biology* 19 (1986): 257–88.

Collopy, Peter Sachs. "Race Relationships: Collegiality and Demarcation in Physical Anthropology." *Journal of the History of the Behavioral Sciences* 51 (2015): 237–60.

Cook, Robert C. *Human Fertility: The Modern Dilemma.* New York: Sloane, 1951.

Cooke, John S. "Sectionalization." In *History of the American Physiological Society: The First Century, 1887–1987*, edited by John R. Brobeck, Orr E. Reynolds, and Tobey A. Appel, 435–61. Bethesda, MD: American Physiological Society, 1987.

Coon, Carleton S. "Chairman's Opening Remarks." *Proceedings of the National Academy of Sciences* 47 (1961): 427.

———. *The Origin of Races.* New York: Knopf, 1962.

Cooper, Gregory J. *The Science of the Struggle for Existence: On the Foundations of Ecology*. Cambridge: Cambridge University Press, 2003.

Cooper, Steven J. "From Claude Bernard to Walter Cannon: Emergence of the Concept of Homeostasis." *Appetite* 51 (2008): 419–27.

Cosgrave, John O'Hara. "With a Spice of Fiction." *Saturday Review of Literature*, April 30, 1932, 32.

Cournand, André. "Dickinson Woodruff Richards 1895–1973." *Biographical Memoirs of the National Academy of Sciences*, 458–87. Washington, DC, 1989.

———. "Some Aspects of the Pulmonary Circulation in Normal Man and in Chronic Cardiopulmonary Diseases." *Circulation* 2 (1950): 641–57.

Cournand, André, and Michael Meyer. *From Roots . . . to Late Budding: The Intellectual Adventures of a Medical Scientist.* New York: Gardner Press, 1986.

Craige, Betty Jean. *Eugene Odum: Ecosystem Ecologist & Environmentalist.* Athens: University of Georgia Press, 2001.

Crispo, Erika. "The Baldwin Effect and Genetic Assimilation: Revisiting Two Mechanisms of Evolutionary Change Mediated by Phenotypic Plasticity." *Evolution* 61 (2007): 2469–79.

Croker, Robert A. *Pioneer Ecologist: The Life and Works of Victor Ernest Shelford, 1877–1968.* Washington, DC: Smithsonian Institution Press, 1991.

Crone, Christian. "The Autoregulation of the Microcirculation." *Acta Medica Scandinavica* 197 (1975): 15–18.

Cross, Stephen J., and William R. Albury. "Walter B. Cannon, L. J. Henderson and the Organic Analogy." *Osiris* 3 (1987): 165–92.

Crowcroft, Peter. *Elton's Ecologists: A History of the Bureau of Animal Population.* Chicago: University of Chicago Press, 1991.

Cuddington, Kim. "The 'Balance of Nature' Metaphor and Equilibrium in Population Ecology." *Biology and Philosophy* 16 (2001): 463–79.

Dantzler, William H., and S. Donald Bradshaw. "Osmotic Regulation in Reptiles." In *Osmotic and Ionic Regulation: Cells and Animals,* edited by David H. Evans, 443–504. Boca Raton, FL: CRC Press, 2009.

Darwin, Charles. *The Voyage of the Beagle.* Project Gutenberg e-book (posted June 24, 2013), chap. 17, http://www.gutenberg.org/files/944/944-h/944-h.htm.

Davis, David E. "The Characteristics of Rat Populations." *Quarterly Review of Biology* 28 (1953): 373–401.

———. "A Comparison of Reproductive Potential of Two Rat Populations." *Ecology* 32 (1951): 469–75.

———. "Early Behavioral Research on Populations." *American Zoologist* 27 (1987): 825–37.

———. *Integral Animal Behavior.* New York: Macmillan, 1966.

Davis, David E., and John J. Christian. "Changes in Norway Rat Populations Induced by Introduction of Rats." *Journal of Wildlife Management* 20 (1956): 378–83.

Dawkins, Richard. *The Selfish Gene.* Oxford: Oxford University Press, 1976.

Dawson, William R. "George A. Bartholomew, 1916–2006." *Biographical Memoirs of the National Academy of Sciences,* 1–33. Washington, DC, 2011.

———. "George A. Bartholomew's Contributions to Integrative and Comparative Biology." *Integrative and Comparative Biology* 45 (2005): 219–30.

———. "Laurence Irving: An Appreciation." *Physiological and Biochemical Zoology* 80 (2007): 9–24.

———. "A Profound Loss to SICB and to Integrative and Comparative Biology Generally: The Passing of Professor George A. Bartholomew (1919–2006)." *SICB Newsletter* 11 (Fall 2006): 1–2.

———. "The Relation of Oxygen Consumption to Temperature in Desert Rodents." *Journal of Mammalogy* 36 (1955): 543–53.

Depew, David J. "Adaptation as Process: The Future of Darwinism and the Legacy of Theodosius Dobzhansky." *Studies in History and Philosophy of Biological and*

Biomedical Sciences 42 (2011): 89–98.

Dethlefsen, Les, Margaret McFall-Ngai, and David A. Relman. "An Ecological and Evolutionary Perspective on Human–Microbe Mutualism and Disease." *Nature* 449 (2007): 811–18.

Dietrich, Michael R. "The Electrophoresis Revolution." *Perspectives on Molecular Evolution*, 2004. http://authors.library.caltech.edu/5456/1/hrst.mit.edu/hrs/evolution/public/techniques/electrophoresis.html.

———. "The Origins of the Neutral Theory of Molecular Evolution." *Journal of the History of Biology* 27 (1994): 21–59.

———. "Paradox and Persuasion: Negotiating the Place of Molecular Evolution within Evolutionary Biology." *Journal of the History of Biology* 31 (1998): 85–111.

Dietrich, Michael R., and Robert A. Skipper. "A Shifting Terrain: A Brief History of the Adaptive Landscape." In *The Adaptive Landscape in Evolutionary Biology*, edited by Erik I. Svensson and Ryan Calsbeek, 3–15. Oxford: Oxford University Press, 2012.

Dill, David Bruce. *Life, Heat and Altitude: Physiological Effects of Hot Climates and Great Heights*. Cambridge: Harvard University Press, 1938.

Dill, David Bruce, F. G. Hall, and H. T. Edwards. "Changes in Composition of Sweat during Acclimatization to Heat." *American Journal of Physiology* 123 (1938): 412–19.

Dobzhansky, Theodosius. *The Biological Basis of Human Freedom*. New York: Columbia University Press, 1956.

———. "Genetic Homeostasis." *Evolution* 9 (1955): 100–101.

———. *Genetics and the Origin of Species*. New York: Columbia University Press, 1937.

———. *Genetics and the Origin of Species*. 1937. 2nd ed. New York: Columbia University Press, 1941.

———. *Genetics and the Origin of Species*. 3rd ed. New York: Columbia University Press, 1951.

———. *Genetics of the Evolutionary Process*. New York: Columbia University Press, 1970.

———. *Mankind Evolving: The Evolution of the Human Species*. New Haven: Yale University Press, 1962.

———. "Nothing in Biology Makes Sense Except in the Light of Evolution." *American Biology Teacher* 35 (1973): 125–29.

Dobzhansky, Theodosius, and Bruce Wallace. "The Genetics of Homeostasis in *Drosophila*." *Proceedings of the National Academy of Sciences* 39 (1953): 162–71.

Drabkin, David L. "Imperfection: Biochemical Phobias and Metabolic Ambivalence." *Perspectives in Biology and Medicine* 2 (1959): 473–517.

———. "Metabolism of the Hemin Chromoproteins." *Physiological Reviews* 31 (1951): 345–430.

Dubos, René J. *The Bacterial Cell in Its Relation to Problems of Virulence, Immunity, and Chemotherapy*. Cambridge: Harvard University Press, 1945.

———. "The Despairing Optimist." *American Scholar* 46 (1977): 152–53, 156–58.

———. "Environmental Biology." *Bioscience* 14 (1964): 11–14.

———. "Gaia and Creative Evolution." *Nature* 282 (1979): 154–55.

———. "Homeostasis, Illness, and Biological Creativity." *Lahey Clinic Foundation Bulletin* 23 (1974): 94–100.

———. "Infection into Disease." *Perspectives in Biology and Medicine* 1 (1958): 425–35.

———. "Influence of Environmental Conditions on the Activities of Cellulose Decomposing Organisms in the Soil." *Ecology* 9 (1928): 12–27.

———. "The Lichen Sermon." In *The World of René Dubos*, edited by Gerard Piel and Osborn Segerberg Jr., 409–15. New York: Henry Holt, 1990.

———. *Man Adapting*. New Haven: Yale University Press, 1965.

———. *The Mirage of Health: Utopias, Progress, and Biological Change*. New York: Harper and Row, 1959.

———. *The Resilience of Ecosystems: An Ecological View of Environmental Restoration*. Boulder: Colorado University Press, 1978.

———. "Symbiosis of the Earth and Humankind." *Science* 193 (1976): 459–62.

Dubos, René J., and Alex Kessler. "Integrative and Disintegrative Factors in Symbiotic Associations." In *Symbiotic Associations*, edited by P. S. Nutman and Barbara Moss, 1 –11. Cambridge: Cambridge University Press, 1963.

Dubos, René J., and Russell W. Schaedler. "The Digestive Tract as an Ecosystem." *American Journal of Medical Sciences* 248 (1964): 267–71.

Dubos, René J., Russell W. Schaedler, and Richard Costello. "The Effect of Antibacterial Drugs on the Weight of Mice." *Journal of Experimental Medicine* 117 (1963): 245–57.

Dubos, René J., Russell W. Schaedler, and Mallory Stephens. "The Effect of Antibacterial Drugs on the Fecal Flora of Mice." *Journal of Experimental Medicine* 117 (1963): 231–43.

Duffuss, R. L. "Taking the Lung-Fish as a Text for Philosophy." *New York Times*, April 17, 1932, 6.

Egerton, Frank. "Changing Concepts of the Balance of Nature." *Quarterly Review of Biology* 48 (1973): 322–50.

Elsner, Robert. "The Irving–Scholander Legacy in Polar Physiology." *Comparative Biochemistry and Physiology—Part A* 126 (2000): 137–42.

Emerson, Alfred E. "Regenerate Behavior and Social Homeostasis of Termites." *Ecology* 37 (1956): 248–58.

———. "Social Coordination and the Superorganism." *American Midland Naturalist* 21 (1939): 182–209.

Emlen, John T., Allen W. Stokes, and David E. Davis. "Methods for Estimating

Populations of Brown Rats in Urban Habitats." *Ecology* 30 (1949): 430–42.

Emlen, John T., Allen W. Stokes, and Charles P. Winsor. "The Rate of Recovery of Decimated Populations of Brown Rats in Nature." *Ecology* 29 (1948): 133–45.

Engelberg, Joseph, and Louis L. Boyarsky. "The Noncybernetic Nature of Ecosystems." *American Naturalist* 114 (1979): 317–24.

Engels, Vincent. "A Snob of Science." *Commonweal* 17 (November 16, 1932): 82–83.

Epstein, Franklin H., ed. *A Laboratory by the Sea: The Mount Desert Island Biological Laboratory*. Rhinebeck, NY: River Press, 1998.

Evans, David H. *Marine Physiology Down East: The History of the Mt. Desert Island Biological Laboratory*. New York: Springer, 2015.

———. "Teleost Osmoregulation: What Have We Learned since August Krogh, Homer Smith, and Ancel Keys." *American Journal of Physiology—Regulatory, Integrative and Comparative Physiology* 295 (2008): R704–13.

Ewel, John J. "Resolution of Respect: Howard Thomas Odum (1924–2002)." *Bulletin of the Ecological Society of America* 84 (2003): 13–15.

Farber, Paul Lawrence. "Dobzhansky and Montagu's Debate on Race: The Aftermath." *Journal of the History of Biology* 49 (2016): 625–39.

———. *Mixing Races: From Scientific Racism to Modern Evolutionary Ideas*. Baltimore: Johns Hopkins University Press, 2011.

———. *The Temptations of Evolutionary Ethics*. Berkeley: University of California Press, 1994.

Farish, Matthew. "Creating Cold War Climates: The Laboratories of American Globalism." In *Environmental Histories of the Cold War*, edited by J. R. McNeill and Corinna R. Unger, 51–84. Cambridge: Cambridge University Press, 2010.

———. "The Lab and the Land: Overcoming the Arctic in Cold War Alaska." *Isis* 104 (2013): 1–29.

Frank, Lawrence K. "Foreword." *Annals of the New York Academy of Sciences* 50 (1948): 189–96.

Fry, Iris. "On the Biological Significance of Properties of Matter: L. J. Henderson's Theory of the Fitness of the Environment." *Journal of the History of Biology* 29 (1996): 155–96.

Geison, Gerald L. "Research Schools and New Directions in the Historiography of Science." *Osiris* 8 (1993): 227–38.

Gilbert, Scott F. "Dobzhansky, Waddington, and Schmalhausen." In *The Evolution of Theodosius Dobzhansky*, edited by Mark B. Adams, 143–54. Princeton: Princeton University Press, 1994.

Gilbert, Scott F., and David Epel. *Ecological Developmental Biology: Integrating Epigenetics, Medicine, and Evolution*. Sunderland, MA: Sinnauer, 2009.

Gilbert, Scott F., Jan Sapp, and Alfred Tauber. "A Symbiotic View of Life: We Have Never Been Individuals." *Quarterly Review of Biology* 87 (2012): 325–41.

Glanville, Ranulph. "Cybernetics: Thinking through Technology." In *Traditions of*

Systems Theory: Major Figures and Contemporary Developments, edited by Darrell P. Arnold, 45–77. New York: Routledge, 2014.

Goldsmith, Maurice. "A New Deal for the World's Arid Lands." *UNESCO Courier* 4, no. 6 (1951): 13.

Gonzalez, Antonio, Jose C. Clemente, Ashley Shade, Jessica L. Metcalfe, Sejin Song, Bharath Prithiviraj, Brent E. Palmer, and Rob Knight. "Our Microbial Selves: What Ecology Can Teach Us." *EMBO Reports* 12 (2011): 775–84.

Gould, S. J., and R. C. Lewontin. "The Spandrels of San Marco and the Panglossian Paradigm: A Critique of the Adaptationist Programme." *Proceedings of the Royal Society of London*, Part B 205 (1979): 581–98.

Granger, Harris J., and Arthur C. Guyton. "Autoregulation of the Total Systemic Circulation Following Destruction of the Central Nervous System in the Dog." *Circulation Research* 25 (1969): 379–88.

Grantham, Dewey W. *Southern Progressivism: The Reconciliation of Progress and Tradition*. Knoxville: University of Tennessee Press, 1983.

Green, R. G., and C. L. Larson. "A Description of Shock Disease in the Snowshoe Hare." *American Journal of Hygiene* 28 (1938): 190–212.

———. "Shock Disease and the Snowshoe Hare Cycle." *Science* 87 (1938): 298–99.

Grote, Mathias. "Petri Dish versus Winogradsky Column: A Long Durée Perspective on Purity and Diversity in Microbiology, 1880s–1980s." *History and Philosophy of the Life Sciences* 40, no. 1 (2018): 11.

Guyton, Arthur C. "An Author's Philosophy of Physiology Textbook Writing." *Advances in Physiology Education* 19 (1998): S1–S5.

———. "Electronic Counting and Size Determination of Particles in Aerosols." *Journal of Industrial Hygiene and Toxicology* 28 (1946): 133–41.

———. "Past-President's Address: Physiology: A Beauty and a Philosophy." *Physiologist* 18 (1975): 495–501.

———. *Textbook of Medical Physiology*. Philadelphia: Saunders, 1956.

———. *Textbook of Medical Physiology*. 2nd ed. Philadelphia: Saunders, 1961.

Guyton, Arthur C., Thomas G. Coleman, and Harris J. Granger. "Circulation: Overall Regulation." *Annual Review of Physiology* 34 (1972): 13–46.

Guyton, Arthur C., John E. Hall, Thomas E. Lohmeier, R. Davis Manning Jr., and Thomas E. Jackson. "Position Paper: The Concept of Whole Body Autoregulation and the Dominant Role of the Kidneys for Long-Term Blood Pressure Regulation." In *Frontiers in Hypertension Research*, edited by John H. Laragh, Fritz R. Bühler, and Donald W. Seldin, 125–34. New York: Springer, 1981.

Guyton, Arthur C., Carl E. Jones, and Thomas G. Coleman. *Circulatory Physiology: Cardiac Output and Its Regulation*. 2nd ed. Philadelphia: Saunders, 1973.

Guyton, Arthur C., Howard T. Milhorn, and Thomas G. Coleman. "Simulation of Physiological Mechanisms Part I." *Simulation* 9 (1967): 15–20.

———. "Simulation of Physiological Mechanisms Part II." *Simulation* 9 (1967): 73–79.

Hadley, Malcolm. "Nature to the Fore: The Early Years of UNESCO's Environmental Program, 1945–1965." In *Sixty Years of Science at UNESCO, 1945–2005,* edited by P. Petitjean, V. Zharov, G. Glaser, J. Richardson, B. de Padirac, and G. Archibald, 201 –232. Paris: UNESCO Publishing, 2006.

Hagen, Joel B. "Bergmann's Rule, Adaptation, and Thermoregulation in Arctic Animals: Conflicting Perspectives from Physiology, Evolutionary Biology, and Physical Anthropology after World War II." *Journal of the History of Biology* 50 (2017): 235–65.

———. "Camels, Cormorants, and Kangaroo Rats: Integration and Synthesis in Organismal Biology after World War II." *Journal of the History of Biology* 48 (2015): 169–99.

———. "The Diving Reflex and Asphyxia: Working across Species in Physiological Ecology." *History and Philosophy of the Life Sciences* 40, no. 1 (2018): 18.

———. *An Entangled Bank: The Origins of Ecosystem Ecology*. New Brunswick, NJ: Rutgers University Press, 1992.

———. "Eugene Odum and the Homeostatic Ecosystem: The Resilience of an Idea." In *Traditions of Systems Theory: Major Figures and Contemporary Developments,* edited by Darrell P. Arnold, 179–93. New York: Routledge, 2014.

———. "Experimentalists and Naturalists in Twentieth-Century Botany: Experimental Taxonomy, 1920–1950." *Journal of the History of Biology* 17 (1984): 249–70.

———. "Naturalists, Molecular Biologists, and the Challenges of Molecular Evolution." *Journal of the History of Biology* 32 (1999): 321–41.

———. "The Statistical Frame of Mind in Systematic Biology from *Quantitative Zoology* to *Biometry*." *Journal of the History of Biology* 36 (2003): 353–84.

———. "Teaching Ecology during the Environmental Age, 1965–1980." *Environmental History* 13 (2008): 704–23.

Hall, Brian K. "Fifty Years Later: I. Michael Lerner's *Genetic Homeostasis* (1954)—a Valiant Attempt to Integrate Genes, Organisms, and Environment." *Journal of Experimental Zoology* 304B (2005): 187–97.

———. "Organic Selection: Proximate Environmental Effects on the Evolution of Morphology and Behavior." *Biology and Philosophy* 16 (2001): 215–37.

———. "Waddington's Legacy in Development and Evolution." *American Zoologist* 32 (1992): 113–22.

Hall, Charles A. S., ed. *Maximum Power: The Ideas and Applications of H. T. Odum.* Niwot: University Press of Colorado, 1995.

Hamilton, William F., and Dickinson W. Richards. "The Output of the Heart." In *Circulation of the Blood: Men and Ideas,* edited by Alfred P. Fishman and Dickinson W. Richards, 71–126. Bethesda: American Physiological Society, 1982.

Hammel, H. T. "Terrestrial Animals in Cold: Recent Studies of Primitive Man." In *Handbook of Physiology. Section 4: Adaptation to the Environment*, edited by David Bruce Dill, Edward F. Adolph, and C. G. Wilber, 413–34. Washington, DC: American Physiological Society, 1964.

Hammond, Debora. "Ecology and Ideology in the General Systems Community." *Environment and History* 3 (1997): 197–207.

———. *The Science of Synthesis: Exploring the Social Implications of General Systems Theory*. Boulder: University Press of Colorado, 2003.

Hardin, Garrett. *Nature and Man's Fate*. New York: Rhinehart, 1959.

Heinrich, Bernd. "A Brief Historical Survey." In *Insect Thermoregulation*, edited by Bernd Heinrich, 7–18. New York: John Wiley, 1981.

———. "Thoracic Temperature Stabilization by Blood Circulation in a Free-Flying Moth." *Science* 168 (1970): 580–82.

Heinrich, Bernd, and George A. Bartholomew. "An Analysis of Pre-flight Warm-Up in the Sphinx Moth, *Manduca sexta*." *Journal of Experimental Biology* 55 (1971): 223–39.

———. "Temperature Control in Flying Moths." *Scientific American* 226, no. 6 (June 1972): 70–77.

Henderson, Lawrence J. *The Fitness of the Environment: An Inquiry into the Biological Significance of the Properties of Matter*. New York: Macmillan, 1913.

———. *The Order of Nature*. Cambridge: Harvard University Press, 1917.

Hillis, David M. "Systematics of the *R. pipiens* Complex: Puzzle and Paradigm." *Annual Review of Ecology and Systematics* 19 (1988): 39–63.

Hillis, David M., and Thomas P. Wilcox. "Phylogeny of the New World True Frogs (*Rana*)." *Molecular Phylogeny and Evolution* 34 (2005): 299–314.

Hiltzik, Michael. *Colossus: Hoover Dam and the Making of the American Century*. New York: Free Press, 2010.

Hirsch, James G., and Carol L. Moberg. "René J. Dubos, February 20, 1901–February 20, 1982." *Biographical Memoirs of the National Academy of Sciences*, 113–16. Washington, DC, 1989.

Hoenig, Melanie P., and Mark L. Zeidel. "Homeostasis, the Milieu Intérieur, and the Wisdom of the Nephron." *Renal Physiology* 9 (2014): 1–10.

Holmes, Frederic L. "Claude Bernard, the Milieu Intérieur, and Regulatory Physiology." *History and Philosophy of the Life Sciences* 8 (1986): 3–25.

———. "The Old Martyr of Science: The Frog in Experimental Physiology." *Journal of the History of Biology* 26 (1993): 311–28.

Honigsbaum, Mark. "René Dubos, Tuberculosis, and the 'Ecological Facets' of Virulence." *History and Philosophy of the Life Sciences* 39 (2017): 15.

Howell, A. Brazier, and I. Gersh. "Conservation of Water by the Rodent *Dipodomys*." *Journal of Mammalogy* 16 (1935): 1–9.

Hubbs, Carl L. "Physiological Regulations." *American Naturalist* 78 (1944): 82–83.

Hudson, Jack W. "The Role of Water in the Biology of the Antelope Ground Squirrel, *Citellus leucurus*." *University of California Publications in Zoology* 64 (1962): 1–56.

Huey, Raymond B., and Albert F. Bennett. "Bart's Familiar Quotations: The Enduring Wisdom of George A. Bartholomew." *Physiological and Biochemical Zoology* 8 (2008): 519–25.

Hughes, David P., Naomi E. Pierce, and Jacobus J. Boomsma. "Social Insect Symbionts: Evolution in Homeostatic Fortresses." *Trends in Ecology and Evolution* 23 (2008): 672–77.

Hutchinson, G. Evelyn. "Circular Causal Systems in Ecology." *Annals of the New York Academy of Sciences* 50 (1948): 221–46.

———. "On Living in the Biosphere." *Scientific Monthly* 67 (1948): 393–97.

Irving, Laurence. "Heterothermous Operation of Warm-Blooded Animals." *Physiologist* 2 (1959): 18–32.

———. "Heterothermy in the Cold Adaptations of Warm Blooded Animals." In *Comparative Physiology of Temperature Regulation*, edited by John P. Hannon, 133–74. Fort Wainwright, AK: Arctic Aeromedical Laboratory, 1962.

———. "Human Adaptation to Cold." *Nature* 185 (1960): 572–74.

———. "The Usefulness of Scholander's Views on Adaptive Insulation of Animals." *Evolution* 11 (1957): 257–59.

Jackson, Donald C., and Knut Schmidt-Nielsen. "Countercurrent Heat Exchange in the Respiratory Passages." *Proceedings of the National Academy of Sciences* 51 (1964): 1192–97.

Jackson, John. "'In Ways Unacademical': The Reception of Carleton S. Coon's *The Origin of Races*." *Journal of the History of Biology* 34 (2001): 247–85.

Jackson, Mark. *The Age of Stress: Science and the Search for Stability*. Oxford: Oxford University Press, 2013.

———. "Evaluating the Role of Hans Selye in the Modern History of Stress." In *Stress, Shock, and Adaptation in the Twentieth Century*, edited by David Cantor and Edmund Ramsden, 21–48. Rochester, NY: University of Rochester Press, 2014.

Jennings, Herbert Spencer. *The Biological Basis of Human Nature*. New York: Norton, 1930.

Johnson, Andi. "'They Sweat for Science': The Harvard Fatigue Laboratory and Self-Experimentation in American Exercise Physiology." *Journal of the History of Biology* 48 (2015): 425–54.

Johnson, Kristin. "The Return of the Phoenix: The 1963 International Congress on Zoology and the American Zoologists in the Twentieth Century." *Journal of the History of Biology* 42 (2009): 417–56.

Johnson, Paul C. "Autoregulation of Blood Flow." *Circulation Research* 59 (1989): 483–95.

Jones, Clive G., John H. Lawton, and Moshe Shachak. "Organisms as Ecosystem Engineers." *Oikos* 69 (1994): 373–86.

Jones, Susan D. "Population Cycles, Disease, and Networks of Ecological Knowledge." *Journal of the History of Biology* 50 (2017): 357–91.

Kaempffert, Waldemar. "The Ostracoderms Lead the Pageant." *New York Times*, November 29, 1953, 22.

Kean, Melissa. *Desegregating Private Higher Education in the South*. Baton Rouge: Louisiana State University Press, 2008.

Keiner, Christine. "Wartime Rat Control, Rodent Ecology, and the Rise of Chemical Rodenticides." *Endeavor* 29 (2005): 119–25.

Kelt, Douglas A. "Comparative Ecology of Desert Small Mammals: A Selective Review of the Past 30 Years." *Journal of Mammalogy* 92 (2011): 1158–78.

Keys, Ancel. "Chloride and Water Secretion and Absorption by the Gills of the Eel." *Zeitschrift für Vergleichende Physiologie* 15 (1931): 364–89.

———. "The Heart-Gill Preparation of the Eel and Its Perfusion for the Study of a Natural Membrane in situ." *Zeitschrift für Vergleichende Physiologie* 15 (1931): 352–63.

Kimler, William C. "Advantage, Adaptiveness, and Evolutionary Ecology." *Journal of the History of Biology* 19 (1986): 215–33.

Kinchy, Abby J. "On the Borders of Post-War Ecology: Struggles over the Ecological Society of America's Preservation Committee, 1917–1945." *Science as Culture* 15 (2006): 23–44.

Kingsland, Sharon E. *The Evolution of American Ecology, 1890–2000*. Baltimore: Johns Hopkins University Press, 2005.

———. "Mathematical Figments, Biological Facts: Population Ecology in the Thirties." *Journal of the History of Biology* 19 (1986): 235–56.

———. *Modeling Nature: Episodes in the History of Population Ecology*. Chicago: University of Chicago Press, 1985.

Kirk, Robert G. W. "Life in a Germ-Free World: Isolating Life from the Laboratory Animal to the Bubble Boy." *Bulletin of the History of Medicine* 86 (2012): 237–75.

Knight, Robert L., and Dennis P. Swaney. "In Defense of Ecosystems." *American Naturalist* 117 (1981): 991–92.

Kohler, Robert E. *Landscapes and Labscapes: Exploring the Lab-Field Border in Biology*. Chicago: University of Chicago Press, 2002.

Krebs, Charles J. *Population Fluctuations in Rodents*. Chicago: University of Chicago Press, 2013.

Kricher, John. *The Balance of Nature: Ecology's Enduring Myth*. Princeton: Princeton University Press, 2009.

Krogh, August. "A Micro-habitat Recorder." *Ecology* 21 (1940): 275–78.

———. *Osmotic Regulation in Aquatic Animals*. Cambridge: Cambridge University Press, 1939.

Laland, Kevin N., Kim Sterelny, John Odling-Smee, William Hoppitt, and Tobias Uller. "Cause and Effect in Biology Revisited: Is Mayr's Proximate–Ultimate Dichotomy Still Useful?" *Science* 334 (2011): 1512–16.

Landauer, Walter. "The Genetics of Dynamic Equilibria." *American Naturalist* 89 (1955): 183–85.

Landry, Stuart O. "John Jermyn Christian: 1917–1997." *Journal of Mammalogy* 79 (1998): 1432–38.

Lanyon, Wesley E., Stephen T. Emlen, and Gordon H. Orians. "In Memoriam: John Thompson Emlen, Jr." *Auk* 117 (2000): 222–27.

Lauson, Henry D., Stanley E. Bradley, and André Cournand. "The Renal Circulation in Shock." *Journal of Clinical Investigation* 23 (1944): 381–402.

Lefèbvre, Pierre. "Early Milestones in Glucagon Research." *Diabetes, Obesity, and Metabolism* 13, Supplement 1 (2011): 1–4.

———. "Glucagon's Golden Jubilee at the University of Liège." *British Journal of Diabetes and Vascular Disease* 12 (2012): 278–84.

Lerner, I. Michael. *Genetic Homeostasis*. New York: John Wiley, 1954

———. *Heredity, Evolution and Society*. San Francisco: Freeman, 1968.

Lewontin, R. C. "Adaptation." *Scientific American* 239, no. 3 (September 1978): 212–31.

———. "The Adaptations of Populations to Varying Environments." *Cold Spring Harbor Symposia on Quantitative Biology* 22 (1956): 395–408.

———. *The Genetic Basis of Evolutionary Change*. New York: Columbia University Press, 1974.

———. "Studies on Heterozygosity and Homeostasis II: Loss of Heterosis in a Constant Environment." *Evolution* 12 (1958): 494–503.

———. "Studies on Homeostasis and Heterozygosity I. General Considerations. Abdominal Bristle Number in Second Chromosome Homozygotes of *Drosophila melanogaster*." *American Naturalist* 90 (1956): 237–55.

———. "Twenty-Five Years in Genetics: Electrophoresis in the Development of Evolutionary Genetics: Milestone or Millstone?" *Genetics* 128 (1991): 657–62.

———. "The Units of Selection." *Annual Review of Ecology and Systematics* 1 (1970): 1–18.

Limburg, Karin E. "The Biogeochemistry of Strontium: A Review of H. T. Odum's Contributions." *Ecological Modeling* 178 (2004): 31–33.

Littlejohn, M. J., and R. S. Oldham. "*Rana pipiens* Complex: Mating Call Structure and Taxonomy." *Science* 162 (1968): 1003–5.

Maienschein, Jane. "History of American Marine Laboratories: Why Do Research at the Seashore?" *American Zoologist* 28 (1988): 15–25.

Maren, Thomas H. "Eli Kennerly Marshall, Jr." *Biographical Memoirs of the National Academy of Sciences*, 313–52. Washington, DC, 1986.

Marshall, E. K. "The Comparative Physiology of the Kidney in Relation to Theories of Renal Secretion." *Physiological Reviews* 14 (1934): 133–59.

Marshall, E. K., and Homer W. Smith. "The Glomerular Development of the Vertebrate Kidney in Relation to Habitat." *Biological Bulletin* 59 (1930): 135–53.

Masani, Pesi R. *Norbert Wiener, 1894–1964*. Basel: Birkhauser, 1990.

Matteson, Sumner W. "John T. Emlen, Jr.: A Naturalist for All Seasons, Part 2: Of Adventure, Innovation, and Conscience." *Passenger Pigeon* 60 (1998): 203–50.

Mayr, Ernst. *Animal Species and Evolution*. Cambridge: Belknap Press, 1963.

———. "Cause and Effect in Biology." *Science* 134 (1961): 1501–6.

———. "Geographical Character Gradients and Climatic Adaptation." *Evolution* 10 (1956): 105–8.

———. *The Growth of Biological Thought: Diversity, Evolution, and Inheritance*. Cambridge: Belknap Press, 1982.

———. "Origin of Human Races." *Science* 138 (1962): 420–22.

———. *Systematics and the Origin of Species*. New York: Columbia University Press, 1942.

Mayr, Ernst, and William B. Provine, eds. *The Evolutionary Synthesis: Perspectives on the Unification of Biology*. Cambridge: Harvard University Press, 1981.

McIntosh, Robert. *The Background of Ecology: Concept and Theory*. Cambridge: Cambridge University Press, 1985.

McNab, Brian K. "Energy Expenditure: A Brief History." In *Mammalian Energetics: Interdisciplinary Views of Metabolism and Reproduction*, edited by Thomas Edward Tomasi and Teresa Helen Horton, 1–15. Ithaca: Cornell University Press, 1992.

———. *Extreme Measures: The Ecological Energetics of Birds and Mammals*. Chicago: University of Chicago Press, 2012.

———. *The Physiological Ecology of Vertebrates: A View from Energetics*. Cornell University Press, 2002.

McNaughton, S. J., and Michael B. Coughenour. "The Cybernetic Nature of Ecosystems." *American Naturalist* 117 (1981): 985–90.

Merrell, David J. "Migration and Gene Dispersal in *Rana pipiens*." *American Zoologist* 10 (1970): 47–52.

Milam, Erika Lorraine. *Creatures of Cain: The Hunt for Human Nature in Cold War America*. Princeton: Princeton University Press, 2018.

———. "The Equally Wonderful Field: Ernst Mayr and Organismic Biology." *Historical Studies in the Natural Sciences* 40 (2010): 279–317.

Milhorn, Howard T. *The Application of Control Theory to Physiological Systems*. Philadelphia: Saunders, 1966.

Mindell, David A. *Between Human and Machine: Feedback, Control, and Computing before Cybernetics*. Baltimore: Johns Hopkins University, 2002.

Mitman, Gregg. *The State of Nature: Ecology, Community, and American Social Thought, 1900–1950*. Chicago: University of Chicago Press, 1992.

Moberg, Carol L. *René Dubos: Friend of the Good Earth*. Washington, DC: ASM Press, 2005.

Moore, John A. "An Embryological and Genetical Study of *Rana burnsi* Weed." *Genetics* 27 (1942): 408–16.

———. "An Embryologist's View of the Species Problem." In *The Species Problem*, edited by Ernst Mayr, 325–38. Washington, DC: American Association for the Advancement of Science, 1957.

———. "Further Studies on *R. pipiens* Racial Hybrids." *American Naturalist* 84 (1950): 247–54.

———. "Geographic Variation in *Rana pipiens* Schreber of the Eastern United States." *Bulletin of the American Museum of Natural History* 82 (1944): 345–70.

———. "Geographic Variation of Adaptive Characters in *Rana pipiens* Schreber." *Evolution* 3 (1949): 1–24.

———. "Incipient Intraspecific Isolation Mechanisms in *R. pipiens.*" *Genetics* 31 (1946): 304–26.

———, ed. *The Physiology of the Amphibia.* New York: Academic Press, 1964.

———. "*R. pipiens*: The Changing Paradigm." *American Zoologist* 15 (1975): 837–49.

———. "The Role of Temperature in the Speciation of Frogs." *Biological Symposia* 6 (1942): 189–213.

———. "Studies on the Development of Frog Hybrids. II. Competence of the Gastrula Ectoderm of *Rana pipiens* x *Rana sylvatica* Hybrids." *Journal of Experimental Zoology* 105 (1947): 349–70.

———. "Temperature Tolerances and Rates of Development in the Eggs of Amphibia." *Ecology* 20 (1939): 459–78.

Muñoz, Mateo Jasmine. "Lawrence Joseph Henderson: Bridging Laboratory and Social Life." PhD diss., Department of the History of Science, Harvard University, 2014.

Nelkin, Dorothy. "Scientists and Professional Responsibility: The Experience of American Ecologists." *Social Studies of Science* 7 (1977): 75–95.

Newman, Marshall T. "Adaptation of Man to Cold Climates." *Evolution* 10 (1956): 101–5.

———. "The Application of Ecological Rules to the Racial Anthropology of the Aboriginal New World." *American Anthropologist* 55 (1953): 311–27.

———. "Review of *Races: A Study of the Problems of Race Formation in Man.*" *Boletín Bibliográfico de Antropología Americana* 13 (1950): 188–92.

Nicholson, Daniel J., and Richard G. Gawne. "Rethinking Woodger's Legacy in the Philosophy of Biology." *Journal of the History of Biology* 47 (2014): 243–92.

Odling-Smee, F. John, Kevin N. Laland, and Marcus W. Feldman. *Niche Construction: The Neglected Process in Evolution.* Princeton: Princeton University Press, 2003.

Odum, Eugene P. *Basic Ecology.* Philadelphia: Saunders, 1983.

———. "The Emergence of Ecology as a New Integrative Discipline." *Science* 195 (1977): 1289–93.

———. "Energy Flow in Ecosystems: A Historical Review." *American Zoologist* 8 (1968): 11–18.

———. *Fundamentals of Ecology.* Philadelphia: Saunders, 1953.

———. *Fundamentals of Ecology*. 3rd ed. Philadelphia: Saunders, 1971.

———. "The New Ecology." *Bioscience* 14 (1964): 14–16.

———. "The Strategy of Ecosystem Development." *Science* 164 (1969): 262–70.

Odum, Howard T. "Ecological Potential and Analogue Circuits for the Ecosystem." *American Scientist* 48 (1960): 1–8.

———. *Environment, Power, and Society*. New York: John Wiley, 1971.

———. "The Stability of the World Strontium Cycle." *Science* 114 (1951): 407–11.

———. "Trophic Structure and Productivity of Silver Springs, Florida." *Ecological Monographs* 27 (1957): 55–112.

Odum, Howard T., and Eugene P. Odum. "Trophic Structure and Productivity of a Windward Coral Reef Community on Eniwetok Atoll." *Ecological Monographs* 25 (1955): 291–320.

Odum, Howard W., Harold D. Meyer, B. S. Holden, and Fred M. Alexander. *American Democracy Anew*. New York: Henry Holt, 1940.

Odum, Howard W., and Harry Estill Moore. *American Regionalism: A Cultural-Historical Approach to National Integration*. New York: Henry Holt, 1938.

Oksanen, Lauri. "Ecosystem Organization: Mutualism and Cybernetics or Plain Darwinian Struggle for Existence?" *American Naturalist* 131 (1988): 424–44.

Orians, Gordon H. "A Diversity of Textbooks: Ecology Comes of Age." *Science* 181 (1973): 1238–39.

Pace, Ann E. *Systematic and Biological Studies of the Leopard Frogs* (Rana pipiens *Complex*) *of the United States*. Miscellaneous Publications of the Museum of Zoology No. 148 Lansing: University of Michigan, 1974.

Parascandola, John. "Organismic and Holistic Concepts in the Thought of L. J. Henderson." *Journal of the History of Biology* 4 (1971): 63–113.

Pastorius, Arlene. "'Ice-Box' Treatment Tests Native Resistance to Cold." *University of Washington Daily*, April 17, 1958.

Patten, Bernard C., and Odum, Eugene P. "The Cybernetic Nature of Ecosystems." *American Naturalist* 118 (1981): 886–95.

Patten, Bernard C. "This Week's Citation Classic." *Current Contents* #5 (February 4, 1985): 22.

Paul, Diane B. "Dobzhansky in the Nature–Nurture Debate." In *The Evolution of Theodosius Dobzhansky*, edited by Mark B. Adams, 219–31. Princeton: Princeton University Press, 1994.

———. *The Politics of Heredity: Essays on Eugenics, Biomedicine, and the Nature–Nurture Debate*. Albany: State University of New York Press, 1998.

Pauly, Philip J. "Summer Resort and Scientific Discipline: Woods Hole and the Structure of American Biology." In *The American Development of Biology*, edited by Ronald Rainger, Keith R. Benson, and Jane Maienschein, 121–50. Philadelphia: University of Pennsylvania Press, 1988.

Peitzman, Steven J. "The Flame Photometer as Engine of Nephrology: A Biography."

American Journal of Kidney Diseases 56 (2010): 379–86.

Pepper, John W., and Simon Rosenfeld. "The Emerging Medical Ecology of the Human Gut Microbiome." *Trends in Ecology and Evolution* 27 (2012): 381–84.

Perlman, Robert. "The Concept of the Organism in Physiology." *Theory in Biosciences* 119 (2000): 174–86.

Pitts, Robert F. "Homer William Smith 1895–1962." *Biographical Memoirs of the National Academy of Sciences*, 445–69. Washington, DC, 1967.

Plutynski, Anya. "The Rise and Fall of the Adaptive Landscape?" *Biology and Philosophy* 23 (2008): 605–23.

Poulson, Thomas L., and George A. Bartholomew. "Salt Balance in the Savannah Sparrow." *Physiological Zoology* 35 (1962): 109–99.

Price, David H. *Threatening Anthropology: McCarthyism and the FBI's Surveillance of Activist Anthropologists*. Durham, NC: Duke University Press, 2004.

Prosser, C. Ladd, ed. *Comparative Animal Physiology*. Philadelphia: Saunders, 1950.

———. "Comparative Physiology in Relation to Evolutionary Theory." In *Evolution after Darwin*, edited by Sol Tax, 569–94. Chicago: University of Chicago Press, 1960.

———. "The Making of a Comparative Physiologist." *Annual Reviews in Physiology* 48 (1986): 1–6.

———. "The 'Origin' after a Century: Prospects for the Future." *American Scientist* 47 (1959): 536–60.

———. "Physiological Variation in Animals." *Biological Reviews* 30 (1955): 229–61.

———. "The Species Problem from the Viewpoint of a Physiologist." In *The Species Problem*, edited by Ernst Mayr, 339–69. Washington, DC: American Association for the Advancement of Science, 1957.

———. "Theory of Physiological Adaptation of Poikilotherms to Heat and Cold." Brody Memorial Lecture 5. *University of Missouri Agricultural Station Special Report #59*, 1965.

Provine, William B. "Ernst Mayr: Genetics and Speciation." *Genetics* 167 (2004): 1041–46.

———. "The Origin of Dobzhansky's *Genetics and the Origin of Species*." In *The Evolution of Theodosius Dobzhansky*, edited by Mark B. Adams, 99–114. Princeton: Princeton University Press, 1994.

———. *Sewall Wright and Evolutionary Biology*. Chicago: University of Chicago Press, 1986.

Rader, Karen A. *Making Mice: Standardizing Animals for American Biomedical Research, 1900–1955*. Princeton: Princeton University Press, 2004.

Ramsden, Edmund. "From Rodent Utopia to Urban Hell: Population, Pathology, and the Crowded Rats of NIMH." *Isis* 102 (2011): 659–88.

———. "Rats, Stress, and the Built Environment." *History of the Human Sciences* 25 (2012): 123–47.

Ramsden, Edmund, and Jon Adams. "Escaping the Laboratory: The Rodent Experiments of John B. Calhoun & Their Cultural Influence." *Journal of Social History* 42 (2009): 761–92.

Redford, Kent H., Julia A. Segre, Nick Salafsky, Carlos Martinez del Rio, and Denise McAloose. "Conservation and the Microbiome." *Conservation Biology* 26 (2012): 195–97.

Richards, Dickinson W. "The Circulation in Traumatic Shock in Man." *Bulletin of the New York Academy of Medicine* 20 (1944): 363–93.

———. "The Effects of Hemorrhage on the Circulation." *Annals of the New York Academy of Sciences* 44 (1948): 534–41.

———. "Homeostasis: Its Dislocations and Perturbations." *Perspectives in Biology and Medicine* 3 (1960): 238–51.

———. "Homeostasis versus Hyperexis: Or Saint George and the Dragon." *Scientific Monthly* 77 (1953): 289–94.

———. *Medical Priesthoods and Other Essays*. Lakeville, CT: Connecticut Printers, 1970.

———. "Right Heart Catheterization: Its Contributions to Physiology and Medicine." *Science* 125 (1957): 1181–85.

———. "The Right Heart and the Lung: With Some Observations on Teleology." *American Review of Respiratory Diseases* 94 (1966): 691–702.

Richards, Dickinson W., and Alvan L. Barach. "Prolonged Residence in High Oxygen Atmospheres: Effects on Normal Individuals and on Patients with Chronic Cardiac and Pulmonary Insufficiency." *Quarterly Journal of Medicine* 3 (1934): 437–66.

Richards, Dickinson W., André Cournand, and Natalie A. Bryan. "Applicability of Rebreathing Method for Determining Mixed Venous CO_2 in Cases of Chronic Pulmonary Disease." *Journal of Clinical Investigation* 14 (1935): 173–80.

Richards, Dickinson W., and Marjorie L. Strauss. "Carbon Dioxide and Oxygen Tensions of the Mixed Venous Blood of Man at Rest." *Journal of Clinical Investigation* 9 (1930): 475–532.

Richter, Curt R. "Rats, Man, and the Welfare State." *American Psychologist* 14 (1959): 18–28.

Robson, G. C., and O. W. Richards. *The Variation of Animals in Nature*. New York: Longmans, 1936.

Romer, Alfred Sherwood. "The Early Evolution of Fishes." *Quarterly Review of Biology* 21 (1946): 33–69.

Rosebury, Theodor. *Microorganisms Indigenous to Man*. New York: McGraw-Hill, 1962.

Rosenblueth, Arturo. "A Critique of Homeostasis." In *Perspectives in Biology: A Collection of Papers Dedicated to Bernardo A. Houssay on the Occasion of His 75th Birthday*, edited by Carl F. Cori, V. G. Foglia, L. F. Leloir, and S. Ochoa, 323–31. New York: Elsevier, 1963.

Rosenblueth, Arturo, Norbert Wiener, and Julian Bigelow. "Behavior, Purpose, and Teleology." *Philosophy of Science* 10 (1943): 18–24.

Ruibel, Rudolfo, Vaughn H. Shoemaker, and Margaret M. Stewart. "Historical Perspective: John A. Moore." *Copeia* 4 (2001): 1155–57.

Ruse, Michael. *The Evolution Wars*. Santa Barbara, CA: ABC-CLIO, 2000.

———. *Monad to Man: The Concept of Progress in Evolutionary Biology*. Cambridge: Harvard University Press, 1996.

Sage, Richard D., and Robert K. Selander. "Hybridization between Species of the *Rana pipiens* Complex in Central Texas." *Evolution* 33 (1979): 1069–88.

Salthe, Stanley N. "Geographic Variation of the Lactate Dehydrogenases of *Rana pipiens* and *Rana palustris*." *Biochemical Genetics* 2 (1969): 271–303.

Sands, Jeff M. "Micropuncture: Unlocking the Secrets of Renal Function." *American Journal of Physiology—Renal Physiology* 287 (2004): F866–87.

Sapp, Jan. *Evolution by Association: A History of Symbiosis*. New York: Oxford University Press, 1994.

Sarkar, Sahotra. "From the *Reaktionsnorm* to the Adaptive Norm: The Norm of Reaction, 1909–1960." *Biology and Philosophy* 14 (1999): 35–252.

Saunders, Joseph F., and Aubrey E. Taylor. "Membership." In *History of the American Physiological Society: The First Century, 1887–1987*, edited by John R. Brobeck, Orr E. Reynolds, and Tobey A. Appel, 301–14. Bethesda, MD: American Physiological Society, 1987.

Savage, Dwayne C. "Microbial Biota of the Human Intestine: A Tribute to Some Pioneering Scientists." *Current Issues in Intestinal Microbiology* 2 (2001): 1–15.

Schaedler, Russell W., René Dubos, and Richard Costello. "Association of Germfree Mice with Bacteria Isolated from Normal Mice." *Journal of Experimental Medicine* 122 (1965): 77–82.

———. "The Development of the Bacterial Flora in the Gastrointestinal Tract of Mice." *Journal of Experimental Medicine* 122 (1965): 59–66.

Scheffler, Robin Wolfe. "The Power of Exercise and the Exercise of Power: The Harvard Fatigue Laboratory, Distance Running, and the Disappearance of Work." *Journal of the History of Biology* 48 (2015): 391–423.

Schickele, Elizabeth. "Environment and Fatal Heat Stroke: An Analysis of 157 Cases Occurring in the Army in the U.S. during World War II." *Military Surgeon* 100 (1947): 235–56.

Schmidt-Nielsen, Bodil. *August and Marie Krogh: Lives in Science*. New York: Oxford University Press, 1995.

———. "The Resourcefulness of Nature in Physiological Adaptation to the Environment." *Physiologist* 1 (1958): 4–20.

———. "Water Conservation in Small Desert Rodents." In *Biology of Deserts*, edited by J. L. Cloudsley-Thompson, 173–81. London: Institute of Biology, 1954.

Schmidt-Nielsen, Bodil, and Roberta O'Dell. "Structure and Concentrating

Mechanism in the Mammalian Kidney." *American Journal of Physiology* 200 (1961): 1119–24.

Schmidt-Nielsen, Bodil, and Knut Schmidt-Nielsen. "The Camel: Facts and Fables." *UNSESCO Courier* 8, no. 8 (1955): 70.

———. "Do Kangaroo Rats Thrive When Drinking Seawater?" *American Journal of Physiology* 160 (1950): 291–94.

———. "The Water Economy of Desert Mammals." *Scientific Monthly* 69 (1949): 180–85.

———. "Water Loss in Desert Rodents in Their Natural Habitat." *Ecology* 31 (1950): 75–85.

Schmidt-Nielsen, Bodil, Knut Schmidt-Nielsen, Adelaide Brokaw, and Howard Scheiderman. "Water Conservation in Desert Rodents." *Journal of Cellular and Comparative Physiology* 32 (1948): 331–60.

Schmidt-Nielsen, Bodil, Knut Schmidt-Nielsen, T. Richard Houpt, and Stig A. Jarnum. "Water Balance of the Camel." *American Journal of Physiology* 185 (1956): 185–94.

Schmidt-Nielsen, Knut. *The Camel's Nose: Memoirs of a Curious Scientist.* Washington, DC: Island Press, 1998.

———. "Comparative Physiology of Desert Mammals." Brody Memorial Lecture 2. *University of Missouri Agricultural Experiment Station Special Report #21*, 1962.

———. "Countercurrent Systems in Animals." *Scientific American* 244, no. 5 (May 1981): 118–28.

———. *Desert Animals: Physiological Problems of Heat and Water.* Oxford: Oxford University Press, 1964.

———. "Heat Regulation in Small and Large Desert Mammals." In *Biology of Deserts*, edited by J. L. Cloudsley-Thompson, 182–87. London: Institute of Biology, 1954.

———. *How Animals Work.* Cambridge: Cambridge University Press, 1972.

———. "Per Scholander 1905–1980." *Biographical Memoirs of the National Academy of Sciences*, 387–412. Washington, DC, 1987.

———. "The Physiology of the Camel." *Scientific American* 201, no. 6 (December 1959): 140–51.

———. "The Salt-Secreting Gland of Marine Birds." *Circulation* 21 (1960): 955–67.

———. *Scaling: Why Is Animal Size So Important?* Cambridge: Cambridge University Press, 1984.

Schmidt-Nielsen, Knut, and Ragnär Fange. "Salt Glands in Marine Reptiles." *Nature* 182 (1958): 783–85.

Schmidt-Nielsen, Knut, C. Barker Jörgensen, and Humio Osaki. "Extrarenal Salt Excretion in Birds." *American Journal of Physiology* 193 (1958): 101–7.

Schmidt-Nielsen, Knut, and Bodil Schmidt-Nielsen. "The Desert Rat." *Scientific American* 189, no. 1 (July 1953): 173–78.

Schmidt-Nielsen, Knut, Bodil Schmidt-Nielsen, and Adelaide Brokaw. "Urea Excretion in Desert Rodents Exposed to High Protein Diets." *Journal of Cellular and Comparative Physiology* 32 (1948): 361–80.

Schmidt-Nielsen, Knut, Bodil Schmidt-Nielsen, T. Richard Houpt, and Stig A. Jarnum. "The Question of Water Storage in the Stomach of the Camel." *Mammalia* 20 (1956): 1–15.

Schmidt-Nielsen, Knut, Bodil Schmidt-Nielsen, Stig A. Jarnum, and T. Richard Houpt. "Body Temperature of the Camel and Its Relation to Water Economy." *American Journal of Physiology* 188 (1956): 103–12.

Schoeb, Trenton R., and Roger P. Orcutt. "Historical Overview." In *Gnotobiotics*, edited by Trenton R. Schoeb and Kathryn A. Eaton, 13–14. San Diego: Academic Press, 2017.

Scholander, P. F. "Climatic Rules." *Evolution* 10 (1956): 339–40.

———. *Enjoying a Life in Science: The Autobiography of P. F. Scholander*. Fairbanks: University of Alaska Press, 1990.

———. "Evolution of Climatic Adaptation in Homeotherms." *Evolution* 9 (1955): 15–26.

———. "The Wonderful Net." *Scientific American* 196, no. 4 (April 1957): 97–107.

Scholander, P. F., Walter Flagg, Vladimir Walters, and Laurence Irving. "Climatic Adaptation in Arctic and Tropical Poikilotherms." *Physiological Zoology* 26 (1953): 67–92.

Scholander, P. F., H. T. Hammel, K. Lange Andersen, and Y. Løyning. "Metabolic Acclimatization to Cold in Man." *Journal of Applied Physiology* 12 (1958): 1–8.

Scholander, P. F., H. T. Hammel, J. S. Hart, D. H. LeMessurier, and J. Steen. "Cold Adaptation in Australian Aborigines." *Journal of Applied Physiology* 13 (1958): 211–18.

Scholander, P. F., Raymond Hock, Vladimir Walters, and Laurence Irving. "Adaptation to Cold in Arctic and Tropical Mammals and Birds in Relation to Body Temperature, Insulation, and Basal Metabolic Rate." *Biological Bulletin* 99 (1950): 259–71.

Scholander, P. F., Raymond Hock, Vladimir Walters, Fred Johnson, and Laurence Irving. "Heat Regulation in Some Arctic Mammals and Birds." *Biological Bulletin* 99 (1950): 237–58.

Scholander, P. F., Vladimir Walters, Raymond Hock, and Laurence Irving. "Body Insulation of Some Arctic and Tropical Mammals and Birds." *Biological Bulletin* 99 (1950): 225–36.

Schulkin, Jay, ed. *Allostasis, Homeostasis, and the Costs of Physiological Adaptation*. Cambridge: Cambridge University Press, 2004.

———. *Curt Richter: A Life in the Laboratory*. Baltimore: Johns Hopkins University Press, 2005.

———. *Rethinking Homeostasis: Allostatic Regulation in Physiology and Pathophysiology*. Cambridge, MA: MIT Press, 2003.

Selye, Hans. "The General Adaptation Syndrome and the Diseases of Adaptation." *Journal of Clinical Endocrinology* 6 (1946): 117–230.

———. *The Stress of Life*. New York: McGraw-Hill, 1956.

———. "A Syndrome Produced by Diverse Nocuous Agents." *Nature* 138 (1936): 32.

Shanahan, Timothy. *The Evolution of Darwinism*. Cambridge: Cambridge University Press, 2004.

Shapiro, Stephanie. "Is There a Doctor in the House?" *Hopkins Medicine Magazine*, Winter 2013. https://www.hopkinsmedicine.org/news/publications/hopkins_medicine_magazine/archives/winter_2013/is_there_a_doctor_in_the_house.

Shulins, Nancy. "Doctor Overcomes Polio to Bring 8—or 10—Children into Medicine: For Arthur Guyton of Mississippi a Handicap Is Not Disabling." *Los Angeles Times*, August 3, 1988. http://articles.latimes.com/1986–08–03/news/mn-971_1_arthur-guyton/4.

Siegal, Mark, and Aviv Bergman. "Waddington's Canalization Revisited: Developmental Stability and Evolution." *Proceedings of the National Academy of Sciences* 99 (2002): 10528–32.

Simpson, George Gaylord. "The Baldwin Effect." *Evolution* 7 (1953): 110–17.

———. *The Meaning of Evolution*. New Haven: Yale University Press, 1949.

Slack, Jonathan M. W. "Conrad Hal Waddington: The Last Renaissance Biologist?" *Nature Reviews Genetics* 3 (2002): 889–94.

Slack, Nancy G. *G. Evelyn Hutchinson and the Invention of Modern Ecology*. New Haven: Yale University Press, 2011.

Smith, Homer W. "The Absorption and Excretion of Water and Salts by Marine Teleosts." *American Journal of Physiology* 93 (1930): 480–505.

———. "The Composition of the Body Fluids of Elasmobranchs." *Journal of Biological Chemistry* 86 (1929): 407–19.

———. "The Excretion of Ammonia and Urea by the Gills of Fish." *Journal of Biological Chemistry* 81 (1929): 727–42.

———. *From Fish to Philosopher*. Boston: Little, Brown, 1953.

———. "The Functional and Structural Evolution of the Vertebrate Kidney." *Sigma Xi Quarterly* 21 (1931): 141–51.

———. *Kamongo*. New York: Viking Press, 1932.

———. "The Kidney." *Scientific American* 188, no. 1 (January 1953): 40–48.

———. *The Kidney: Structure and Function in Health and Disease*. New York: Oxford University Press, 1950.

———. *Lectures on the Kidney*. Lawrence: University Extension Division, University of Kansas, 1943.

———. "Lung-Fish." *Scientific Monthly* 31 (1930): 467–70.

———. *Man and His Gods*. New York: Little, Brown, 1952.

———. "Metabolism of the Lung-Fish, *Protopterus aethiopicus*." *Journal of Biological Chemistry* 88 (1930): 97–130.

———. "Observations on the African Lung-Fish, *Protopterus aethiopicus*, and on

Evolution from Water to Land Environments." *Ecology* 12 (1931): 164–81.
———. "Organism and Environment: Dynamic Oppositions." In *Adaptation*, edited by John Romano, 25–52. Ithaca: Cornell University Press, 1949.
———. *Principles of Renal Physiology*. New York: Oxford University Press, 1956.
———. "Renal Physiology." In *Circulation of the Blood: Men and Ideas*, edited by Alfred P. Fishman and Dickinson W. Richards, 545–606. Bethesda, MD: American Physiological Society, 1982.
———. "The Retention and Physiological Role of Urea in the Elasmobranchii." *Biological Reviews* 49 (1936): 49–81.
———. "Water Regulation and Its Evolution in the Fishes." *Quarterly Review of Biology* 7 (1932): 1–26.
Smocovitis, Vassiliki Betty. "Disciplining Biology: Ernst Mayr and the Founding of the Society for the Study of Evolution and *Evolution* (1939–1950)." *Evolution* 48 (1994): 1–8.
———. "The 1959 Darwin Centennial Celebration in America." *Osiris* 14 (1999): 274–323.
———. "Unifying Biology: The Evolutionary Synthesis and Evolutionary Biology." *Journal of the History of Biology* 25 (1992): 1–65.
———. *Unifying Biology: The Evolutionary Synthesis and Evolutionary Biology*. Princeton: Princeton University Press, 1996.
Somero, George M. "Clifford Ladd Prosser 1907–2002." *Biographical Memoirs of the National Academy of Sciences*, 1–17. Washington, DC, 2008.
Stap, Don. *Birdsong: A Natural History*. New York: Scribner, 2005.
Steffes, David M. "Panpsychic Organicism: Sewall Wright's Philosophy for Understanding Complex Genetic Systems." *Journal of the History of Biology* 40 (2007): 327–61.
Sterelny, Kim. *Dawkins vs. Gould: Survival of the Fittest*. Cambridge, UK: Icon Books, 2007.
Sterling, Peter. "Allostasis: A Model of Predictive Regulation." *Physiology & Behavior* 106 (2011): 5–15.
Sturdy, Steve. "Biology as Social Theory: John Scott Haldane and Physiological Regulation." *British Journal for the History of Science* 21 (1988): 315–40.
Sunderland, Mary E. "Modernizing Natural History: Berkeley's Museum of Vertebrate Zoology in Transition." *Journal of the History of Biology* 46 (2013): 369–400.
Swanson, Kara W. *Banking on the Body: The Market in Blood, Milk, and Sperm in Modern America*. Cambridge: Harvard University Press, 2014.
Swetlitz, Marc. "Julian Huxley and the End of Evolution." *Journal of the History of Biology* 28 (1995): 181–217.
Tasnim, Nishat, Nijiati Abulizi, Jason Pither, Miranda M. Hart, and Deanna L. Gibson. "Linking the Gut Microbial Ecosystem with the Environment: Does Gut Health Depend on Where We Live?" *Frontiers in Microbiology* 8 (2017): 1–8.
Taylor, Peter J. "Technocratic Optimism, H. T. Odum, and the Partial

Transformation of Ecological Metaphor after World War II." *Journal of the History of Biology* 21 (1988): 213–44.

Tracy, Randall L., and Glenn E. Walsberg. "Kangaroo Rats Revisited: Re-evaluating a Classic Case of Desert Survival." *Oecologia* 133 (2002): 449–57.

Tracy, Sarah W. "The Physiology of Extremes: Ancel Keys and the International High Altitude Expedition of 1935." *Bulletin of the History of Medicine* 86 (2012): 627–60.

Turner, J. Scott. "Biology's Second Law: Homeostasis, Purpose, and Desire." In *Beyond Mechanism: Putting Life Back into Biology*, edited by Brian G. Henning and Adam C. Scarfe, 183–204. Lanham, MD: Lexington Books, 2013.

Vize, Peter D. "A Homeric View of Kidney Evolution: A Reprint of H. W. Smith's Classic Essay with a New Introduction." *Anatomical Record* 277A (2004): 345–54.

Vogel, Steven. "Knut Schmidt-Nielsen. 24 September 1915–25 January 2007." *Biographical Memoirs of the Fellows of the Royal Society* 54 (2008): 319–31.

Waddington, C. H. "Canalization of Development and the Inheritance of an Acquired Trait." *Nature* 150 (1942): 563–65.

———. *The Ethical Animal.* London: George Allen and Unwin, 1960.

———. "The Resistance to Evolutionary Change." *Nature* 175 (1955): 51–52.

———. *The Strategy of the Genes: A Discussion of Some Aspects of Theoretical Biology.* London: George Allen and Unwin, 1957.

———, ed. *Towards a Theoretical Biology*. Chicago: Aldine, 1968.

Wallace, Bruce. "Genetic Homeostasis." *Science* 121 (1955): 558.

Walsberg, Glenn E. "Small Mammals in Hot Deserts: Some Generalizations Revisited." *BioScience* 50 (2000): 109–20.

Weber, Bruce H. "Lawrence Henderson's Natural Teleology." In *Water and Life: The Unique Properties of H_2O*, edited by Ruth M. Lynden-Bell, Simon Conway Morris, John D. Barlow, John L. Finney, and Charles L. Harper Jr., 327–44. Boca Raton, FL: CRC Press, 2010.

Weed, Alfred C. "New Frogs from Minnesota." *Proceedings of the Biological Society of Washington* 35 (1922): 107–10.

West, John B. *High Life: A History of High-Altitude Physiology and Medicine.* New York: Oxford University Press, 1998.

Wiener, Norbert. *Cybernetics: Or Control and Communication in the Animal and the Machine.* Cambridge, MA: MIT Press, 1948.

Williams, George C. *Adaptation and Natural Selection.* Princeton: Princeton University Press, 1966.

Williams, Michael Vinson. *Medgar Evers: Mississippi Martyr.* Fayetteville: University of Arkansas Press, 2011.

Wolfe, Eline L., A. Clifford Barger, and Saul Benison. *Walter B. Cannon, Science and Society*. Cambridge: Harvard University Press, 2000.

Wynne-Edwards, V. C. *Animal Dispersion in Relation to Social Behaviour*. London: Oliver and Boyd, 1962.

Zimmer, Carl. "Tending the Body's Microbial Garden." *New York Times*, June 18, 2012. http://www.nytimes.com/2012/06/19/science/studies-of-human-microbiome-yield-new-insights.html?pagewanted=all.

Zimmer, Heinz-Gerd. "August Krogh." *Clinical Cardiology* 29 (2006): 231–33.

Zupanc, G. K. H., and M. M. Zupanc. "Theodore H. Bullock: Pioneer of Integrative and Comparative Neurobiology." *Journal of Comparative Physiology A* 194 (2008): 119–34.

Index

Page numbers in italics indicate figures.